Annals of Mathematics Studies

Number 112

+M

BEIJING LECTURES IN HARMONIC ANALYSIS

EDITED BY

E. M. STEIN

PRINCETON UNIVERSITY PRESS

PRINCETON, NEW JERSEY

1986

Clothbound editions of Princeton University Press books are printed on acid-free paper, and binding materials are chosen for strength and durability. Paperbacks, while satisfactory for personal collections, are not usually suitable for library rebinding

ISBN 0-691-08418-1 (cloth)
ISBN 0-691-08419-X (paper)

Printed in the United States of America
by Princeton University Press, 41 William Street
Princeton, New Jersey

☆

Library of Congress Cataloging in Publication data will be found on the last printed page of this book

TABLE OF CONTENTS

PREFACE

In September 1984 a summer school in analysis was held at Peking University. The subjects dealt with were topics of current interest in the closely interrelated areas of Fourier analysis, pseudo-differential and singular integral operators, partial differential equations, real-variable theory, and several complex variables. Entitled the "Summer Symposium of Analysis in China," the conference was organized around seven series of expository lectures whose purpose was to give both an introduction of the basic material as well as a description of the most recent results in these areas. Our objective was to facilitate further scientific exchanges between the mathematicians of our two countries and to bring the students of the summer school to the level of current research in those important fields.

On behalf of all the visiting lecturers I would like to acknowledge our great appreciation to the organizing committee of the conference: Professors M. T. Cheng and D. G. Deng of Peking University, S. Kung of the University of Science and Technology of China, S. L. Wang of Hangzhou University, and R. Long of the Institute of Mathematics of the Academia Sinica. Their efforts helped to make this a most fruitful and enjoyable meeting.

E. M. STEIN

Beijing Lectures in

Harmonic Analysis

NON-LINEAR HARMONIC ANALYSIS, OPERATOR THEORY AND P.D.E.

R. R. Coifman and Yves Meyer

Our purpose is to describe a certain number of results involving the study of non-linear analytic dependence of some functionals arising naturally in P.D.E. or operator theory.

To be more specific we will consider functionals i.e., functions defined on a Banach space of functions (usually on R^n) with values in another Banach space of functions or operators.

Such a functional $F: B_1 \to B_2$ is said to be real analytic around 0 in B_1 if we can expand it in a power series around 0 i.e.

$$F(f) = \sum_{k=0}^{\infty} \Lambda_k(f)$$

where $\Lambda_k(f)$ is a "homogeneous polynomial" of degree k in f. This means that there is a k multilinear function

$$\Lambda_k(f_1 \cdots f_k) : B_1 \times B_1 \cdots \times B_1 \to B_2$$

(linear in each argument) such that $\Lambda_k(f) = \Lambda_k(f, f, \cdots f)$ and

$$\|\Lambda_k(f_1 \cdots f_k)\|_{B_2} \le C^k \prod_{j=1}^{k} \|f_j\|_{B_1} \tag{1}$$

for some constant C. (This last estimate guarantees the convergence of the series in the ball $\|f\|_{B_1} < \frac{1}{C}$.)

Certain facts can be easily verified. In particular if F is analytic it can be extended to a ball in B_1^c (the complexification of B_1) and the extension is holomorphic from B_1^c to B_2^c i.e., $F(f+zg)$ is a holomorphic (vector valued) function of $z \in \mathbf{C}$, $|z|<1$, $\forall f, g$ sufficiently small. The converse is also true. Any such holomorphic function can be expanded in a power series, (where Λ_k is $\frac{1}{k!} \times$ the kth Frechet differential at 0).

We will concentrate our attention on very concrete functionals arising in connection with differential equations or complex analysis, and would like to prove that they depend analytically on certain functional parameters.

As you know there are two ways to proceed.

1. Expand in a power series and show that one has estimates (1).

2. Extend the functional to the complexification as "formally holomorphic" and prove some boundedness estimates.

Let L denote a differential operator like

$$a(x)\frac{d}{dx} \qquad x \in \mathbf{R},$$

$$a(z)\frac{\partial}{\partial z} \qquad z \in \mathbf{C}$$

$$\sum \frac{\partial}{\partial x_i} a_{ij}(x) \frac{\partial}{\partial x_j} = \operatorname{div} A(x) \operatorname{grad}, \quad A=(a_{ij}) \qquad x \in \mathbf{R}^n$$

$$\sum a_{ij}(x) \frac{\partial}{\partial x_i \partial x_j} \qquad x \in \mathbf{R}^n$$

the coefficients $a(x)$ (or $a_{ij}(x)$) will be assumed to belong to some Banach space B_1 of functions (for example L^∞). It is natural to ask when such objects as:

$$L^{-1}, \ \sqrt{L}, \ \operatorname{sgn} L, \ e^{-tL}, \ e^{-t\sqrt{L}}$$

or more generally, $\phi(L)$ (where $\phi: \mathbf{C} \to \mathbf{C}$), can be defined as a bounded operator (say on L^2 or some Soboleff space), and a functional calculus developed i.e., $\phi_1(L)\phi_2(L) = \phi_1\phi_2(L)$.

Many questions arise:

a) Does $F(a) = \phi(L)$ viewed as an operator valued function of a depend analytically on a?

This is equivalent to asking whether we can consider complex valued coefficients in L and still have estimates on $\phi(L)$.

b) What is the largest domain of coefficients a for which we have estimates for $\phi(L)$? This question is the same as asking what is the *largest* B_1 for which (1) holds, and what is the domain of holomorphy of $F(a)$ in this space.

The answer to question a) will require first that we understand methods for expanding functionals in a power series, and second, that the nature of the multilinear operators Λ_k be sufficiently well understood to provide estimates (1). As for question b) we will see that the largest spaces possible for the coefficients involve rough coefficients and leads us to work with coefficients in L^∞, B.M.O. and other "exotic spaces."

We now start with a fundamental example related to the Cauchy integral. We let

$$L_a = \frac{1}{1+a}\,\frac{1}{i}\frac{d}{dx} \quad \text{with } \|a\|_\infty < 1 \quad a(x), \text{ real valued.}$$

If we define $h(x) = x + A(x)$, $A'(x) = a$. We then have

$$L_a f = \left(\frac{1}{i}\frac{d}{dx} f \circ h^{-1}\right) \circ h = \frac{1}{i}\, U_h \frac{d}{dx} U_h^{-1} f$$

where

$$U_h f = f \circ h\,.$$

Of course, in this case, if we use the Fourier transform we can define

$$\phi\left(\frac{1}{i}\frac{d}{dx}\right) f = \int_{-\infty}^{\infty} e^{ix\xi}\phi(\xi)\hat{f}(\xi)\,d\xi\,.$$

This gives, for example

$$\operatorname{sgn}\left(\frac{1}{i}\frac{d}{dx}\right)f = \int e^{ix\xi}\operatorname{sgn}\xi\,\hat{f}(\xi)\,d\xi = \frac{1}{\pi}\ \text{p.v.}\int_{-\infty}^{\infty}\frac{f(t)}{x-t}\,dt = H(f)\,.$$

Thus we can define

$$\pi\operatorname{sgn}(L_a)f = \pi U_h \operatorname{sgn}\frac{1}{i}\frac{d}{dx}\,U_h^{-1}f = \text{p.v.}\int_{-\infty}^{\infty}\frac{f(t)\,(1+a(t))}{x-t+A(x)-A(t)}\,dt$$

(where we used the observation that $\phi(ULU^{-1}) = U\phi(L)\,U^{-1}$).

We view

$$F(a) = \operatorname{sgn} L_a \quad \text{as an operator on } L^2(R)$$

and wish to know whether it is analytic on L^∞ or if we can replace a by complex a and still have a bounded operator.

If we do this, writing $a = \alpha + i\beta$ $\|a\|_\infty < 1$, we find

$$F(a)f = \int\frac{f(t)\ \ (1+i\alpha+i\beta)}{x-t+A(x)+iB(x)-A(t)-iB(t)}\,dt$$

$$= \int\frac{f(t)[(1+a)/(1+\alpha)](1+\alpha)}{x+A(x)-t-A(t)+i(B(x)-B(t))}\,dt$$

$$= U_h C U_h^{-1} f_1$$

where

$$h = x+A(x),\ Cf = \int_{-\infty}^{\infty}\frac{f(t)\ \ (1+B_1'(t))}{x-t+iB_1(x)-iB_1(t)}\,dt$$

$$f_1(t) = f(t)\,\frac{1+a}{1+\alpha}\,\frac{1}{B_1'(t)} \qquad B_1 = B\circ h^{-1}\,.$$

Since U_h is bounded on L^2 it would suffice to prove that C is bounded on L^2 for all B such that B′ is small.

We could also try to prove this by expanding

$$-i\pi\,\mathrm{sgn}(L_a)f = \int \frac{f(t)\ (1+a(t))}{(x-t)+A(x)-A(t)}\,dt = \sum (-1)^k \int_{-\infty}^{\infty} \left(\frac{A(x)-A(t)}{x-t}\right)^k \frac{f(1+a)}{x-t}\,dt\,.$$

Observe that the operators are of the form

$$T(f) = \int \Psi\left(\frac{A(x)-A(t)}{x-t}\right)\frac{f(t)}{x-t}\,dt = \int k(x,t)\,f(t)\,dt\,.$$

We will prove Theorem I: Let $\Psi \in C^\infty(C)$ and $A(x)$ such that

$$\left|\frac{A(x)-A(t)}{x-t}\right| \le M \quad\text{and}\quad T(f) = \text{p.v.}\int \Psi\left(\frac{A(x)-A(y)}{x-y}\right)\frac{f(y)}{x-y}\,dy\,.$$

Then the operator T is bounded on $L^2(R)$ (and L^p $1 < p < \infty$). This result will then be extended to R^n and other settings.

We now return to the interpretation of C as the Cauchy integral for the curve $z(t) = t + iA(t)$ where A is Lipschitz

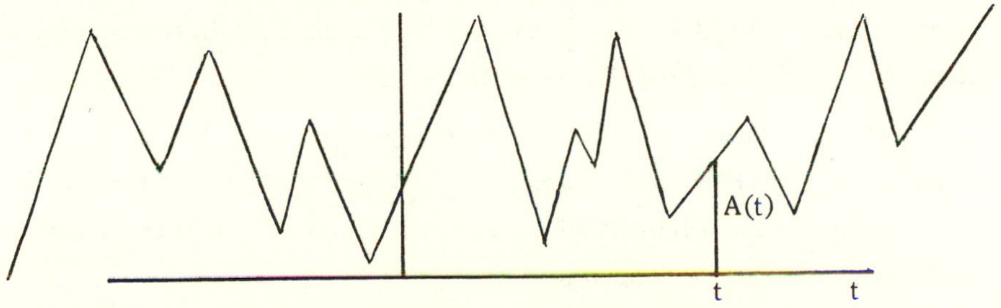

as we can see its boundedness in L^2 is equivalent to the analytic dependence of $C(a)f$ on the curve a. This now is related to the lectures by C. Kenig (to which we shall return later).

Let us consider a more general version of the Cauchy integral.

Let Γ be a rectifiable curve through 0, s be the arc length parameter

$$z'(s) = e^{i\alpha(s)} \text{ i.e., } z(s) = \int_0^s e^{i\alpha(t)}dt\,.$$

The Cauchy integral on Γ is given as:

$$C_\Gamma(f) = \text{p.v.} \int_{-\infty}^{\infty} \frac{f(t)z'(t)}{z(s)-z(t)}\,dt$$

$$= \int_{-\infty}^{\infty} \frac{1}{\frac{z(s)-z(t)}{s-t}} \frac{f(t)z'(t)}{s-t}\,dt$$

$$= \int_{-\infty}^{\infty} \psi\left(\frac{z(s)-z(t)}{s-t}\right) \frac{f_1(t)}{s-t}\,dt$$

if we assume

$$* \qquad 0 < \delta < \left|\frac{z(s)-z(t)}{s-t}\right| \le 1$$

we can take $\psi \in C_0^\infty(\mathbb{C})\psi(z) = \frac{1}{z}$ on $\delta < |z| \le 1$ and obtain the boundedness on L^2 of C_Γ (from Theorem I).

Condition * is the so-called chord arc condition and * for δ small is equivalent to $\alpha \in \text{BMO}$ with $\|\alpha\|_{\text{BMO}}$ small (see [3]). If we think of C as an operator valued functional of α, we will see that B.M.O. is the space of analyticity or holomorphy of C_α.

§2.

All the operators which we encountered previously had the form

$$T(f) = \text{p.v.} \int K(x,y) f(y)\, dy$$

where $|K(x,y)| < \frac{C}{|x-y|}$ $|\partial_x K| + |\partial_y K| \leq \frac{C}{|x-y|^2}$, moreover they were also antisymmetric i.e.,

$$K(x,y) = -K(y,x) .$$

(For example, $K(x,y) = \phi\left(\frac{A(x)-A(y)}{x-y}\right)\frac{1}{x-y}$ $\phi \in C^4$, $A' \in L^\infty$.) Recently G. David and J.-L. Journé found a necessary and sufficient condition for such operators to be bounded on L^2 (or L^p). This condition is simply that $T(1)$ must be of bounded mean oscillation.

We now would like to state certain facts concerning B.M.O. and prove their theorem.

Recall that $b \in BMO(R)$ if

$$\|b\|_* = \left(\sup_I \frac{1}{|I|} \int_I |b - m_I(b)|^2 dx\right)^{1/2} < \infty, \quad \text{where } m_I(b) = \frac{1}{|I|} \int_I b(x)\, dx$$

and I is an interval (or a cube in R^n), and that this norm is equivalent to the following "Carleson" norm

$$\sup_I \left(\frac{1}{|I|} \int_I \int_0^{|I|} |\psi_t * b|^2 \frac{dx\, dt}{t}\right)^{1/2}$$

where $\psi_t = \frac{1}{t}\psi\left(\frac{x}{t}\right)$ $\psi \in C_0^\infty$, $\int \psi dx = 0$ $(\psi \neq 0)$ (see [5]).

A basic reason for the frequent occurrence of functions in B.M.O. is the following simple fact.

PROPOSITION. *If* T *is as above and* T *is bounded on* L^2 *then* T *maps* L^∞ *into* B.M.O.

Proof. Let $b \in L^\infty$ and let I be given. Consider $\overline{I} = 2I$ and write $\tilde{b} = Tb = T(b\chi_{\overline{I}}) + T(b(1-\chi_{\overline{I}})) = \tilde{b}_1 + \tilde{b}_2$.

Clearly

$$\left(\frac{1}{|I|}\int_I |\tilde{b}-m_I(\tilde{b})|^2\right)^{1/2} \le \left(\frac{1}{|I|}\int_I |\tilde{b}_1-m_I(\tilde{b}_1)|^2\right)^{1/2}$$

$$+\left(\frac{1}{|I|}\int_I |\tilde{b}_2-m_I(\tilde{b}_2)|^2\right)^{1/2}.$$

The first term is dominated by

$$2\left(\frac{1}{|I|}\int_I |\tilde{b}_1|^2 dx\right)^{1/2} \le C\left(\frac{1}{|I|}\int_I |b_1|^2\right)^{1/2} \le C\|b\|_\infty .$$

For the second we observe that

$$T(b_2)(x) - T(b_2)(u) = \left|\int [K(x,y)-K(u,y)]b_2(y)dy\right|$$

$$\le \int_{|x-y|>|I|} \frac{|I|}{|x-y|^2}\, dy\|b\|_\infty \le C\|b\|_\infty .$$

Integrating in y we get

$$|T(b_2)(x) - m_I(T(b_2))| < C\|b\|_\infty$$

which shows that second term is bounded by $C\|b\|_\infty$. We have thus shown the necessity of the condition $T(1) \in BMO$. Before stating the theorem precisely we would like to reformulate it somewhat.

Let $\phi \in C_0^\infty(R^1)$ with $\int \phi dx = 1$ and $\psi = \phi(x)$. Let $\phi_t^x(u) = \frac{1}{t}\phi\left(\frac{u-x}{t}\right)$ and similarly for $\psi_t^x(u)$. We claim that under the preceding assumptions on T we have

$$|<T\psi_t^x, \phi_t^y>| \le CP_t(x-y) = C\frac{1}{t}\frac{1}{1+\left(\frac{x-y}{t}\right)^2}.$$

In fact, assume for simplicity, that ϕ is supported in $(-1,1)$. Since $\int \psi du = 0$, if we assume $|x-y| > 3t$

$$|<T\psi_t^x, \phi_t^y>| = \left|\int \phi_t^y(z)\left\{\int [K(z,u) - K(z,x)]\psi_t^x(u)\,du\right\}dz\right|$$

$$\le \int |\phi_t^y(z)| \frac{t}{|y-x|^2}|\psi_t^x(u)|\,du\,dz \le C\frac{t}{|y-x|^2}$$

(where we used the fact that $|y-z| < t$, $|x-u| < t$, $|x-y| > 3t$ and the hypothesis $|\partial_y K(x,y)| \le |x-y|^{-2}$).

If $|x-y| < 3t$ we use the antisymmetry of $k(x,y)$ to write

$$|<T\psi_t^x, \phi_t^y>| = \left|\frac{1}{2}\iint K(z,u)\,(\psi_t^x(u)\,\phi_t^y(z) - \psi_t^x(z)\,\phi_t^y(u))\,dz\,du\right|$$

but $|\psi_t^x(u)\,\phi_t^y(z) - \phi_t^y(u)\,\psi_t^x(z)| \le \frac{|u-z|}{t^3}$ and the fact that $|u-x| < t$, $|u-z| < t$, $|x-y| < 3t$ and $|K(z,u)| \le \frac{1}{|z-u|}$ imply

$$|<T\psi_t^x, \phi_t^y>| \le \frac{C}{t}.$$

Combining these estimates proves our claim.

We can now state

THEOREM (G. David, J. L. Journe) [7]. *Let* $T:\mathcal{D} \to \mathcal{D}'$ *such that for some* $\varepsilon > 0$

$$|<T\psi_t^v, \phi_t^u>| \leq \frac{1}{t} \cdot \frac{1}{1 + |\frac{u-v}{t}|^{1+\varepsilon}} = p_t(u-v)$$

and

$$* \qquad |<T^*\psi_t^v, \phi_t^u>| \leq \frac{1}{t} \frac{1}{1 + |\frac{u-v}{t}|^{1+\varepsilon}} = p_t(u-v) \; .$$

Then the necessary and sufficient condition for T to extend to a bounded operator on L^2 into L^2 is that

$T(1)$ and $T^*(1)$ be in B.M.O.

We would like to make a few comments concerning the conditions *.

We have just seen that if

$$T(f) = \text{p.v.} \int k(x,y) f(y) dy$$

where

$$1° \qquad |k(x,y)| < \frac{1}{|x-y|}$$

$$2° \qquad |k(x,y^1) - k(x,y)| \leq \frac{|y-y^1|^\varepsilon}{|x-y|^{1+\varepsilon}} \quad \text{for } |x-y| > 2|y-y^1|$$

and

$$|k(x^1,y) - k(x,y)| \leq \frac{|x-x^1|^\varepsilon}{|x-y|^{1+\varepsilon}} \quad \text{for } |x-y| > 2|x-x^1|$$

for some $\varepsilon > 0$.

$$3° \qquad |<T(\psi_t^x), \phi_t^y>| + |<T^*(\psi_t^x), \phi_t^y>| \leq \frac{c}{t} \; .$$

(This last condition followed from $k(x,y) = -k(x,y)$ and 1°, 2°) then the conditions $*$ are verified. It can be shown that if $*$ is valid then T can be represented as a limit of integral operators whose kernels satisfy the conditions 1°, 2°, 3°. We will refer to 1°, 2° as standard estimates, and to 3° as the weak cancellation (or boundedness) property. This last condition is independent of 1°, 2°, and can be proved in a variety of ways, as we shall see.

To see how the theorem can be applied to reduce the degree of non-linearity a of a "polynomial" and to obtain estimates consider

$$C_n(a,f) = \int \left(\frac{A(x)-A(y)}{x-y}\right)^n \frac{f(y)}{x-y}\, dy \, .$$

We check by a simple integration by parts that

$$C_n(a,1)(x) = C_{n-1}(a,a)(x) \, .$$

If we make the induction hypothesis that $C_{n-1}(a,f)$ is bounded on L^2 it would follow by the preceding remarks that $C_{n-1}(a,f)$ maps L^∞ to B.M.O. from which we deduce that $C_n(a,1) \in$ B.M.O. and that $\exists C > 0$ such that

$$\|C_n(a,f)\|_{L^2} \leq C^n \|a\|^n_{L^\infty} \|f\|_{L^2} \, .$$

We will see later that this reduction of C_n to C_{n-1} is not a trick but can be accomplished in general to reduce the study of $n+1$ multilinear operators to an n-multilinear.

Proof of the theorem. For simplicity of notation we'll denote

$$P_t f = \phi_t * f \qquad Q_t f = \psi_t * f \, .$$

We start with the fundamental theorem of calculus writing for $f \in C_0^\infty(R)$

$$Tf = \lim_{t\to 0} P_t^2 T P_t^2 f = -\lim_{\varepsilon\to 0} \int_{\varepsilon}^{1/\varepsilon} t\,\frac{\partial}{\partial t}(P_t^2 T P_t^2 f)\,\frac{dt}{t}$$

$$= -\int\left\{\left(t\,\frac{\partial}{\partial t}P_t^2\right) T P_t^2 f + P_t^2 T\left(t\,\frac{\partial}{\partial t}P_t^2\right) f\right\}\frac{dt}{t}\,.$$

We'll study only the first term (since the second is its adjoint).

We start by observing that

$$t\,\frac{\partial}{\partial t}P_t^2 g\;\hat{}(\xi) = t\,\frac{\partial}{\partial t}\hat{\phi}^2(t\xi)\hat{g}(\xi) = 2t\xi\,\hat{\phi}(t\xi)\,\hat{\phi}'(t\xi)\,\hat{g}(\xi)$$

$$= 2\hat{\psi}(t\xi)\,\hat{\psi}_1(t\xi)\,\hat{g}(\xi) = (Q_{1,t}Q_t g)\hat{}$$

where $\psi_1(x) = x\phi(x)$ and $Q_{1,t}g = \psi_{1,t} * g$. This permits us to rewrite the first term as

$$-2\int_0^{\infty} Q_{1,t}L_t P_t f\,\frac{dt}{t}$$

where

$$L_t g = Q_t T P_t g = \int \ell_t(x,y)\,g(y)\,dy$$

and

$$\ell_t(x,y) = \langle T^*\psi_t^x, \phi_t^y\rangle\,.$$

Also observe that $L_t(1) = \int \ell_t(x,y)\,dy = \langle T^*\psi_t^x, 1\rangle = \psi_t * T(1) = \psi_t * b$ (here all integrals converge absolutely since $|\ell_t(x,y)| \le p_t(x-y)$).

Before proceeding we remark that since $|\ell_t(x,y)| \le p_t(x-y)$ we can think of L_t as an averaging operator on the scale t, and of $P_t f$ as being essentially constant on that scale i.e., $L_t(P_t f)$ would look like $L_t(1)\,P_t(f)$. (This would be exact if P_t is a conditional expectation and

L_t had the correct measurability.) We are thus led to write

$$\int_0^\infty Q_{1,t} L_t P_t f \frac{dt}{t} = \int_0^\infty Q_{1,t}\{L_t(1) P_t(f)\} \frac{dt}{t}$$

$$+ \int_0^\infty Q_{1,t}\{L_t P_t f - L_t(1) P_t(f)\} \frac{dt}{t}$$

$$= \int_0^\infty Q_{1,t}\{\psi_t * b(x) \cdot \phi_t * f\} \frac{dt}{t} + E(f) .$$

We consider $E(f)$ as an error term and prove that

$$\int |E(f)|^2 dx \leq e \int |f|^2 dx$$

(this is valid using only $*$). In fact, let $g \in L^2$

$$|\langle g, E(f) \rangle| = \left| \iint Q_{1,t}(g) \{(L_t P_t) f - L_t(1) P_t(f)\} \frac{dx\,dt}{t} \right|$$

$$\leq \left(\iint |Q_{1,t}(g)|^2 \frac{dx\,dt}{t} \right)^{1/2} \left(\iint \left| \int \ell_t(x,y)(P_t(f)(y) \right.\right.$$

$$\left.\left. - P_t(f)(x))\, dy \right|^2 \frac{dx\,dt}{t} \right)^{1/2} .$$

A simple application of Plancherel's theorem permits us to estimate the first term. For the second we use $*$ and Minkowsky's inequality to dominate it by

$$\left(\iiint P_t(x-y)|P_t(f)(y)-P_t(f)(x)|^2 \frac{dx\,dy\,dt}{t}\right)^{1/2}$$

$$=\left(\iiint P_t(u)|P_t(f)(y)-P_t(f)(y+u)|^2 \frac{du\,dy\,dt}{t}\right)^{1/2}$$

$$=\left(\iiint P_t(u)|\hat{\phi}(t\xi)|^2|e^{iu\xi}-1|^2 \hat{f}(\xi)^2 \frac{du\,d\xi\,dt}{t}\right)^{1/2}$$

$$\leq\left(\iiint P_t(u)|\tfrac{u}{t}|^\delta|\hat{\phi}(t\xi)|^2|t\xi|^\delta|\hat{f}(\xi)|^2 \frac{du\,d\xi\,dt}{t}\right)^{1/2}$$

$$=\left(\int p(u)\,u^\delta\right)^{1/2}\left(\int_0^\infty |\phi(t)|^2 t^\delta \frac{dt}{t}\right)^{1/2}\left(\int |\hat{f}(\xi)|^2 d\xi\right)^{1/2}.$$

As for the first term, we proceed as above and are led to estimate

$$\left(\int |\psi_t * b(x)|^2 |\phi_t * f|^2 \frac{dx\,dt}{t}\right)^{1/2}.$$

This is dominated by $\|f\|_2$ if and only if $(\psi_t * b(x))^2 \frac{dx\,dt}{t}$ is a Carleson measure i.e., b is in B.M.O.

In fact, recall that $d\nu(x,t)$ on $\mathbf{R}^2_+$ is a Carleson measure if for each interval I

$$\nu(\hat{I}) \leq C|I|$$

$$\hat{I} = \{(x,t) \in \mathbf{R}^2_+ : (x-t, x+t) \leq I\}.$$

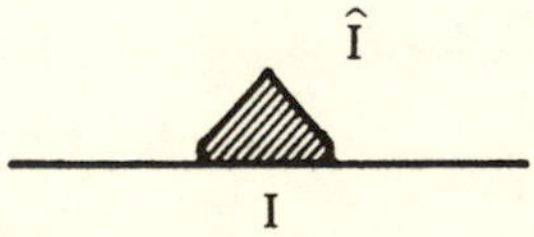

Carleson's lemma states that

$$\int F^p(x,t)\,d\nu(x,t) \le c \int F^*(x)^p\,dx ,$$

where $F^*(x) = \sup_{|y|<t} |F(x-y,t)|$.

This is proved by observing that if $O_\lambda = \{x : F^*(x) > \lambda\}$ then outside $\hat{O}_\lambda$ $F(x,t) < \lambda$, thus

$$\nu(\{F > \lambda\}) \le \nu(\hat{O}_\lambda) \le c\,|O_\lambda|$$

and the L^p result follows by integration.

We now would like to make a few comments concerning the principal term:

$$\int_0^\infty Q_t((\psi_t * b)(\phi_t * f))\,\frac{dt}{t} = \pi(b; f) .$$

This is essentially equivalent to a simpler looking version

$$\int_0^\infty (\psi_t * b)(\phi_t * f)\,\frac{dt}{t}$$

which is the basic bilinear operation in (b, f) commuting with translations and dilations i.e.:

If $b^\tau(x) = b(x-\tau)$ and $b_\lambda(x) = b(\lambda x)$ then

$$\pi(b^\tau_\lambda; f^\tau_\lambda) = \pi(b; f)^\tau_\lambda .$$

Since all the problems we will be considering are invariant under such transformations, it is not surprising that the bilinear operations which arise look like π.

One cannot conclude this section without mentioning that π is also the principal term in the Bony linearization formula and is called para-product by him i.e., given $F \in C^\infty$, $f \in \Lambda^\alpha$, $0 < \alpha < 1/2$ one has the remarkable formula (Bony)

$$F(f) = \pi(f; F'(f)) + e$$

where

$$e \in \Lambda^{2\alpha} \; (f \in \Lambda^\alpha \;\text{ iff }\; |f(x) - f(y)| < c|x-y|^\alpha) \, .$$

The proof of this linearization formula is the "same" as for the T(1) theorem and is achieved as follows.

$$F(f) = \lim_{t \to 0} F(f * \phi_t) = -\int_0^\infty t \frac{\partial}{\partial t} F(f * \phi_t) \frac{dt}{t}$$

$$= -\int_0^\infty F'(f * \phi_t) f * \psi_t \frac{dt}{t} \quad \left(\text{here } \psi_t = t \frac{\partial}{\partial t} \phi_t\right)$$

$$= -\int_0^\infty F'(f) * \phi_t f * \psi_t \frac{dt}{t} + e = \pi(f; F'(f)) + e$$

and it is quite easy to verify the $e \in \Lambda^{2\alpha}$ if $f \in \Lambda^\alpha$. (This method should be explored further in connection with non-linear estimates [15].)

§3. *Multilinear Fourier Analysis*

We would like to understand various explicit representations for "homogeneous polynomials" i.e. multilinear functionals. For simplicity and to avoid technicalities we consider Λ_k a multilinear operator defined on trigonometric polynomials on T' with values in periodic continuous functions i.e.

$$\Lambda_k(t_1, \cdots, t_k)(\theta) \text{ is multilinear in the } t_i .$$

We will assume that Λ_k commutes with simultaneous translation of t_i i.e.

$$* \qquad \Lambda_k(t_1^h, t_2^h, \cdots, t_k^h)(\theta) = \Lambda_k(t_1 \cdots t_k)(\theta - h) ,$$

where

$$t_i^h(\psi) = t_i(\psi - h) .$$

Clearly if we take $t_i(\theta) = e^{ik_i\theta}$, $t_i^h(\theta) = e^{-ihk_i} t_i(\theta)$ we find using the linearity in each argument that

$$e^{-i\, h(\sum_1^k k_i)} \Lambda_k(e^{ik_1\theta}, \cdots, e^{ik_k\theta})(\theta) = \Lambda_k(e^{ik_1\theta} \cdots e^{ik_k\theta})(\theta - h)$$

taking $h = \theta$ we get

$$\Lambda_k(e^{ik_1\psi}, \cdots, e^{ik_k\psi})(\theta) = \lambda(k_1 \cdots k_k) e^{i\theta(k_1 + k_2 \cdots + k_k)} .$$

More generally if $t_i(\theta) = \Sigma\, \hat{t}_i(j) e^{ij\theta}$ we find

$$\Lambda(t_1 t_2 \cdots t_k)(\theta) = \sum_{j_k} \cdots \sum_{j_1} \lambda(j_1, \cdots, j_k) \hat{t}_1(j_1) \hat{t}_2(j_2) \hat{t}_k(j_k) e^{i\theta(\sum_1^k j_i)}$$

where

$$\lambda(j_1 \cdots j_k) = \Lambda(e^{ij_1\theta}, \cdots, e^{ij_k\theta})(0) .$$

Before continuing let us observe that for $k = 1$, $e^{ik\theta}$ form a basis diagonalizing all linear operators commuting with translation, in fact, all such operators are normal. For $k \geq 2$ it isn't as simple.

Consider for a moment a finite dimensional vector space V and a bilinear transformation $\Lambda : V \times V \to V$. We can ask when is it true that there exists a basis in V $e_1 \cdots e_N$ a linear operator $\widetilde{\Lambda}$ defined on $V \otimes V$ diagonalized by $e_i \otimes e_j$

$$\widetilde{\Lambda}\,(e_i \otimes e_j) = \lambda_{i,j}\; e_i \otimes e_j$$

such that

$$\Lambda(e_i, e_j) = \lambda_{i,j}\; e_{\pi(i,j)} \quad \text{for some map } \pi(i,j) \text{ of}$$

$$(1, \cdots, N)^2 \to (1, \cdots, N)$$

i.e. we want the diagram to commute

$$\begin{array}{ccc} \widetilde{\Lambda} \quad V \otimes V & \longrightarrow & V \otimes V \\ \uparrow \nu & & \downarrow \pi \\ \Lambda : V \times V & \longrightarrow & V \end{array}$$

where

$$\nu(e_i, e_j) = e_i \otimes e_j \quad \pi(e_i \otimes e_j) = e_{\pi(i,j)}\,.$$

This is clearly the situation for periodic functions. It would be interesting to understand the bilinear operators admitting such a diagonizable lift to the tensor product.

This observation indicates that the hypothesis concerning commutation with translations, imposes severe restrictions on the nature of a multilinear operation.

We now return to the line on which we realize multilinear operations as

$$\Lambda_k(f_1 \cdots f_k)(x) = \iint_{R^k} e^{ix(\xi_1 + \cdots \xi_k)} \lambda_k(\xi_1 \cdots \xi_k)\hat{f}(\xi_1)\hat{f}(\xi_2) \cdots \hat{f}(\xi_k)\, d\xi_1 d\xi_k$$

and

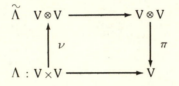

$$\Lambda_k(e^{i\xi_1 x}, \dots, e^{i\xi_k x})(x) = e^{ix(\xi_1+\cdots\xi_k)}\lambda_k(\xi_1 \cdots \lambda_k)\,.$$

(Note that this realization is only valid for multilinear operations verifying some mild continuity conditions.) This realization permits us for example to show that the study of $\int_0^\infty \psi_t * b\, \phi_t * f \frac{dt}{t}$ can easily be converted to the study of

$$\int_0^\infty \psi_t^1 * (\psi_t * b \cdot \phi_t * f)\frac{dt}{t}$$

which we considered previously in the proof of Theorem [II].

In fact

$$\int_0^\infty \psi_t * b\, \phi_t * f \frac{dt}{t} = \int e^{ix(\xi_1+\xi_2)}\left(\int_0^\infty \hat{\psi}(t\xi_1)\hat{\phi}(t\xi_2)\frac{dt}{t}\right)\hat{b}(\xi_1)\hat{f}(\xi_2)d\xi_1 d\xi_2$$

$$= \int e^{ix(\xi_1+\xi_2)}\lambda(\xi_1,\xi_2)\hat{b}(\xi_1)\hat{f}(\xi_2)d\xi_1\, d\xi_2$$

if we take $\hat{\psi}$ with support in $1 < |\xi| < 2$ and $\hat{\phi}$ supported in $|\xi| < \frac{1}{10}$ we have

$$\hat{\psi}(\xi_1 t)\hat{\phi}(t\xi_2) = \hat{\psi}^1(t(\xi_1+\xi_2))\hat{\psi}(t\xi_1)\hat{\phi}(t\xi_2)$$

where $\psi^1(\xi)$ is any function equal to 1 on $1.9 < |\xi| < 2.1$. In this case we see that the two expressions are equal. This sort of analysis comparing frequencies and interaction of various functions can be carried out permitting one to understand the structure of some multilinear operations. We are thus led to consider the general question of studying multilinear multipliers. We state two known results.

PROPOSITION 1. *Let* $|\lambda(\xi_1 \cdots \xi_k)| < \dfrac{1}{(\max|\xi_i|)^{\frac{k-1}{2}}}$ *then*

$$\Lambda(f_1 \cdots f_k) = \int e^{ix\cdot\sum_1^k \xi_i} \lambda(\xi_1 \cdots \xi_k)\hat{f}(\xi_1)\cdots\hat{f}(\xi_k)d\xi_1\cdots d\xi_k$$

maps $L^2 \times \cdots \times L^2 \to L^2$ *continuously.*

This is an elementary result which we leave as an exercise. It would be desirable to obtain more subtle results for L^2.

THEOREM [5]. *Let* $\lambda(\xi_1 \cdots \xi_k)$ *be such that*

$$\left|\frac{\partial^\alpha \lambda}{\partial \xi^\alpha}\right| \le \frac{C_\alpha}{|\xi|^{|\alpha|}} \quad |\alpha| \le \left[\frac{n}{2}\right] + 1 \qquad \alpha = (\alpha_1 \cdots \alpha_k) \; |\alpha| = \sum \alpha_i \, ,$$

then

$$\Lambda(f_1 \cdots f_k) : L^{p_1} \times L^{p_2} \cdots \times L^{p_k} \to L^q$$

$$\frac{1}{q} = \sum_1^k \frac{1}{p_i} \quad \infty \ge p_i > 1 \quad \infty > p_1 \, .$$

More generally one can take operators of the form

$$\sigma(f_1 \cdots f_k) = \int e^{ix(\Sigma\xi_i)} \sigma(x, \xi_1 \cdots \xi_k)\hat{f}(\xi_1)\cdots\hat{f}(\xi_k)\, d\xi_1 \cdot d\xi_k$$

with $\sigma(x, \xi)$ such that

$$|\partial_\xi^\alpha \partial_x^\beta \sigma(x, \xi)| \le \frac{C_{\alpha_1\beta}}{1 + |\xi|^{|\alpha|}} \quad \text{for all } \alpha, \beta \text{ sufficiently large}$$

and the same conclusion is valid.

We now prove the result by induction on k; for simplicity we consider only $k = 2$.

We take $f \in L^p$, $f_2 = a \in L^\infty$ and write

$$\Lambda(f,a)(x) = \iint e^{ix(\xi_1+\xi_2)} \lambda(\xi_1,\xi_2)\hat{a}(\xi_1)\hat{f}(\xi_2)d\xi_1 d\xi_2$$

$$= \left[\int k(x-t, x-y)a(t)\right] f(y)dy$$

where $k(x_1,x_2) = \hat{\lambda}(x_1 x_2)$ satisfies $|k(x_1,x_2)| \le \frac{C}{x_1^2 + x_2^2}$, $|\nabla k| \le \frac{C}{|x|^3}$ thus $k(x,y) = \int k(x-t, x-y)a(t)dt$ verifies conditions $1°, 2°$, and $3°$ of the T(1) theorem: thus to verify the boundedness in L^2 (for a fixed) it suffices to calculate $T(1)$ and $T^*(1)$. These can easily be calculated, in fact

$$T(1) = \int e^{ix\xi_1} \lambda(\xi_1, 0)\hat{a}(\xi_1)d\xi_1$$

which is a linear Calderon Zygmund operator applied to the bounded function a. Thus it is in BMO.

Since T^*, (in f, for a fixed) is given by the symbol $\lambda^*(\xi_1,\xi_2) = \lambda(\xi_1, -\xi_1-\xi_2)$, verifying the same hypothesis the result follows for $T^*(1)$. Property $3°$ can be verified once we observe that the same induction shows that

$$\|T(e^{ix\xi})\|_{BMO} \le C .$$

and that this condition implies

$$\int_{|I|} |T(\phi_t)|\,dx \le c \quad \text{on any interval of length } t .$$

In fact

$$\phi_t(x) = \int e^{ix\xi}\,\hat{\phi}(\xi t)\,d\xi \Longrightarrow \|T(\phi_t)(x)\|_{BMO} \leq \quad |\hat{\phi}(\xi t)|dt \leq \frac{c}{t} \ ,$$

now assume $\phi_t(y)$ is supported in I.

$$|T(\phi_t)(x)| \leq \int \frac{|\phi_t(y)|}{|x-y|}\,dy \leq \frac{c}{t}$$

if $|x| > 2|t|$ thus

$$\int_{\bar{I}} |T\,\phi_t(x)|\,dx \leq c \quad \text{if} \quad \text{dist}(\bar{I},I) > t\ .$$

Since $T(\phi_t)$ is in B.M.O. it follows that

$$\frac{c}{t} \geq \frac{1}{|I|} \int_I |T(\phi_t)(x) - m_{\bar{I}}(T(\phi_t))| \Longrightarrow \int_I |T\phi_t|\,dx \leq c\ ,$$

thus we obtain

$$\int |T(\phi_t)\cdot\psi_t| \leq \frac{c}{t}\ ,$$

provided $\psi_t\,\phi_t$ are supported in I.

Again we conclude with a few general observations. As you all know the Fourier transform in a basic tool in linear P.D.E. permitting the reduction for example, of initial value problems to simple O.D.E. We claim that this can be achieved also for non-linear partial differential equations. Before illustrating on an example recall also that a common method for solving differential equations (linear and non-linear) is to plug into the equation a formal power series whose coefficients are determined recursively using the differential equation. We now sketch how these two ideas can be combined to "formally" solve the Korteweg-de Vries equation (KdV)

$$\frac{\partial u}{\partial t} + \frac{\partial^3 u}{\partial x^3} = -6\, u \frac{\partial u}{\partial x} \qquad x \in R,\ t > 0\ .$$

The set of solutions is clearly invariant under x translations and can be parametrized by functions of x (for example by specifying $u(x,0) = f(x)$). Let us write u as a power series

$$u(x,t) = \sum \Lambda_k(\nu) = \sum u_k(x,t)$$

where

$$u_k(x,t) = \int e^{ix(\sum_1^k \xi_i)} \sigma_k(\xi,t)\,\hat{\nu}(\xi_1)\cdots\hat{\nu}(\xi_k)\,d\xi_1\cdots d\xi_k$$

(here ν is some unspecified functional parameter) plugging into the equation we obtain

$$\frac{\partial u_k}{\partial t} + \frac{\partial^3 u_k}{\partial x^3} = -3\frac{\partial}{\partial x}\sum_{j=1}^{n-1} u_j u_{k-j}$$

we find first that $\sigma_k(\xi, t) = e^{it\sum_1^k \xi_j^3}\sigma_k(\xi)$, and that

$$\sigma_k(\xi)\left(\sum_1^k \xi_i^3 - \left(\sum_1^k \xi_i\right)^3\right) = -3\left(\sum_1^k \xi_i\right)\sum_{j=1}^{k-1}\sigma_j(\xi_1\cdots\xi_j)\,\sigma_{k-j}(\xi_{j+1}\cdots\xi_k)\ .$$

(Strictly speaking σ_k is determined only if we assume that it is invariant under permutations of ξ_k, so that the equality should be true only after symmetrization.)

It can be shown (not so simply) that

$$\sigma_k = \frac{\xi_1 + \cdots + \xi_k}{(\xi_1+\xi_2)(\xi_2+\xi_3)\cdots(\xi_{k-1}+\xi_k)}$$

solves the recurrence above.

Thus

$$u_k(x,t) = \frac{1}{i}\frac{\partial}{\partial x}\int \frac{1}{\xi_1+\xi_2}\,\frac{1}{\xi_2+\xi_3}\cdots\frac{1}{\xi_{k-1}+\xi_k}\prod_{i=1}^{k}\hat{\nu}_{x,t}(\xi_i)\,d\xi_1\cdots d\xi_k$$

$$\hat{\nu}_{x,t} = e^{ix\xi + t\xi^3}\,\hat{\nu}(\xi)\,.$$

If we let $C_\nu(f)(\xi) = \text{p.v.}\int \frac{1}{\xi+\eta}\,\hat{\nu}_{x,t}(\eta) f(\eta)\,d\eta$, we get

$$u_k(x,t) = \frac{\partial}{i\partial x}\int \hat{\nu}_{x,t}(\eta)\,C_\nu^{k-1}(1)(\eta)\,d\eta$$

and

$$u(x,t) = \frac{\partial}{\partial x}\int \nu(I-C_\nu)^{-1}(1)\,d\eta = \frac{\partial}{\partial x}\int \nu\,\psi\,d\eta$$

where ψ satisfies the integral equation

$$\psi = 1 + \int \frac{1}{\eta+\xi}\,\nu(\eta)\,\psi\,d\eta\,.$$

This set of remarkable formulas constitute the so-called method of "inverse scattering" for solving K deV see [9], and was shown to us by B. Dahlberg.

§4. *Functional calculus, resolvent expansions*

We start by considering a small perturbation of $-\Delta$

$$L = -\Delta + \sum_{ij} b_{ij}(x) \frac{\partial^2}{\partial x_i \partial x_j} \qquad \|b\|_\infty < \varepsilon_0 .$$

The $b_{ij}(x)$ are complex valued functions. We would like to prove L^2 estimates for such functions as

$$e^{-tL}, e^{itL}, e^{it\sqrt{L}}, e^{-t\sqrt{L}} .$$

These provide us with estimates for solutions of the initial value or Dirichlet problem for

$$\frac{\partial}{\partial t} u = -Lu , \quad \frac{1}{i} \frac{\partial}{\partial t} u = Lu , \quad \frac{\partial^2}{\partial t^2} u = -Lu$$

$$\left(\frac{\partial^2}{\partial t^2} - L\right) u = 0$$

or more generally for $F(L)$ where F is a bounded holomorphic function in a sector $|\arg z| < \delta$ containing the spectrum of L (this can be shown to be true if ε_0 is sufficiently small).

We define formally

$$F(L) = \frac{1}{2\pi i} \int_\Gamma (\zeta - L)^{-1} F(\zeta)\, d\zeta$$

where Γ is the boundary of the sector.

We let

$$R_\zeta^{i,j} = +(\zeta+\Delta)^{-1} \frac{\partial^2}{\partial x_i \partial x_j} \qquad \hat{R} = \frac{-\xi_i \xi_j}{\zeta - |\xi|^2}$$

$$\zeta - L = \left(\zeta + \Delta - \sum b_{ij} \frac{\partial^2}{\partial x_i \partial x_j}\right) = (I - BR)(\zeta + \Delta)$$

$$(\zeta - L)^{-1} = (\zeta+\Delta)^{-1}(I-BR)^{-1}$$

$$= \sum (\zeta+\Delta)^{-1}(BR)^k .$$

We thus find

$$F(L) = \sum_0^\infty \int_\Gamma (\zeta+\Delta)^{-1}(BR_\zeta)^k F(\zeta)\,d\zeta = \sum_0^\infty \Lambda_k(B) .$$

To better understand such expressions we consider the first two terms

$$(\Lambda_0(B)f)\hat{} = \left(\int_\Gamma \frac{1}{\zeta+\Delta} F(\zeta)\,d\zeta\right)\hat{} = \frac{1}{2\pi i} \int_\Gamma \frac{1}{\zeta - |\zeta|^2} F(\zeta)\,d\zeta\,\hat{f}(\zeta)$$

$$= F(|\zeta|^2)\hat{f}(\zeta) = m(\zeta)\hat{f}(\zeta) .$$

Since F is assumed to be bounded holomorphic in a wedge containing the real axis we obtain (using Cauchy's theorem) that $|(m^{(k)}(\zeta)| \le \frac{c_k}{|\zeta|^k}$ and thus Λ_0 is a Calderon-Zygmund kernel operator mapping $L^2 \to L^2$, $L^\infty \to BMO$. We now consider $\Lambda_1(B)$ and to simplify the exposition we will assume that $\Gamma = iR$ (i.e., F is bounded holomorphic in the half plane $\mathrm{Re}\, z > 0$). This is not going to affect the argument.

$$\Lambda_1(B) = \int_{-\infty}^{\infty} \frac{t}{it+\Delta} B \frac{\partial^2}{\partial x_i \partial x_j} \frac{1}{it+\Delta} F(it) \frac{dt}{t}$$

and consider separately $t > 0$, $t < 0$. Change variables $t = \pm \frac{1}{s^2}$, to obtain 2 terms of the form

$$\int_0^{\infty} \frac{1}{i+s^2\Delta} B\, s^2 D_i D_j \frac{1}{i+s^2\Delta} \mu(s) \frac{ds}{s} \qquad \mu(s) = F(1/s^2)\,.$$

Now write

$$\frac{s^2 D_i D_j}{i+s^2\Delta} = \frac{s^2\Delta}{i+s^2\Delta} \frac{D_i D_j}{\Delta} = \frac{D_i D_j}{\Delta} - \frac{i}{i+s^2\Delta} \frac{D_i D_j}{\Delta}\,.$$

This gives $\int_0^{\infty} \frac{1}{1+s^2\Delta} B \frac{D_i D_j}{\Delta} \mu(s) \frac{ds}{s}$ which we recombine with the other term corresponding to $t < 0$ to get

$$\Lambda_0(B)\left(B \frac{D_i D_j}{\Delta} f\right)\,.$$

The second term is of the form

$$\left\{\int_0^{\infty} \frac{1}{1+s^2\Delta} B \frac{1}{1+s^2\Delta} \mu(j) \frac{ds}{s}\right\} \frac{D_i D_j}{\Delta}\,.$$

The operator in curly brackets has a kernel satisfying $*$. Thus, to check boundedness in L^2 we need to calculate $T(1)$, $T*(1)$. This again reduces to the linear case when the terms are recombined.

The higher order case is much more complicated since terms appearing lack regularity and need to be replaced by more regular terms complicating the induction. The main idea is the same, see [7], leading to L^2 estimates for $F(L)$ for ε_0 sufficiently small.

In general a functional calculus is obtained by considering functions $F(L)$ where F is holomorphic in a neighborhood of the spectrum of L, and using the Cauchy formula. This becomes impossible for an operator like $\frac{\partial}{\partial z}$, $z \in \mathbb{C}^1$, $z = x_1 + ix_2$ since

$$\frac{\partial}{\partial z} e^{i\xi_1 x_1 + \xi_2 x_2} = \frac{i}{2}(\xi_1 - i\xi_2) e^{i\xi \cdot x}$$

and the spectrum is the whole complex plane. We are thus led to consider for $\phi \in C_0^\infty(\mathbb{C}^1)$ expressions of the form

$$\phi(L) = \frac{1}{2\pi i} \int_{\mathbb{C}} \frac{\partial \phi}{\partial \bar{\lambda}} \frac{1}{\lambda - L}\, d\lambda \wedge d\bar{\lambda} = -\frac{1}{2\pi i} \int_{\mathbb{C}} \phi(\lambda) \frac{\partial}{\partial \bar{\lambda}} \left(\frac{1}{\lambda - L}\right) d\lambda \wedge d\bar{\lambda} .$$

These formal expressions need to be given sense and one should prove that $\phi(L)\psi(L) = (\phi\psi)(L)$. In the case $L_0 = \frac{2}{i} \frac{\partial}{\partial z}$ such an expression is easily justified using the Fourier transform. Since

$$\frac{1}{2\pi i}\left[\left(\int_{\mathbb{C}} \frac{\partial \phi}{\partial \bar{\lambda}} \frac{1}{\lambda - L}\, d\lambda \wedge d\bar{\lambda}\right) f\right]^{\wedge}(\xi) = \frac{1}{2\pi i} \int_{\mathbb{C}} \frac{\partial \phi(\lambda)}{\partial \bar{\lambda}} \frac{1}{\lambda - \bar{\xi}}\, d\lambda \wedge d\bar{\lambda}\, \hat{f}(\xi)$$

$$= \phi(\bar{\xi}) \hat{f}(\xi)$$

(here we used the fact that $\frac{1}{2\pi i} \partial_{\bar{z}} \frac{1}{z} = \delta(z)$, where δ is the Dirac mass at 0).

We see that

$$\phi(L_0) = \int e^{ix\xi} \phi(\bar{\xi}) \hat{f}(\xi)\, d\xi = \int \phi^{\vee}(\bar{x} - \bar{y}) f(y)\, dy .$$

If we now consider $L = \frac{1}{1+a} \frac{\partial}{\partial z}$ where $a \in L^\infty$, $\|a\| \le \delta_0$ we would like to understand the nature of

$$\phi(L) = \frac{1}{2\pi i} \int \phi(\lambda) \frac{\partial}{\partial \bar{\lambda}} \frac{1}{L-\lambda} \, d\lambda \wedge d\bar{\lambda}$$

for functions ϕ homogeneous of $d^0 0$. For example, we would like $\frac{\bar{L}}{L}$ (this will give us $\bar{L} = \frac{\bar{L}}{L} \cdot L$ which will turn out to be very important).

We can, in this case, calculate explicitly

$$(L-\lambda)^{-1}\text{, in fact, } L^{-1}f = \int \frac{i}{\pi(\bar{z}-\bar{w})} f(1+a) f dv(w) .$$

We now observe that if $h(z)$ is such

$$\frac{\partial}{\partial z} h = 1+a \quad \text{and} \quad h(z) - z \in L^{\infty}$$

then

$$L\chi_{\zeta} = L(e^{i(h(z)\zeta + \bar{z}\bar{\zeta})}) = \zeta \chi_{\zeta}$$

from which we find

$$(L-\zeta) f = \chi L \frac{1}{\chi} f \quad \text{and} \quad (L-\zeta)^{-1} = \chi L^{-1} \frac{1}{\chi} f ,$$

or more precisely

$$(L-\zeta)^{-1} f = \frac{i}{\pi} \int \frac{1}{\bar{z}-\bar{w}} e^{i\{(h(z)-h(w))\zeta + (\bar{z}-\bar{w})\bar{\zeta}\}} \frac{\partial h}{\partial w} f(w) dw$$

$$\frac{\partial}{\partial \bar{\zeta}} (L-\zeta)^{-1} f = \int e^{i\{(h(z)-h(w))\zeta + (\bar{z}-\bar{w})\bar{f}\}} \frac{\partial h}{\partial w} f(w) dw$$

and

$$\phi(L)f = \int\left[\int \phi(\zeta)\, e^{i\{(h(z)-h(w))\zeta+(\bar{z}-\bar{w})\bar{\zeta}\}} d\tau\right] \frac{\partial h}{\partial w}\, f(w)\,dw$$

$$= \int k(h(z)-h(w), \bar{z}-\bar{w})\, \frac{\partial h}{\partial w}\, f(w)\,dw$$

where

$$k(u,v) = \int_{\mathbf{C}} \phi(\zeta)\, e^{iu\zeta+v\bar{\zeta}}\, d\zeta$$

is the Laplace transform of ϕ. This gives us an explicit kernel realization of the calculus.

It can be shown independently that

$$\frac{\bar{L}}{L} f = -\frac{1}{\pi}\int \frac{f(w)}{(\bar{z}-\bar{w})^2}\,dw - \frac{\partial h}{\partial \bar{z}} f\,.$$

Since $\frac{\partial h}{\partial z} = 1+a$, for this to be bounded we must have $\frac{\partial h}{\partial \bar{z}} \in L^\infty$ i.e., $a * \frac{1}{z^2} \in L^\infty$. In other words, in order to have a functional calculus in $\frac{1}{1+a}\frac{\partial}{\partial z}$ (consisting of bounded operators in $L^2(\mathbf{C}^1)$) it is necessary to assume that $a \in L^\infty$ and the 2-dimensional Hilbert transform (or Ahlfors Beurling transform) of a is also bounded. This version of H^∞ is quite interesting and should be better understood.

§5.

Until now we have only shown that small perturbations of certain operators are bounded, we would like to describe an extension method due to G. David.

We proved earlier that the commutators

$$C_k(a,f) = \int \left(\frac{A(x)-A(y)}{x-y}\right)^k \frac{f(y)}{x-y}\,dy \qquad A'(x) = a(x)$$

satisfy

$$\|C_k(a,f)\|_1 \leq C^k \|a\|_\infty^k \|f\|_2 \,.$$

(This is A. P. Calderon's theorem [1].)

If we recombine these terms in a series we obtain, for example, that

$$\int \frac{f(y)}{x-y+iA(x)-iA(y)}\,dy$$

is a bounded operator on L^2 if $\|A'\|_\infty < \frac{1}{c}$ and that

$$\left\| \int \frac{e^{i\frac{A(x)-A(y)}{x-y}\xi}}{x-y} f(y)\,dy \right\|_{L^2} \leq e^{c|\xi|\,\|A'\|_\infty} \|f\|_2 \,.$$

It was proved in [4] by a careful analysis of the functional calculus interpretation of the Cauchy integral that in fact,

$$\|C_k(a,f)\|_{L^2} \leq c(1+k^4)\,\|a\|_\infty^k \|f\|_2 \,.$$

This more precise result implied a much stronger statement.

THEOREM 1 [4]. *Let* $\phi \in C^{11}(R)$. *A real valued* $\|A'\|_\infty < \infty$ *then*

$$\left\| \int \phi\left(\frac{A(x)-A(y)}{x-y}\right) \frac{f(y)}{x-y}\,dy \right\|_{L^2} \leq c\|f\|_2 \,.$$

It is our purpose to describe a direct real variable method to obtain this result. The main idea, due to G. David, is that on each interval a Lipschitz function with a given norm is, in fact, equal (modulo an additive constant) on a substantial subset to a Lipschitz function with a fraction of the norm. By iteration this permits estimates for large norms in terms of smaller ones.

To get results like Theorem 1 it is enough to show that for A real

$$\|T_\xi(f)\|_2 = \left\| \int e^{i\xi \frac{A(x)-A(y)}{x-y}} \frac{f(y)}{x-y}\, dy \right\|_2 \le C(\|\xi a\|_\infty + 1)^N \|f\|_2$$

for some $N > 0$, $C > 0$.

In fact, this will imply that for $\phi(x)$ such that $\int |\hat{\phi}(\xi)|(1+|\xi|^N)\,d\xi < \infty$ we have

$$\left\| \int \phi\left(\frac{A(x)-A(y)}{x-y}\right) \frac{f(y)}{x-y}\, dy \right\|_2 = \left\| \int \hat{\phi}(\xi) \left\{ \int e^{i\xi \frac{A(x)-A(y)}{x-y}} \frac{f(y)}{x-y}\, dy \right\} d\xi \right\|_2$$

$$\le \int |\hat{\phi}(\xi)|\, \|T_\xi f\|_2\, d\xi \le c\|f\|_2 .$$

To show that the norm of

$$T_A(f) = \int e^{i \frac{A(x)-A(y)}{x-y}} \frac{f(y)}{x-y}\, dy$$

grows like $\|A'\|_\infty^N$ we start with a number of easy observations. The kernel of T satisfies the estimates (*) and is antisymmetric, thus it suffices to estimate the B.M.O. norm of $T(1)$.

The operator norm of T is unchanged if we add a constant to a, or replace a by $-a$ (this in fact will imply that the result is true for $a \in \mathrm{BMO}$).

We also need a lemma indicating that the space B.M.O. can be characterized by a weak local estimate.

LEMMA (Stromberg). *Let* b *be measurable and assume that there exist* $\alpha > 0$, $N > 0$ *such that for each interval* I *there is a constant* C(I) *(depending continuously on* I *) for which*

$$|x \in I : |b(x) - C(I)| \leq \alpha| > \frac{1}{N} |I| .$$

Then $b \in \mathrm{BMO}$ *and* $\|b\|_{\mathrm{BMO}} \leq c_N \alpha$.

The main idea to estimate the BMO norm of T(1) is to replace inside each interval I, T_A by an operator T_{A_I} where $A_I = A$ on a large fraction of I and A_I' has a smaller Lipschitz norm, and then compare $T_{A_I}(1)$ to $T_A(1)$. This is achieved via the following lemma, the first of which is the rising sun lemma (or the one-dimensional version of the Calderon-Zygmund decomposition).

LEMMA 1. *Let* A *be such that*

$$C - M < A'(x) < C + M$$

then for each I *there exists a function* A_I *and a constant* C_I *such that*

$$A_I = A \quad \textit{on a set } E \quad |E| > \frac{1}{4} |I|$$

and

$$C_I - \frac{2}{3} M < A_I'(x) < C_I + \frac{2}{3} M .$$

Proof. We can assume $C = M$ i.e., $0 < A' < M2$. There are two cases:

a) $m_I(A') \geq M$.

In that case consider the smallest function $A_I \geq A$ with $A'(y) \geq \frac{2M}{3}$. We

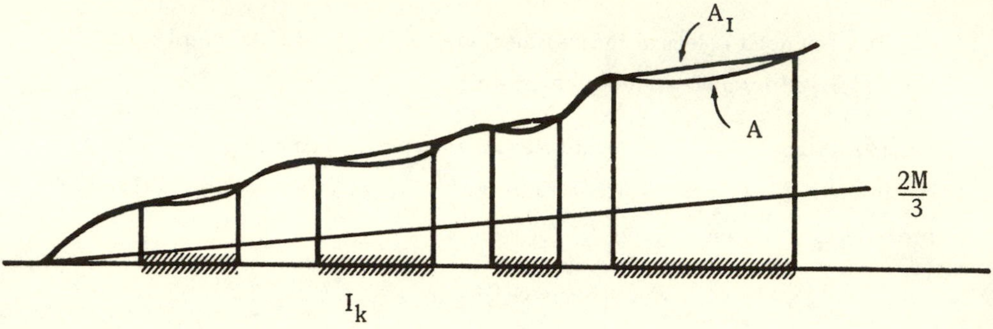

then have $A_I(y) = A(y)$ except for disjoint intervals I_k on which $A_I(y) = \frac{2M}{3}$. But

$$\int_I A_I'(y)\,dy = \int_{I-\cup I_k} + \int_{\cup I_k} \leq 2M\,|I-\cup I_k| + \frac{2M}{3}\,|\cup I_k|$$

$$= 2M|I| - \frac{4}{3}M\,|\cup I_k|$$

$$\frac{4}{3}M\,|\cup I_k| \leq (2M - m_I(A_I'))\,|I| \leq (2M-M)\,|I| = |I|\,M$$

i.e.,

$$|\cup I_k| < \frac{3}{4}\,|I|\,.$$

Since $\frac{2M}{3} \leq A_I'(y) \leq M2$, we have

$$\frac{4}{3}M - \frac{2}{3}M \leq A_I'(y) \leq \frac{4M}{3} + \frac{2}{3}M\,.$$

b) $m_I(A') < M$.

We consider the function

$$2M - A'(x) = A_1' \qquad |A_1'| \leq M2 .$$

Then we have $m_I(A_1') > M$ and we construct A_{1I} as above.

$$A_{1I} = 2M(x-a) - A(x) \text{ except on a set of meas } \leq \frac{3}{4}|I|$$

$$A(x) = 2M(x-a) - A_{1I} = A_I$$

$$A'(x) = 2M - A_I' = A_I'$$

$$\frac{4}{3}M - \frac{2}{3}M \leq A_{1I}' \leq \frac{4M}{3} + \frac{2}{3}M$$

$$2M - \frac{4}{3}M - \frac{2}{3}M \leq 2M - A_{1I}' \leq \frac{21}{3}M + \frac{2}{3}M + 2M$$

$$\frac{2}{3}M - \frac{2}{3}M \leq A_I' \leq \frac{2}{3}M + \frac{2}{3}M .$$

The main result is the following.

THEOREM (G. David). *Assuming that there exist* $\delta > 0$, $c > 0$ *such that for each* I *there exists* $K_I(x,y)$ *satisfying standard estimates uniformly in* I *with*

$$T_I(f) = \int K_I(x,y)\, f(y)\, dy$$

satisfying

$$\|T_1\|_{L^2(L),L^2(I)} \leq C_0$$

and there is a subset $E \subseteq I$ *with* $|E| > \delta|I|$ *such that*

$$\forall x \in E , \quad \forall y \in E , \qquad K_I(x,y) = K(x,y) .$$

Then T *maps* L^∞ *to* BMO *with*

$$\|T\|_{L^\infty, BMO} \le C_\eta C_0 .$$

If we let $\sigma(M) = \sup_{\|A'\|_\infty \le 1} \left\| \int \frac{e^{iM\frac{A(x)-A(y)}{x-y}}}{x-y}\, dy \right\|_{BMO}$ *a direct application of this theorem choosing for each interval* I

$$K_I(x,y) = \frac{e^{im\frac{A_I(x)-A_I(y)}{x-y}}}{x-y}$$

shows that

$$\sigma(M) \le C\sigma\left(\frac{2}{3}M\right)$$

i.e., $\sigma(M) \le c_1(1+M)^N$ *for some* N *or*

$$\|Tf\|_{L^2} = \left\| \int \frac{e^{i\frac{A(x)-A(y)}{x-y}}}{x-y} f(y)\,dy \right\|_{L^2} \le C(1+\|A'\|_\infty^N \|f\|_{L^2} .$$

Proof. We consider $f \in L^\infty$, $f = f\chi_I + f(1-\chi_I) = f_1 + f_2$, $\|f\|_\infty < 1$. One checks that

$$\int_I |T(f_2)(x) - T(f_2)(x_0)| \le c|I|$$

from which it follows $|x \in I: |Tf_2(x) - c_I| > c| < \eta_0|I|$ if c is large enough.

We now claim $T(f_1)$ is well approximated by $T_I(f_1)$, which we control, so that we want now to estimate

$$\int_{E'} |T(f_1) - T_I(f_1)|dx$$

where $E = I - \cup I_i$, $E' = I - \cup \bar{I}_i$ where $\bar{I}_i = (1+\delta) I_i$

$$\leq \int_{x \in E'} \int_{y \in I} |K(x,y) - K_I(x,y)| \, dy$$

$$= \sum \int_{x \in E'} \int_{I_i} |K(x,y) - K(x,y_i) + K_I(x,y_i) - K_I(x,y)| dy$$

where we have used the fact that $K(x,y) = K_I(x,y)$ outside I_i and y_i are endpoints of I_i, $(K(x,y_i) = K_I(x,y_i))$. This integrand is dominated by the Marcenkiewitz function of $\cup I_i$. Consequently the integral is bounded by $c|I|$ enabling us to apply Lemma 1.

All of these results can be extended to R^n by various methods. The easiest is the so-called method of rotation based on a general transference principle, valid for multilinear operators commuting with translations, and some nonlinear operators. We state the result in general although for the case of R^n this is an easy application of Fubini's theorem.

TRANSFERENCE THEOREM. *Let* U_t *be* 1*-parameter group of measure preserving transformations on a measure space* (X,dx) *and* $\Lambda(a_1 \cdots a_k, f)$ *a* $k+1$ *multilinear operation on* R *given by*

$$\Lambda(a;f) = \iint_{R^n} k(x-t, x-t_2 \cdots x-t_k, x-y) \prod a_i(t_i) f(y) \, dt \, dy$$

$$= \iint k(t_1, t_2, \cdots, t_k, s) \prod a_i(x-t_i) f(x-s) \, dt \, dy \, ds$$

satisfying

$$\|\Lambda(a;f)\|_{L^P} \le \prod_{i=1}^{k} \|a_i\|_\infty \|f\|_L .$$

Then the multilinear operator $\tilde{\Lambda}$ *on* $L^\infty(X)^k \times L^P(X) \to L^P(X)$ defined by

$$\Lambda(A,F) = \iint k(t_1,t_2 \cdots t_k,s) \prod A_i(U_{t_i}x)\,F(U_s x)\,dt\,ds$$

satisfies

$$\|\Lambda(A,F)\|_{L^P(X)} \le \prod \|A_i\|_{L^\infty(X)} \|F\|_{L^P(X)}$$

(*note that the constants are the same*).

The examples we have in mind are the following. $X = R^n$, dx Lebesque measure, e a unit vector

$$U_t x = x - te .$$

If we take $Hf = \int_R \frac{f(x-t)}{t}\,dt$, $\tilde{H}f = \int \frac{f(x-te)}{t}\,dt$. Or consider

$$X = T^1 \qquad U_t(e^{i\theta}) = e^{i(\theta-t)}$$

$$\tilde{H}f = c \int f(\theta-t) \frac{1}{\operatorname{ctg} \frac{t}{2}}\,dt .$$

If we take

$$\Lambda_k(a,f) = \int \left(\frac{A(x)-A(y)}{x-y}\right)^k \frac{f(y)}{x-y}\,dy$$

$$= \int \left(\frac{A(x)-A(x-t)}{t}\right)^k \frac{f(x-t)}{t}\,dt$$

then

$$\tilde{\Lambda}_{k,e}(A,f) = \int \left(\frac{A(x)-A(x-te)}{t}\right)^k f(x-te)\,\frac{dt}{t}\,.$$

Multiplying by $\Omega(e)$ and integrating on $|e| = 1$ in R^n we get

$$\int \Omega(e)\left(\frac{A(x)-A(x-te)}{t}\right)^k f(x-te)\,\frac{dt}{t}$$

$$= \int \left(\frac{A(x)-A(y)}{x-y}\right)^k K(x-y)f(y)dy \quad \text{where} \quad k(y) = \frac{\Omega(y|y|)}{|y|^n}$$

is odd for K even

even for K odd.

§6.

We now are in a position to recall the various ingredients which we discussed previously and recast them in a general setting. We considered operators L which are "small" perturbations of an operator L_0 for example, we took $L^0 = -\Delta$ and $L = -\Delta + \Sigma\, b_{ij} \frac{\partial^2}{\partial x_i \partial x_j}$ $\|b_{ij}\|_\infty < \varepsilon_0$ and then proceeded to expand functions of L, $F(L)$ as a power series in terms of the coefficients of $L-L_0$ i.e., we wrote

$$F(L) = F(L_0) + \sum_{k=1}^{\infty} \Lambda_k(b)$$

where the $\Lambda_k(b)$ was an operator valued homogeneous polynomial of d^0k in b. Of course such perturbations could be shown to converge only in a little ball around 0 (in the space of coefficients). It is then appropriate to ask how far can one extend the results and what is the natural domain of analyticity or holomorphy of the function $F(L)$. This question is meaningless if the Banach space in which we prove analyticity (in a neighborhood of 0) is not the largest possible space for which such estimates can be proved. A first task is to identify this largest space, which we will call the space of holomorphy. Once this space has been found it is natural to ask for the domain of holomorphy of the corresponding functional.

Let us return now to our first example where $F(a) = \operatorname{sgn}\left(\frac{i}{1-a}\frac{d}{dx}\right)$ where a is a function on R, and $F(a)$ is an operator on $L^2(R)$. The series expansion was

$$F(a)f = \sum_0^\infty \int \frac{f(y)}{x-y}\left(\frac{A(x)-A(y)}{x-y}\right)^k (1-a(y))\,dy\,, \qquad A' = a\,.$$

Let us consider for simplicity the commutator series

$$F_0(a)f = \sum_0^\infty \int \left(\frac{A(x)-A(y)}{x-y}\right)^k \frac{f(y)}{x-y}\,dy = \qquad \Lambda_k(a)(f)$$

We wish to find the smallest norm $|||\ \ |||$ (i.e., the largest Banach space) for which

$$* \qquad \|\Lambda_k(a)\|_{L^2,L^2} \le C^k\, |||a|||^k$$

in particular for $k = 1$ we must have

$$\|\Lambda_1(a)\|_{L^2,L^2} \le C\,|||a|||\,.$$

If we decide to define $|||a||| = \|\Lambda_1(a)\|_{(L^2,L^2)}$ it certainly defines a seminorm, if it actually is a norm for which the estimate $*$ can be proved, we would have identified the space of holomorphy. But

$$\Lambda_1(a)f = \int \frac{A(x)-A(y)}{x-y} \frac{f(y)}{x-y}\, dy .$$

We have already shown that $\|\Lambda_1(a)\|_{L^2 L^2} \le c\|a\|_\infty$ and it can easily be shown that $\|a\|_\infty$ is dominated by $c\|\Lambda_1(a)\|_{L^2 L^2}$. (In fact it suffices to consider the L^2 norm of

$$(x-x_0)^2 \Lambda_1(a)(\chi_I) - 2(x-x_0)\Lambda_1(a)((y-y_0)\chi_I) + \Lambda_1(a)((y-y_0)^2\chi_I)$$

on the interval I whose center is x_0. By letting I shrink to x_0 and using the L^2 estimate on $\Lambda_1(a)$ one obtains an estimate for a, for simplicity one can assume that a is smooth since the estimate does not depend on this condition.) Thus our norm $|||\ \ |||$ is equivalent to the L^∞ norm. The estimates * have already been proved in this norm.

We may also find the domain of holomorphy of $F_0(a)$, in fact

$$F_0(a)f = \int \frac{1}{x-y-(A(x)-A(y))} f(y)\,dy = \int \frac{1}{1-\frac{A(x)-A(y)}{x-y}} \frac{f(y)\,dy}{x-y} .$$

We have seen by the method of boosting for the Lipschitz constant that this will be bounded on L^2 as long as

$$\frac{1}{1-\frac{A(x)-A(y)}{x-y}} = \phi\left(\frac{A(x)-A(y)}{x-y}\right) \quad \text{for some } \phi \in C^\infty$$

clearly the case whenever $\inf_{x,y} \left|\frac{A(x)-A(y)}{x-y} - 1\right| > 0$ and we conjecture that this condition gives the domain of holomorphy. [Before discussing other examples we urge the reader to identify the space of holomorphy for $F(a) = \operatorname{sgn}\left(\frac{1}{1-a}\frac{d}{dx}\right)$. [Caution: It is not enough to consider all Λ_k separately.]

A more interesting example arises if we reconsider the Cauchy integral on rectifiable curves, which we chose to parametrize by arc length (and not as graphs).

We write $z(s) = \int_0^s e^{i(\alpha(t))}dt$ and consider the L^2 operator valued functional

$$C(\alpha)f = \frac{1}{2\pi i}\ \mathrm{p.v.} \int_{-\infty}^{\infty} \frac{f(t)z'(t)}{z(s)-z(t)}\,dt = \sum \Lambda_k(\alpha)f$$

here again one can define the norm

$$|||\alpha||| = \|\Lambda_1(\alpha)\|_{L^2,L^2}$$

and one finds that this norm is equivalent to the BMO norm of α. On the other hand it is easy to show that if $\|\alpha\|_{BMO} < \delta_0$ then $1-\delta_0 < \left|\frac{z(s)-z(t)}{s-t}\right| \leq 1$, and this is precisely the condition permitting us to write $C(\alpha) = \int \phi\left(\frac{z(s)-z(t)}{s-t}\right)\frac{f(t)}{s-t}\,dt$ giving rise to a bounded operator on L^2. Thus BMO is the natural space of holomorphy. As you recall from S. Krantz's lecture one can express the Szego projection $S(\alpha)$, projecting $L^2(\Gamma, ds) \simeq L^2(R, ds)$ onto $H^2_+(\Gamma, ds)$ (H^2_+ is the space of functions in L^2 admitting a holomorphic H^2 extension to the "left" of Γ) in terms of the Cauchy operator. This representation yields the result that the $S(\alpha)$ has BMO as its natural space of holomorphy and is an entire function on the manifold of chord-arc curves. (A similar result is true for the Riemann mapping function.)

Another remarkable example leading to an "exotic" space of holomorphy and to interesting geometry involves the functional calculus in $L = \frac{1}{1+a}\frac{\partial}{\partial z}$. As already seen it was necessary to assume that $a \in L^\infty$ and $a * \frac{1}{z^2} \in L^\infty$ with small norm. (This was obtained by considering the example $T(a) = \frac{\bar{L}}{L}$.) It turns out that $T(a)$ is analytic in a, relative to the norm $|||a||| = \|a\|_\infty + \|a * \frac{1}{z^2}\|_\infty$, and the condition really means that there exists a bilipschitz map $h(z): C \to C$ such that $\frac{\partial}{\partial z}h = a(z) + 1$.

It is clear from these and other examples that the identification of the space of holomorphy, or a detailed study of the first bilinear operation $\Lambda_1(a)f$ is basic to the understanding of these functionals.

R. R. COIFMAN
DEPARTMENT OF MATHEMATICS
YALE UNIVERSITY
NEW HAVEN, CONN.

YVES MEYER
CENTRE de MATHEMATIQUES
ECOLE POLYTECHNIQUE
PALAISEAU, FRANCE

REFERENCES

[1] A. P. Calderon, Cauchy Integrals on Lipschitz curves and related operators. Proc. Nat. Acad. Sci., U.S.A. 75 (1977, 1324-1327.

[2] R.R. Coifman, D.G. Deng and Y. Meyer. Domaine de la racine carée de certains opérateurs différentiels accretifs. Ann. Inst. Fourier 33, 2 (1983), 123-134.

[3] R.R. Coifman, Y. Meyer, Lavrentiev's curves and conformal mappings, Rep 5. 1983, Mittag Leffler Inst., Sweden.

[4] R.R. Coifman, A. McIntosh and Y. Meyer. L'intégrale de Cauchy définit un opérateur borné sur L^2 pour les courbes lipschitziennes. Annals of Math. 116 (1982), 361-387.

[5] R.R. Coifman and Y. Meyer, Au delà des opérateurs pseudodifférentiels. Asterisque 57. Societé Mathématique de France (1978).

[6] G. David, Opérateurs intégraux singuliers sur certaines courbes du plan complexe. Ann. Scient. Ec. Norm. Sup. 4° série, 17 (1984), 157-189.

[7] G. David, J. L. Journé, "A boundedness Criterion for Calderon Zygmund operators," Annals of Math. 120 (1984), 371-397.

[8] E. Fabes, D. Jerison and C. Kenig, Multilinear Littlewood-Paley estimates with applications to partial differential equations. Proc. Nat. Acad. Sci. U.S.A. 79 (1982), 5746-5750.

[9] R. Rosales, Exact Solutions of Some Nonlinear Evolution Equations, Studies in Applied Math. 59, 117-151.

[10] E. M. Stein, Singular Integrals and Differentiability Properties of Function, Princeton University Press (1970).

MULTIPARAMETER FOURIER ANALYSIS

Robert Fefferman

Introduction

The article which follows is an attempt to give an exposition of some of the recent progress in that part of Fourier Analysis which deals with classes of operators commuting with multiparameter families of dilations. In some sense, this field is not that new, since already in the early 1930's the properties of the strong maximal function were being investigated by Saks, Zygmund, and others. However, for many of the problems in this area which seem quite classical, answers have either not been found at all, or only quite a short time ago, so that our knowledge of the area is still fragmentary at this time.

The article is divided into six sections. The first treats some basic issues in the classical one-parameter theory whose multiparameter theory is then discussed in the remaining sections. Since the reader is no doubt quite familiar with the main elements of the classical theory, we have omitted references to the materials in section one. The book "Singular Integrals and Differentiability Properties of Functions" by E. M. Stein is an excellent reference for virtually all of the material there.

Finally, it is a pleasure to thank Professors M. T. Cheng and E. M. Stein for all of their hard work in organizing the Summer Symposium in Analysis in China, as well as many others whose generous hospitality made the visit to China such a very enjoyable one.

1. *The maximal function, Calderón-Zygmund decomposition, and Littlewood-Paley-Stein theory*

We hope here to review briefly some aspects of the classical 1-parameter theory of these topics. The three are inseparable and we hope to stress this.

We begin with the fundamental

CALDERÓN-ZYGMUND LEMMA. *Let* $f(x) \geq 0$, $f \in L^1(R^n)$, and $\alpha > 0$. *Then there exist disjoint cubes* Q_k *such that*

$$(1) \quad \alpha < \frac{1}{|Q_k|} \int_{Q_k} f(x)\,dx \leq 2^n \alpha$$

$$(2) \quad f(x) \leq \alpha \text{ a.e. } \textit{for } x \notin \cup Q_k$$

and

$$(3) \quad |\cup Q_k| \leq \frac{1}{\alpha} \int_{R^n} |f(x)|\,dx\ .$$

Proof. Let R^n be subdivided into a grid of congruent cubes so large that $\frac{1}{|Q|} \int_Q f \leq \alpha$ for all of them. Subdivide each cube in this collection into 2^n congruent subcubes. Select from these the cubes Q' such that $\frac{1}{|Q'|} \int_{Q'} f > \alpha$. For these Q' we stop the bisection process. For the rest, we continue until we first arrive at a cube Q' such that $\frac{1}{|Q'|} \int_{Q'} f > \alpha$, at which point we stop.

The cubes Q' at which we stop are then our Q_k. By construction $\frac{1}{|Q_k|} \int_{Q_k} f > \alpha$. Let $\tilde{Q}_k$ be the cube containing Q_k which was bisected to produce Q_k. Then $|\tilde{Q}_k| = 2^n |Q_k|$ and since we did not stop at $\tilde{Q}_k$, $\frac{1}{|\tilde{Q}_k|} \int_{\tilde{Q}_k} f \leq \alpha$. It follows that

$$\frac{1}{|Q_k|}\int_{Q_k} f \leq \frac{|\tilde{Q}_k|}{|Q_k|}\frac{1}{|\tilde{Q}_k|}\int_{\tilde{Q}_k} f \leq 2^n\alpha ,$$

proving (1). Notice that (3) follows from (1) because $|Q_k| \leq \frac{1}{\alpha}\int_{Q_k} f$ so summing on k, we have

$$|\cup Q_k| \leq \frac{1}{\alpha}\sum \int_{Q_k} f \leq \frac{1}{\alpha}\int_{R^n} f .$$

Finally, (2) follows, since for each $x \notin \cup Q_k$, x belongs to a sequence of cubes C_k whose diameters converge to zero and such that $\frac{1}{C_k}\int_{C_k} f \leq \alpha$. It follows from Lebesques theorem on differentiation of integrals that $f(x) \leq \alpha$ a.e. for such x.

We all know how important the maximal operator of Hardy-Littlewood is in the subject of Fourier analysis. This is the operator given by

$$Mf(x) = \sup_{r>0}\frac{1}{|B(x;r)|}\int_{B(x;r)} |f(t)|\,dt .$$

Going along with this we also define

$$M_\delta f(x) = \sup_{x\varepsilon Q \text{ dyadic cube}} \frac{1}{|Q|}\int_Q |f(t)|\,dt .$$

(Recall that a dyadic interval of R^1 is one of the form $[j2^k,(j+1)2^k]$ $j,k \varepsilon Z$ and a dyadic cube is a product of dyadic intervals of equal length; recall also the basic property of dyadic cubes—if Q_1,Q_2 are dyadic either $Q_1 \cap Q_2 = \emptyset$, $Q_1 \subseteq Q_2$ or $Q_2 \subseteq Q_1$.) Then the following simple lemma sheds some light on the relationship of the Calderón-Zygmund lemma to the Maximal Operator.

LEMMA. *Let* $f \geq 0 \in L^1(R^n)$ *and* $\alpha > 0$. *Let* Q_k *be Calderón-Zygmund cubes as above, and let* $\tilde{Q}_k$ *denote the double of* Q_k. *Then there exist positive numbers* c *and* C *such that*

(1) $\cup Q_k \subseteq \{Mf > c\alpha\}$

(2) $\cup \tilde{Q}_k \supseteq \{Mf > C\alpha\}$

(3) *Furthermore, if* Q_k *are dyadic, then* $\cup Q_k \supseteq \{M_\delta f > C\alpha\}$.

Proof. (1) Let $x \in Q_k$. Then there exists a ball $B(x;r)$ such that $x \in Q_k \subseteq B(x;r)$ and $\frac{|B(x;r)|}{|Q_k|} \leq C_n$. Then

$$Mf(x) \geq \frac{1}{|B(x;r)|} \int_{B(x;r)} |f| \geq \left(\frac{|Q_k|}{|B(x;r)|}\right) \frac{1}{|Q_k|} \int_{Q_k} |f| \geq \frac{1}{C_n} \alpha .$$

(2) Let $x \notin \cup \tilde{Q}_k$. Let $r > 0$. Then we estimate

$$\int_{B(x;r)} f = \int_{B(x;r) \cap^c [\cup Q_k]} f + \sum_{Q_j \cap B \neq \emptyset} \int_{B(x;r) \cap Q_j} f$$

$$\leq \alpha |B(x;r)| + \sum_{Q_j \cap B \neq \emptyset} \int_{Q_j} f$$

$$\leq \alpha |B(x;r)| + 2^n \alpha \sum_{Q_j \cap B \neq \emptyset} |Q_j| .$$

Key point: if $Q_j \cap B(x;r) \neq \emptyset$ then $Q_j \subseteq B(x;10r)$ so that

$$\sum_{Q_j \cap B \neq \emptyset} |Q_j| \leq C|B(x;r)| \quad \text{and} \quad \int_{B(x;r)} f \leq C_n \alpha |B(x;r)| .$$

That is, $Mf(x) \leq C_n \alpha$.

The use of the Calderón-Zygmund lemma is apparent in the Calderón-Zygmund Theorem on Singular Integrals: Let X and N be Banach spaces and let B(X,N) denote the bounded operators from X to N. Let $K : R^n \times R^n/\{x=y\} \to B(X,N)$ satisfy

(1) $$|K(x,y+h) - K(x,y)|_{B(X,N)} \le \frac{|h|^\delta}{|x-y|^{n+\delta}} \text{ for } |h| < \frac{|x-y|}{2}$$

and for some $\delta > 0$.

(2) if $Tf(x) = \int_{R^n} K(x,y) f(y) dy$ for $f \in L^P(X)$, and suppose for some

$p_0 > 1$ $\|Tf\|_{L^{p_0}(N)} \le C\|f\|_{L^{p_0}(X)}$.

Then, for T we have

$$|\{x| \ |Tf(x)|_N > \alpha\}| \le \frac{C}{\alpha} \|f\|_{L^1(X)}$$

and

$$\|Tf\|_{L^p(N)} \le C_p \|f\|_{L^p(X)} \text{ for } 1 < p < p_0 .$$

Proof. Let $\alpha > 0$ and $f \in L^1(X)$. Set

$$g(x) = \begin{cases} \dfrac{1}{|Q_k|} \displaystyle\int_{Q_k} f\, dt & \text{if } x \in Q_k , \\ f(x) & \text{if } x \notin Q_k . \end{cases}$$

and $b(x) = f(x) - g(x)$. Then

$$|\{|Tg(x)|_N > \alpha\}| \le \frac{C}{p_0} \|g\|^{p_0}_{L^{p_0}(X)} \le \frac{C'}{\alpha} \|g\|_{L^1(X)} \le \frac{C'}{\alpha} \|f\|_{L^1(X)} .$$

As for $Tb(x)$, suppose $x \notin \cup \widetilde{Q}_k$.

Let $b_k(x) = \chi_{Q_k}(x)\, b(x)$; $\int_{Q_k} b_k(x)\, dx = 0$. Then

$$Tb_k(x) = \int_{Q_k} K(x,y)\, b_k(y)\, dy\ .$$

Let $\bar{y}_k$ be the center of Q_k; then

$$\int_{Q_k} K(x,\bar{y}_k)\, b_k(y)\, dy = K(x,\bar{y}_k) \int_{Q_k} b_k(y)\, dy = 0$$

so

$$Tb_k(x) = \int_{Q_k} \{K(x,y) - K(x,\bar{y}_k)\}\, b_k(y)\, dy$$

and

$$|Tb_k(x)|_N \le \frac{\operatorname{diam}(Q_k)^\delta}{|x-\bar{y}_k|^{n+\delta}} \|b_k\|_{L^1(X)}\ ;$$

summing over k we have

$$\int_{x \notin \cup \tilde{Q}_k} |T(b)(x)|_N \le \sum_k \int_{x \notin \tilde{Q}_k} \frac{\operatorname{diam}(Q_k)^\delta}{\operatorname{dist}(x,Q_k)^{n+\delta}} \|b_k\|_{L^1}\, dx \le C\|f\|_{L^1(X)}\ .$$

Thus

$$|\{Tb(x)|_N > \alpha\}| \le |\cup \tilde{Q}_k| + \frac{1}{\alpha} \|f\|_1 \le \frac{C'}{\alpha} \|f\|_1\ .$$

From this weak (1,1) estimate, interpolate to get the L^p result.

We now quote some important examples:

1. *Classical Calderon-Zygmund Convolution Operators.* Here $Tf = f * K$ where $K(x)$ is a complex-valued function satisfying

(α) $|K(x)| \le C/|x|^n$;

(β) $$\int_{\rho_1 \le |x| \le \rho_2} K(x)\,dx = 0 \quad \text{for all } 0 < \rho_1 < \rho_2 ;$$

and

(γ) $$|K(x+h) - K(x)| \le C \frac{|h|^\delta}{|x|^{n+\delta}} \qquad |h| < \frac{1}{2}|x| .$$

The Riesz transforms $R_j f = f * x_j/|x|^{n+1}$ are especially important since they are related to H^p spaces and analytic functions.

For a Calderón-Zygmund singular integral T, it is easily seen to be bounded on $L^2(R^n)$, since $\hat{K}(\xi) \in L^\infty$. Also, since T^* is also a Calderón-Zygmund singular integral, T is bounded on the full range of $L^p(R^n)$, $1 < \rho < \infty$.

2. *Littlewood-Paley-Stein Functions.* The most basic, simplest of these are the g-function and S function defined as follows: Let $\psi \in C_c^\infty(R^n)$, $\int_{R^n} \psi = 0$. Let $\psi_t(x) = t^{-n}\psi\left(\frac{x}{t}\right)$ for $t > 0$. Then

$$g^2(f)(x) = \int_0^\infty |f * \psi_t(x)|^2 \frac{dt}{t}$$

$$S^2(f)(x) = \iint_{\Gamma(x)} |f * \psi_t(y)|^2 \frac{dt\,dy}{t^{n+1}}$$

where $\Gamma(x) = \{(y,t) |\ |y-x| < t\}$. Then it is a basic fact that $\|S(f)\|_{L^p} \le C_p\|f\|_{L^p}$ and $\|g(f)\|_{L^p} \le C_p\|f\|_{L^p}$, when $1 < p < \infty$. If, say, ψ is suitably non-trivial, (radial, non-zero is good enough) then the reverse inequalities hold:

$$\|S(f)\|_{L^p} \ge c_p\|f\|_{L^p} \quad \text{and} \quad \|g(f)\|_{L^p} \ge c_p\|f\|_{L^p} .$$

Now take $S(f)$. We want to point out here that S is a singular integral. In fact define $K: R^n \mapsto L^2(\Gamma(0); dydt)$ by $K(x)(y,t) = \psi_t(x-y)$. Then

$$S(f)(x) = |f * K(x)|_{L^2(\Gamma; dtdy/t^{n+1})}$$

and K satisfies

$$|K(x+h) - K(x)| \leq C \frac{|h|}{|x|^{n+1}} \qquad |h| < \frac{1}{2}|x| .$$

Also by a Fourier transform argument $\|Sf\|_{L^2(R^n)} \leq c\|f\|_{L^2(R^n)}$ so the Calderón-Zygmund theorem applies. In fact, the adjoint operator also maps $L^p(L^2(\Gamma)) \to L^p(R^n)$, $1 < p < 2$, because it is also C-Z, so again this explains why we get boundedness of S on the full range $1 < p < \infty$.

3. *The Hardy-Littlewood Maximal Operator as a Singular Integral.* Let $\phi(x) \in C^\infty(R^n)$ and suppose for $|x| < 1$, $\phi(x) = 1$ and for $|x| > 2$, $\phi(x) = 0$. Then define $K: R^n \mapsto L^\infty((0,\infty); dt)$ by $K(x)(t) = \phi_t(x) = t^{-n}\phi(x/t)$. Then

$$|\nabla_x K(x)(t)| = |t^{-(n+1)}\nabla\phi \ \tfrac{x}{t}| \leq C\|\nabla\phi\|_{L^\infty(R^n)} \frac{1}{|x|^{n+1}}$$

(since if $t > |x|/2$, $\nabla\phi(x/t) = 0$).

Again

$$|K(x+h) - K(x)|_{L^\infty} \leq C \frac{|h|}{|x|^{n+1}} \text{ if } |h| < \frac{1}{2}|x| ,$$

and we also have $\|f * K\|_{L^\infty(L^\infty)} \leq C\|f\|_{L^\infty}$ since $|f * \phi_t(x)| \leq \|\phi\|_1 \|f\|_\infty$. Then $Mf(x) \sim |f * K(x)|_{L^\infty}$ so M is bounded on $L^p(R^n)$, $p > 1$ and weak 1-1.

4. *The Estimates for Pointwise Convergence of Singular Integrals on* $L^1(R^n)$. Suppose that $K(x)$ is a classical Calderón-Zygmund kernel and let $K_\varepsilon(x) = K(x) \cdot \chi_{|x|>\varepsilon}(x)$, for $\varepsilon > 0$. We are interested in the existence a.e. of $\lim_{\varepsilon \to 0} f * K_\varepsilon(x)$ for $f \in L^1(R^n)$. In order to know this, it

is enough to show that $T^*f(x) = \sup_{\varepsilon>0} |f * K_\varepsilon(x)|$ satisfies the weak type estimate $|\{T^*f(x) > \alpha\}| \le \frac{C}{\alpha} \int_{R^n} |f|$. It turns out that by using the Hardy Littlewood maximal operator it is not difficult to prove $T^*f(x) \le C\{M(Tf)(x) + Mf(x)\}$ which immediately gives the boundedness of T^* on $L^p(R^n)$ for $p > 1$. However, it fails to give the weak type inequality for functions on $L^1(R^n)$. This inequality follows easily from the observation that T^* is a singular integral.

Let

$$\phi(x) \in C_c^\infty(R^n), \phi(x) = \begin{cases} 1 & \text{if } |x| \le 1 \\ 0 & \text{if } |x| > 2 \end{cases}$$

and set $\tilde{K}_\varepsilon(x) = K(x)\left[1 - \phi\left(\frac{x}{\varepsilon}\right)\right]$. Then $|\tilde{K}_\varepsilon(x) - K(x)| \le \frac{C}{|x|^n} \chi_{\varepsilon \le |x| \le 2\varepsilon}(x)$ so that $T^*f(x) \le \sup_{\varepsilon>0} |f * \tilde{K}_\varepsilon(x)| + Mf(x)$, and so we need only show that $\sup_{\varepsilon>0} |f * \tilde{K}_\varepsilon|$ is weak type (1,1). In order to do this let $H: R^n \to L^\infty((0,\infty); d\varepsilon)$ be given by $H(x)(\varepsilon) = \tilde{K}_\varepsilon(x)$. Then $|H(x) - H(x+h)|_{L^\infty} \le \frac{C|h|}{|x|^{n+1}}$ and H is bounded from $L^2 \to L^2(L^\infty)$, so H is weak 1-1.

5. *The Maximal Function as a Littlewood-Paley-Stein Function.* Let $f \in L^2(R^n)$, $f(x) \ge 0$ for all x. Use the Calderón-Zygmund decomposition with $\alpha = C^j$, $j \in Z$ for some $C > 0$ sufficiently large, to get (dyadic) cubes Q_k^j where $\frac{1}{|Q_k^j|} \int_{Q_k^j} f \sim C^j$. Define f_j as in the Calderón-Zygmund decomposition,

$$\begin{cases} \dfrac{1}{|Q_k^j|} \displaystyle\int_{Q_k^j} f & \text{if } x \in Q_k^j \\ f(x) & \text{if } x \notin \bigcup_k Q_k^j \end{cases}$$

and $\Delta_j f = f_{j+1} - f_j$, then observe that:

(1) $\Delta_j f$ lives on $\bigcup_k Q_k^j$ and has mean value 0 on each Q_k^j.

(2) $\Delta_i f$ is constant on every Q_k^j for $i < j$.

(3) $f_j \to 0$ as $j \to -\infty$ and $f_j \to f$ as $j \to +\infty$ so $f = \sum_{j=-\infty}^{+\infty} \Delta_j f$.

From (1) and (2) it is clear that the $\Delta_j f$ are orthogonal so that

$$\|f\|_{L^2(R^n)} = \left(\sum_j \|\Delta_j f\|_{L^2}^2\right)^{1/2} = \left\| \left(\sum_j |\Delta_j f(x)|^2\right)^{1/2} \right\|_{L^2} .$$

Finally, observe that the square function $(\sum_j |\Delta_j f(x)|^2)^{1/2}$ is essentially just the dyadic maximal function. In fact, if $C^j << M_\delta(f)(x)$ then $x \in Q_k^j$ for some k and $\Delta_j(f)(x) \sim C^j$. It follows that

$$\left(\sum_j |\Delta_j f(x)|^2\right)^{1/2} \geq cM_\delta f(x) .$$

Before finishing this section, we shall need estimates near L^1 for the maximal function.

If Q denotes the unit cube in R^n then for k a positive integer

$$\int_Q Mf(\log^+ Mf)^{k-1} dx < \infty \text{ if and only if } \int_Q |f|(\log^+ |f|)^k dx < \infty .$$

The proof runs as follows: If $f \in L(\log^+ L)^k$ then

$$|\{x | Mf(x) > \alpha\}| \leq \frac{C}{\alpha} \int_{|f(x)| > \alpha/2} f(x)\, dx$$

and so

$$\|Mf\|_{L(\log L)^{k-1}} \le$$

$$\int_1^\infty (\log \alpha)^{k-1} \frac{1}{\alpha} \int_{|f(x)|>\alpha/2} |f(x)|dx\, d\alpha \le \int_{Q_0} |f(x)| \int_1^{2|f(x)|} \frac{1}{\alpha} (\log \alpha)^{k-1} d\alpha\, dx$$

$$\le \|f\|_{L(\log L)^k} .$$

Conversely, (Stein) Calderón-Zygmund decompose R^n at height $\alpha > 0$. We have

$$\int_{M_\delta f(x)>C\alpha} f(x)dx \le \int_{\cup Q_k} f(x)dx \le \sum_k \int_{Q_k} f \le C\alpha \sum_k |Q_k| \le C\alpha|\{Mf > c\alpha\}| .$$

This yields

$$\int_{Q_k} |f(x)|(\log^+ M_\delta f(x))^k dx \le \int_1^\infty \frac{1}{\alpha} \int_{M_\delta f(x)>C_n\alpha} |f(x)|dx(\log \alpha)^{k-1} d\alpha$$

$$\le \int_1^\infty \frac{1}{\alpha} \int_{M_\delta f(x)>C_n\alpha} |f(x)|dx \cdot (\log \alpha)^{k-1} d\alpha$$

$$\le \int_{Q_k} Mf(x)\, [\log^+ Mf(x)]^{k-1} dx .$$

2. *Multi-parameter differentiation theory*

During the first lecture we discussed some fundamentally important operators of classical (and sometimes, not so classical) harmonic analysis: the maximal operator, singular integrals, and Littlewood-Paley-

Stein operator. These operators all had one thing in common. They all commute in some sense with the one-parameter family of dilations on R^n, $x \to \delta x$, $\delta > 0$. The nature of the real variable theory involved does not seem to depend at all on the dimension n. In marked contrast, it turns out that a study of the analogous operators commuting with a multi-parameter family of dilations reveals that the number of parameters is enormously important, and changes in the number of parameters drastically change the results.

Let us begin by giving the most basic example, which dates back to Jessen, Marcinkiewicz, and Zygmund. We are referring to a maximal operator on R^n which commutes with the full n-parameter group of dilations $(x_1, x_2, \cdots, x_n) \to (\delta_1 x_1, \delta_2 x_2, \cdots, \delta_n x_n)$, where $\delta_i > 0$ is arbitrary. This is the "strong maximal operator," $M^{(n)}$, defined by

$$M^{(n)}f(x) = \sup_{x \epsilon R} \frac{1}{|R|} \int_R |f(t)|\, dt$$

where R is a rectangle in R^n whose sides are parallel to the axes. Unlike the case of the Hardy-Littlewood operator, $M^{(n)}$ does not satisfy

$$|\{x | M^{(n)}(f)(x) > a\}| \leq \frac{C}{a} \|f\|_{L^1(R^n)} .$$

For instance when $n = 2$ and when $f_\delta = \delta^{-2} \chi_{(|x_1| < \delta/2) \times (|x_2| < \delta/2)}$, then for $|x_1|$, $|x_2| > 2\delta$,

$$M^{(2)}(f_\delta)(x_1, x_2) = M^{(1)}(\chi_{|x_1| < \delta/2})(x_1) M^{(1)}(\chi_{|x_2| < \delta/2})(x_2) \sim \frac{1}{|x_1|} \frac{1}{|x_2|}$$

and

$$|\{x \epsilon Q_0 | M^{(2)}(f_\delta) > a\}| \geq |\{x| \, |x_1| \, |x_2| < \tfrac{1}{a}, \text{ and } \delta < |x_i| < 1\}| \sim \frac{1}{a} \log \frac{1}{a} ,$$

if $a = 1/\delta$.

If we have a weak type inequality, we must have

$$|\{M^{(2)}(f_\delta) > \alpha\}| \leq \frac{1}{\alpha} \|f_\delta\|$$

so that $\|f_\delta\| \geq c \log \frac{1}{\delta}$ and the smallest Orlicz norm for which this holds is the $L(\log L)$ norm. A similar computation in R^n reveals that for $M^{(n)}$ to map L_Φ boundedly to Weak L^1 we must have $L_\Phi \subseteq L(\log L)^{n-1}$. The next theorem shows that indeed $M^{(n)}$ does indeed map $L(\log L)^{n-1}$ boundedly into Weak L^1.

THEOREM OF JESSEN-MARCINKIEWICZ-ZYGMUND (1935) [1]. *For functions* $f(x)$ *in the unit cube of* R^n *we have*

$$|\{x \in Q_0, M^{(n)}(f)(x) > \alpha\}| \leq \frac{C}{\alpha} \|f\|_{L(\log L)^{n-1}(Q_0)} .$$

The proof is strikingly simple. Define M_{x_i} to be the 1-dimensional maximal function in the ith coordinate direction. Consider the case $n = 2$, which is already entirely typical. Let R be a rectangle containing the point $(\overline{x}_1, \overline{x}_2)$, say $R = I \times J$. Then

$$\frac{1}{R} \iint_R |f(x_1,x_2)|\, dx_1 dx_2 = \frac{1}{|I|} \int_I \left(\frac{1}{|J|} \int_J f(x_1,x_2) dx_2 \right) dx_1 \tag{2.1}$$

$$\frac{1}{|J|} \int_J |f(x_1,x_2)| dx_2 \leq M_{x_2} f(x_1, \overline{x}_2)$$

so (2.1) is

$$\leq \frac{1}{|I|} \int_I M_2 f(x_1, \overline{x}_2)\, dx_1 \leq M_{x_1}(M_{x_2} f)(\overline{x}_1, \overline{x}_2) .$$

Thus, for all $(x_1,x_2) \in Q_0$,

$$M^{(2)}f(x_1,x_2) \leq M_{x_1} \circ M_{x_2}(f)(x_1,x_2)\ .$$

We have seen that M_{x_2} maps $L(\log L)(Q_0)$ boundedly into $L^1(Q_0)$ so that

$$\|M_{x_2} f\|_{L^1(Q_0)} \leq C\|f\|_{L(\log L)(Q_0)}\ ,$$

and finally

$$|\{x \in Q_0 | M_{x_1} M_{x_2} f(x) > \alpha\}| \leq \frac{C'}{\alpha} \|M_{x_2} f\|_{L^1(Q_0)} \leq \frac{C''}{\alpha} \|f\|_{L(\log L)(Q_0)}\ .$$

Now, this method of iteration in the proof above gives sharp estimates for $M^{(n)}$, and it may be suspected that the whole story of the harmonic analysis of several parameters can be told by applying this iteration technique. That this is not the case should become clear as this lecture proceeds. We want to describe some of the multi-parameter theory and to do this, let us begin with maximal functions. For many years after the Jessen-Marcinkiewicz-Zygmund theorem, there was no machinery around to treat problems here, and then, only fairly recently, two such machines were created. The one we describe here proceeds by means of covering lemmas while the other, due to Nagel, Stein, and Wainger, which Wainger has described in detail, uses the Fourier transform [2]. Though the two methods seem totally different on the surface, they are really quite closely related and have in common the main theme of reducing higher parameter, complicated operators to lower parameter simpler ones, which are already well understood.

The model for our method is the following, where the operator $M^{(n)}$ is controlled by $M^{(n-1)}$.

COVERING LEMMA FOR RECTANGLES OF THE STONG MAXIMAL OPERATOR [3]. *Let* $\{R_k\}$ *be a given sequence of rectangles in* $R^n \subseteq B(0,1)$ *whose sides are parallel to the axes. Then there is a subsequence* $\{\widetilde{R}_k\}$ of $\{R_k\}$ *so that*

(1) $|\cup\widetilde{R}_k| \geq c_n |\cup R_k|$

(2) $\left\| \exp\left(\sum \chi_{\widetilde{R}_k}\right)^{\frac{1}{n-1}} \right\|_{L^1(B)} \leq C$.

Before we prove this theorem, let us show that it implies the Jessen-Marcinkiewicz-Zygmund result. Let $\alpha > 0$, and for each point $x \in \{M^{(n)}f(x) > \alpha\}$ there is a rectangle R_x containing x with

$$\frac{1}{|R_x|}\int_{R_x} |f| > \alpha . \tag{2.2}$$

Without loss of generality we assume $\cup R_x = \cup R_k$ where R_k are certain if the R_x's. Apply the covering lemma to get $\widetilde{R}_k$ with properties (1) and (2) above. Then by virtue of (1) we need only show that

$$|\cup\widetilde{R}_k| \leq \frac{C}{\alpha} \|f\|_{L(\log L)^{n-1}(Q_0)} .$$

By (2.2), $|\widetilde{R}_k| \leq \frac{1}{\alpha}\int_{\widetilde{R}_k} |f|$ and summing we have

$$|\cup\widetilde{R}_k| \leq \frac{1}{\alpha}\int_{Q_0} |f| \sum \chi_{\widetilde{R}_k} \leq \frac{1}{\alpha} \|f\|_{L(\log L)^{n-1}} \left\| \sum \chi_{\widetilde{R}_k} \right\|_{\exp(L^{1/(n-1)})} .$$

Now, let us prove the covering theorem. We shall proceed by induction on n. Assume the case $n-1$. Let $R_1, R_2, \cdots, R_k, \cdots$ be ordered such that the x_n side length decreases. For a rectangle R, let R_d denote the rectangle whose center and x_i side lengths, $i < n$, are the same as those of R, but whose x_n side length is multiplied by 5. Then we describe the procedure for selecting the $\widetilde{R}_k$ from the R_k: Let $\widetilde{R}_1 = R_1$. Suppose $\widetilde{R}_1, \widetilde{R}_2, \cdots, \widetilde{R}_k$ have already been chosen. We continue along the list, and each time we consider the rectangle R we ask whether or not

$$|R \cap [\cup (\tilde{R}_j)_d]| < \frac{1}{2} |R|$$

where the above union is taken over all the $\tilde{R}_j$, $j \leq k$ for which $\tilde{R}_j \cap R \neq \emptyset$. If the answer is no, we move on to consider the next rectangle on the list. If the answer is yes, we make the rectangle $R = \tilde{R}_{k+1}$, and start the process over again.

Now, we prove (1) as follows: If R is an unselected rectangle, then

$$\left| R \cap \left[\bigcup_{\substack{\tilde{R}_j \cap R = \emptyset \\ \tilde{R}_j \text{ before } R}} (\tilde{R}_j)_d \right] \right| \geq \frac{1}{2} |R| .$$

Let us slice all rectangles with a hyperplane perpendicular to the x_n axis. Then if slices are indicated by using S's instead of R's,

$$|S \cap [\cup (\tilde{S}_j)_d]| \geq \frac{1}{2} |S|$$

so that $M^{(n-1)}(\chi_{\cup(\tilde{R}_j)_d}) > \frac{1}{2}$ on $\cup R_j$, where $M^{(n-1)}$ is acting in the $x_1, x_2, \cdots, x_{n-1}$ coordinates. By the boundedness of $M^{(n-1)}$ on, say, L^2 (by induction) we have

$$|\cup R_j| \leq C |\cup (\tilde{R}_k)_d| \leq C' \sum |\tilde{R}_k| \leq C' |\cup \tilde{R}_k| .$$

To obtain (2), notice that the $\tilde{R}_j$'s satisfy

$$\left| \tilde{R}_k \cap \left[\bigcup_{\substack{\tilde{R}_j \cap \tilde{R}_k \neq \emptyset \\ j<k}} (\tilde{R}_j)_d \right] \right| < \frac{1}{2} |\tilde{R}_k| .$$

If we again slice with a hyperplane perpendicular to the x_n axis,

$$|\tilde{S}_k \cap \bigcup_{j<k} \tilde{S}_j| < \frac{1}{2} |\tilde{S}_k|$$

so that, if $\tilde{E}_k = \tilde{S}_k - \bigcup_{j<k} \tilde{S}_j$ we have $|\tilde{E}_k| > \frac{1}{2} |\tilde{S}_k|$ and if $\phi \in L(\log L)^{n-1}(S_0)$ we shall show that

$$\int_{S_0} \sum \chi_{\tilde{S}_k} \phi \leq C \|\phi\|_{L(\log L)^{n-1}} \tag{2.3}$$

which will give

$$\int_{S_0} \exp\left[c\left(\sum \chi_{\tilde{S}_k}\right)^{1/(n-1)}\right] dx_1 \cdots dx_{n-1} \leq C .$$

Integrating this estimate in x_n finishes things.

To obtain (2.3) we write

$$\int_{S_0} \sum \chi_{\tilde{S}_k} \phi dx_1 \cdots dx_{n-1} \leq \sum_k \int_{\tilde{S}_k} \phi \leq C \sum_k |\tilde{E}_k| \frac{1}{|\tilde{S}_k|} \int \phi \leq$$

$$\leq C \int_{S_0} M^{(n-1)}(\phi) dx_1 dx_2 \cdots dx_{n-1} \leq C' \|\phi\|_{L(\log L)^{n-1}} .$$

Notice that in our argument above the slicing was the most important idea. If you try the proof without it, you will not wind up with the estimate you want, on the $\exp(\ )^{1/(n-1)}$norm, but rather on the $\exp(\ )^{1/n}$ norm instead. Also the slicing is the mechanism by which we control $M^{(n)}$ by the lower parameter operator $M^{(n-1)}$ and here this enables us to proceed by induction. Of course, in the end the theory of the boundedness of $M^{(n)}$ had been known for some 40 years before the covering lemma. But the lowering of the number of parameters, and the induction procedure will be used in what follows as the key ingredient to prove new theorems.

To illustrate the method of the machine, we consider a maximal operator whose relation to multiplier theory will be studied below. Suppose in R^2 we consider the class $\mathfrak{A}$ of all rectangles of arbitrary side lengths which are oriented in one of the directions $\theta_k = 2^{-k}$, $k = 1,2,\cdots$, measured from some fixed direction, say the positive x-direction. Define the maximal operator $\mathfrak{m}$ by

$$\mathfrak{m}f(x) = \sup_{x \in R \in \mathfrak{A}} \frac{1}{|R|} \int_R |f(t)|\,dt \ .$$

This operator was considered following the ideas of the covering approach outlined above by Stromberg [4] and Cordoba-Fefferman [5]. Somewhat later Nagel, Stein, and Wainger [6] used Fourier transform methods to extend the result we shall discuss below.

What we prove is that

$$|\{\mathfrak{m}f(x) > \alpha\}| \leq \frac{C}{\alpha^2} \|f\|^2_{L^2(R^2)} \ .$$

The proof consists of showing that, given a sequence of rectangles $\{R_k\}$ belonging to $\mathfrak{A}$, there exists a subfamily $\{\widetilde{R}_k\}$ such that

(1) $|\cup \widetilde{R}_k| \geq c|\cup R_k|$

and

(2) $\|\sum \chi_{\cup \widetilde{R}_k}\|_{L^2(R^2)} \leq |\cup R_k|^{1/2}$.

To prove this we give a rule for selecting $\widetilde{R}_k$, given that we have already selected $\widetilde{R}_j$ for $j < k$. Assume that the R_k all have their longest side in a direction in the 1st quadrant, and are ordered so that their longer side lengths are decreasing. Then consider the rectangle R following $\widetilde{R}_{k-1}$. Consider in particular

$$\frac{1}{|R|} \sum_{j<k} |R \cup \widetilde{R}_j| = \frac{1}{|R|} \int_R \sum_{j<k} \chi_{\widetilde{R}_j}\,dx \ .$$

If this is less than 1/2, select R as $\tilde{R}_k$. If not go to the next rectangle on the list, and apply the same test to it. In this way we obtain the desired $\tilde{R}_k$. Notice that

$$\begin{aligned}\Big\|\sum \chi_{\tilde{R}_k}\Big\|_2^2 &= \int \sum_{j,k} \chi_{\tilde{R}_j}\chi_{\tilde{R}_k} = 2\int_j \sum_{j<k} \chi_{\tilde{R}_k} + \sum_k |\tilde{R}_k| \\ &= 2\sum_k \int_{\tilde{R}_k} \sum_{j<k} \chi_{\tilde{R}_j} + \sum_k |\tilde{R}_k| \le 2\sum |\tilde{R}_k| \le C|\cup \tilde{R}_k| .\end{aligned} \tag{2.4}$$

This is (2). To show (1), we let R be some rectangle which was unselected. This implies that

$$\frac{1}{|R|} \sum_{\tilde{R}_j \text{ before } R} |R \cap \tilde{R}_j| \ge \frac{1}{2} .$$

Now draw the following picture. Let S be the envelope rectangle to R whose sides are parallel to the coordinate axes.

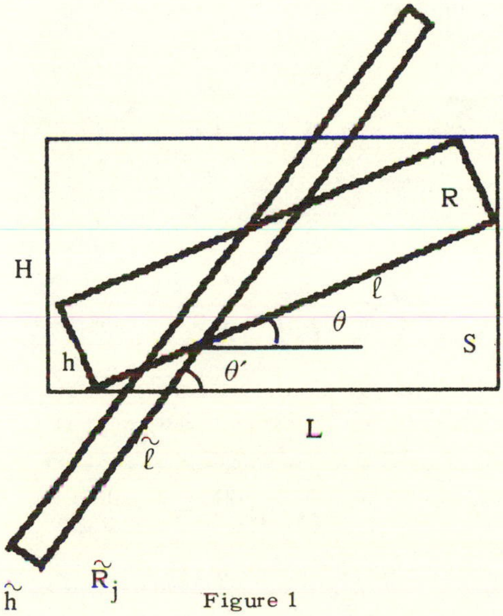

Figure 1

Then by dilating S if necessary, we can assume that $\tilde{R}_j$ is centered at the same center as S. Now the point is that

$$(2.5)\qquad \frac{|\tilde{R}_j \cap S|}{|S|} \geq c\, \frac{|\tilde{R}_j \cap R|}{|R|}\,.$$

In fact,

$$\frac{|\tilde{R}_j \cap S|}{|S|} \sim \frac{\tilde{h} H/\theta'}{hL} = \frac{\tilde{h}}{L\theta'}$$

and

$$\frac{|\tilde{R}_j \cap R|}{|R|} \sim \frac{\tilde{h}\, h(\theta'-\theta)}{h\ell} = \frac{\tilde{h}}{\ell(\theta'-\theta)}$$

and our inequality (2.5), taking into account that $\ell \sim L$, is

$$\frac{1}{\theta'} \geq \frac{1}{(\theta'-\theta)} \quad \text{or} \quad \theta'-\theta \geq c\theta'$$

and this in our case of $\theta_k = 2^{-k}$ is valid with $c = 1/2$. (If $\theta_k = (1-\varepsilon)^k$, $c = \varepsilon$.) Then summing over j in (2.5) we have

$$\frac{1}{|S|}\int_S \sum_{\tilde{R}_j \text{ before } R} \chi_{\tilde{R}_j} \geq \frac{1}{|R|}\int_R \sum_{\tilde{R} \text{ before } R} \chi_{\tilde{R}_j} \geq \frac{1}{2}\, c$$

in other words

$$M^{(2)}\Big(\sum \chi_{\tilde{R}_j}\Big) > \frac{1}{2}\, c \quad \text{on} \quad \cup R_j$$

and by the boundedness of $M^{(2)}$ we see that

$$|\cup R_j| \leq C \Big\|\sum \chi_{\tilde{R}_j}\Big\|_2^2 \leq C'|\cup \tilde{R}_j|$$

by (2.5). We have controlled $\mathfrak{m}$ by $M^{(2)}$ here in just exactly the way $M^{(n)}$ was previously controlled by $M^{(n-1)}$. And while $\mathfrak{m}$ is a 3-parameter maximal operator, $M^{(2)}$ is a 2-parameter one.

Finally, we should note that, as before, this covering lemma implies the maximal theorem for $\mathfrak{m}$ as claimed. In fact, suppose we have shown that given $\{R_k\}$ there exists $\{\tilde{R}_k\}$ a subsequence so that

(1) $|\cup\tilde{R}_k| \geq c|\cup R_k|$

(2) $\|\sum \chi_{R_k}\|_{p'} \leq c|\cup R_k|^{1/p'}$, (here $\frac{1}{p}+\frac{1}{p'}=1$).

Then

$$|\{\mathfrak{m}f>\alpha\}| \leq \left(C\,\frac{\|f\|_p}{\alpha}\right)^p .$$

In fact we have, by definition $\{R_k\}$ so that $\{\mathfrak{m}f>\alpha\}\subseteq \cup R_k$ and $\frac{1}{|R_k|}\int_{R_k}|f|>\alpha$ for all k. By the covering lemma, select the class $\{\tilde{R}_k\}$. Then $|\tilde{R}_k| \leq \frac{1}{\alpha}\int_{\tilde{R}_k}|f|$ and so

$$|\cup\tilde{R}_k| \leq \frac{1}{\alpha}\int f\sum \chi_{\tilde{R}_k} \leq \frac{1}{\alpha}\|f\|_p\|\sum\chi_{\tilde{R}_k}\|_{p'}$$

$$\leq \frac{C}{\alpha}\|f\|_p|\cup R_k|^{1/p'} \leq C'\|f\|_p|\cup\tilde{R}_k|^{1/p'}$$

and the estimate on $|\{\mathfrak{m}f>\alpha\}|$ follows by a division of both sides by $|\cup\tilde{R}_k|^{1/p'}$.

Our last topic for this lecture will be the so-called Zygmund Conjecture. I believe it was Zygmund who was the first to realize the difference between the one-parameter and several parameter harmonic analysis. Particularly, he remarked that in differentiation theory a "big picture" was evolving. He considered n functions $\phi_1,\phi_2,\cdots,\phi_n$ of the positive real variable t, with each $\phi_i(t)$ increasing and the family of rectangles

$\{R_t\}_{t \geq 0}$ given by $R_t = \prod_{i=1}^{n} \left[-\frac{\phi_i(t)}{2}, \frac{\phi_i(t)}{2}\right]$. Form a maximal operator M defined by

$$M(f)(x) = \sup_{t>0} \frac{1}{|R_t|} \int_{R_t} |f(x+y)|\, dy \,.$$

Then M is of weak type 1-1, just as in the special case of the Hardy-Littlewood operator where $\phi_i(t) = t$. Zygmund noticed that the proof of this was virtually the same as the Hardy-Littlewood theorem. All one had to do was to prove a Vitali-type covering lemma for R_t's using the fact that if $\mathfrak{P}$ is the class of all translates of the R_t and if $R, S \in \mathfrak{P}$ and $R \cap S \neq \emptyset$ and if R corresponds to a bigger value of t than does S, then $S \subseteq \widetilde{R}$, the 5-fold dilation of R. Next, he considered the collection of rectangles $R_{s,t}$, $s,t > 0$ where

$$R_{s,t} = \left[-\frac{s}{2}, \frac{s}{2}\right] \times \left[-\frac{t}{2}, \frac{t}{2}\right] \times \left[-\frac{\phi(s,t)}{2}, \frac{\phi(s,t)}{2}\right]$$

where ϕ is a function increasing in each variable separately, fixing the other variable. In other words, Zygmund next conjectured that since $\mathfrak{R}$ is a 2-parameter family of rectangles in R^3, the corresponding maximal operator, which we shall call M_Z should behave like the model 2-parameter operator $M^{(2)}$ in R^2:

$$|\{M_Z f(x) > \alpha, |x| < 1\}| \leq \frac{C}{\alpha} \|f\|_{L(\log L)(|x|<1)} \,.$$

Not long ago, using the methods we have just discussed, Cordoba was able to prove this [7]. Let us give the proof. Suppose $\{R_k\}$ is a sequence of rectangles with side lengths s,t, and $\phi(s,t)$ in R^3. We must show there exists a subcollection $\{\widetilde{R}_k\}$ such that

(1) $|\cup \widetilde{R}_k| > c|\cup R_k|$

(2) $\|\sum \chi_{\widetilde{R}_k}\|_{\exp(L)} \leq C$.

To prove this, order the R_k so that the z side lengths are decreasing. With no loss of generality, we may assume that $|R_k \cap [\bigcup_{j<k} R_j]| < \frac{1}{2}|R_k|$, that there are finitely many R_k and that the R_k are dyadic. (In fact, we may assume this because if $\frac{1}{|R|}\int_R |f| > \alpha$ for some $R \in \mathfrak{R}$ containing x, then there exists a dyadic R_1 whose $\tilde{R}_1$ (double) contains x such that $\frac{1}{|R_1|}\int_{R_1} |f| > \frac{\alpha}{C}$.) Now let $\tilde{R}_1 = R$ and, given $\tilde{R}_1, \cdots, \tilde{R}_k$, select $\tilde{R}_{k+1}$ as follows: Let $\tilde{R}_{k+1}$ be the first R on the list of R_k so that

$$\frac{1}{|R|}\int_R \exp\left(\sum_{j\le k} \chi_{\tilde{R}_j}\right) dx \le C .$$

We claim that the $\tilde{R}_k$ satisfy $\int_{|x|\le 1} \exp(\Sigma\, \chi_{\tilde{R}_k}) \le C'$. To see this let the $\tilde{R}_k$ be $\tilde{R}_1, \cdots, \tilde{R}_N$ and let $\tilde{E}_j = \tilde{R}_j - \bigcup_{k>j} \tilde{R}_k$. Then

$$\int_{\cup\tilde{R}_j} \exp\left(\sum_{k=1}^{N} \chi_{\tilde{R}_k}\right) = \int_{\tilde{E}_N} \exp\left(\sum_{k=1}^{N} \chi_{\tilde{R}_k}\right) dx + \int_{\tilde{E}_{N-1}} \exp\left(\sum_{k=1}^{N-1} \chi_{\tilde{R}_n}\right) + \cdots + \int_{\tilde{E}_1} \exp(\chi_{\tilde{R}_1})$$

and

$$\int_{\tilde{E}_j} \exp\left(\sum_{k=1}^{j} \chi_{\tilde{R}_k}\right) \le C \int_{\tilde{R}_j} \exp\left(\sum_{k<j} \chi_{\tilde{R}_k}\right) \le C|\tilde{R}_j| ,$$

so we have

$$\exp\left(\sum \chi_{\tilde{R}_k}\right) \le C \sum |\tilde{R}_j| \le C' |\cup\tilde{R}_j| .$$

Now let us show that $|\cup\tilde{R}_j| > c|\cup R_j|$. Let R be an unselected rectangle. Then

$$\frac{1}{|R|}\int_R \exp\left(\sum \chi_{\tilde{R}_k}\right)dx \geq C$$

where the sum extends only over those chosen $\tilde{R}_k$ which precede R. Let us slice R with a hyperplane in the x_1,x_2 direction. Call $S,\tilde{S}_j$ the slices of R and $\tilde{R}_j$. Then

$$\frac{1}{|S|}\int_S \exp\left(\sum \chi_{\tilde{S}_j}\right)dx_1dx_2 \geq C\ .$$

(Again we sum only over those $\tilde{S}_j$ which appear before S.) Now, each $\tilde{R}_j$ appearing before R has the property that its x_3 (or z) side length exceeds that of R. It follows that each corresponding $\tilde{S}_j$ has either its x_1 or x_2 side length longer than that of S. Call those $\tilde{S}_j$ having longer x_1 side length than x_1 length of S *of type I*, and the other of type II. Put $S = I\times J$. Then

$$C < \frac{1}{|I\times J|}\iint_{I\times J}\exp\left(\sum \chi_{\tilde{S}_j}\right)dx_1dx_2 = \frac{1}{|I|\,|J|}\int_I\int_J \exp\left(\sum_{I}\chi_{\tilde{S}_j}+\sum_{II}\chi_{\tilde{S}_j}\right)dx_1dx_2$$

$$= \frac{1}{|I|}\int_I \exp\left(\sum_{II}\chi_{\tilde{R}_j}\right)dx_1\ \frac{1}{|J|}\int_J \exp\left(\sum_{I}\chi_{\tilde{R}_j}\right)\cdot$$

so it follows that

$$M_{x_1}\left[\exp\left(\sum \chi_{\tilde{R}_j}\right)\right]M_{x_2}\left[\exp\left(\sum \chi_{\tilde{R}_j}\right)\right] > C$$

on $\cup R_j$; hence

$$\cup R_j \subseteq \left\{M_{x_1}\left[\exp\left(\sum \chi_{\tilde{R}_j}\right)\right] > \sqrt{C}\right\} \cup \left\{M_{x_2}\left[\exp\left(\sum \chi_{\tilde{R}_j}\right)\right] > \sqrt{C}\right\}$$

so

$$|UR_j| \leq C' \left\|\exp\left(\sum \chi_{\tilde{R}_j}\right)\right\|_{L^1} \leq C''|U\tilde{R}_j| .$$

So far what has been done suggests the following general conjecture of Zygmund which says: Let $\{\phi_i(t_1, t_2, \cdots, t_k)\} = \Phi$, $i = 1,2,\cdots,n$ be functions which are increasing in each of the variables $t_i > 0$ separately. Define a k-parameter family of rectangles $R_{t_1 t_2,\cdots,t_k}$ by

$$R_{t_1,t_2,\cdots,t_k} = \prod_{i=1}^{n} \left[\frac{-\phi(t_1,\cdots,t_k)}{2}, \frac{+\phi(t_1,\cdots,t_k)}{2}\right],$$

and a maximal operator on R^n by

$$M_\Phi(f)(x) = \sup_{t_1,t_2,\cdots,t_k>0} \frac{1}{|R_{t_1,\cdots,t_k}|} \int_{R_{t_1,\cdots,t_k}} |f(x+y)|\, dy .$$

Then

$$|\{x \mid |x| < 1, M_\Phi(f)(x) > \alpha\}| \leq \frac{C}{\alpha} \|f\|_{L(\log L)^{k-1}} .$$

Quite recently a beautifully simple counterexample to the general conjecture was given by Fernando Soria of the University of Chicago [8]. Soria's counterexample was: in R^3, consider all rectangles $\mathfrak{S}$ of the form $s \times t\phi_1(s) \times t\phi_2(s)$ where ϕ_i are increasing on $[0,1]$. In fact we claim that given any small rectangle R with side length x,y and z such that $x < y < z \leq 1$ in R^3, there exists a rectangle $S \in \mathfrak{S}$ such that $R \subset S$ and $|S| \leq C|R|$. Clearly, then $M_{\mathfrak{S}}$ cannot be any better than the 3-parameter operator $M^{(3)}$. To do this let ϕ_1 and ϕ_2 be increasing, $\gneq 0$, and continuous on $[0,1]$ so that on each interval $2^{-(k+1)} < s < 2^{-k}$, $\frac{\phi_2(s)}{\phi_1(s)}$ takes every value between 1 and $C \cdot 2^{(k+1)}$. Then given three side lengths $\rho_1, \rho_2, \rho_3 \leq 1$ of R, simply choose s according to the following: say $2^{-k-1} < \rho_1 \le 2^{-k}$. Choose $s \in (2^{-k-1}, 2^{-k})$ satisfying $\frac{\phi_2(s)}{\phi_1(s)} = \frac{\rho_3}{\rho_2}$ (since $1 \le \frac{\rho_3}{\rho_2} \le \rho_1 < 2^{k+1}$ we can do this). Then choose t so that $t\phi_1(s) = \rho_2$. This guarantees $t\phi_2(s) = \rho_3$, and we are finished.

We can make $\frac{\phi_2(s)}{\phi_1(s)}$ assume every value between 1 and $C2^k$ on $[2^{-k-1}, 2^{-k}]$ bt letting $\phi_1(s) = e^{-1/s}$ and $\phi_2(s) = \phi_1(s)$ on $\left[\frac{3}{2} \cdot 2^{-k-1}, 2^{-k}\right]$ and

$$\phi_2(s) = \phi_1\left(\frac{3}{2} \cdot 2^{-k-1}\right) \text{ for all } s \in \left[2^{-k-1}, \frac{3}{2} \cdot 2^{-k}\right].$$

3. *Multiparameter weight-norm inequalities and applications to multipliers*

In this lecture we want to describe further applications of the ideas centering around the covering lemma for rectangles previously described. We shall begin with more about maximal operators, and then move on to multiparameter multiplier operators, and the connection they have with our maximal functions.

The first topic we take up is that of classical weight norm inequalities, which have proven of enormous importance throughout Fourier analysis. Here, we want to know which locally integrable non-negative weight functions $w(x)$ on R^n have the property that some operator T is bounded on $L^p w(x)dx$. The most basic examples are the Hardy-Littlewood maximal operator, and Calderón-Zygmund singular integrals $Tf = f * K$. The theory was developed in R^1 by Muckenhoupt [9] and Hunt, Muckenhoupt and Wheeden [10], and in R^n by Coifman and C. Fefferman [11]. We will present only a small segment of that theory now and list some relevant facts for which the interested reader should see the Studia article of Coifman-C. Fefferman [11].

It is no coincidence that the class of weights w for which the Hardy-Littlewood maximal operator is bounded on $L^p(w)$ is exactly the same as the class of w for which all Calderon-Zygmund operators are bounded on $L^p(w)$. This is the so-called A^p class of Muckenhoupt.

A nonnegative locally integrable function $w(x)$ on R^n is said to belong to A^p if and only if for each cube $Q \subseteq R^n$

$$\left(\frac{1}{|Q|}\int_Q w\,dx\right)\left(\frac{1}{|Q|}\int_Q w^{-1/(p-1)}dx\right)^{p-1} \leq C\,.$$

The smallest such C is called the A^p norm. We say that $w \in A^\infty$ if and only if, whenever Q is a cube and $E \subset Q$, if $|E|/|Q| > 1/2$ then $w(E)/w(Q) > \eta$ for some $\eta > 0$.

Let us list some properties of A^p classes:

(α) If $\rho > 0$ and $w \in A^p$ then $\rho w \in A^p$ with the same norm as w.

(β) If $w \in A^p$ and $\delta > 0$ then $w(\delta x) \in A^p$ with the same norm as w.

(γ) If $w \in A^p$ then $w^{-1/(p-1)} \in A^{p'}$ where $\frac{1}{p} + \frac{1}{p'} = 1$.

(δ) If $w \in A^p$ then $w \in A^\infty$. In fact, if $w \in A^p$ by (α) and (β) it is enough to show that if $|Q| = 1$, and $\int_Q w = 1$ then $|E| > 1/2$ implies $w(E) > \eta$.

(For, in general if Q is arbitrary of side δ and $|E| > 1/2|Q|$, consider $w(\delta x)$ on Q/δ and multiply $w(\delta x)$ by the right constant ρ to have $\int_{1/\delta} \rho w(\delta x) = 1$. Then $\rho w(\delta x)$ on E/δ would have

$$\frac{[\rho w(\delta x)](E/\delta)}{[\rho w(\delta x)](Q/\delta)} > \eta \iff \frac{w(E)}{w(Q)} > \eta\,.)$$

But by the A^p condition,

$$c\left(\int_E w\right)^{-1} \leq \left(\int_E w^{-1/(p-1)}\right)^{p-1} \leq \left(\int_Q w^{-1/(p-1)}\right)^{p-1} \leq C$$

and so

$$\int_E w \geq \frac{c}{C}\,.$$

So far all the properties of A^p weights listed are obvious and follow straight from the definitions. There are some deeper properties which though not difficult to prove are not immediate.:

(ε) If $w \in A^p$ then w satisfies a Reverse Hölder Inequality:

$$\left(\frac{1}{|Q|}\int_Q w^{1+\delta}\right)^{1/(1+\delta)} \leq C_\delta\left(\frac{1}{|Q|}\int_Q w\right)$$

for all cubes Q with $\delta > 0$. The constant C_δ may be taken arbitrarily close to 1 as $\delta > 0$.

(ζ) From (ε) it is immediate (see also (γ)) that $w \in A^p$ implies $w \in A^q$ for some $q < p$.

(η) If f is a locally integrable function in some L^p space and $0 < \alpha < 1$ then $(Mf)^\alpha \in A^1$, i.e., $M((Mf)^\alpha)(x) \leq C(Mf)^\alpha(x)$ (for $w \in A^1$ implies $w \in A^p$ for all $p > 1$).

To prove this let $f \in L^p(R^n)$ be given and $\alpha \in (0,1)$. Let Q be a cube centered at $\bar{x}$, and $\tilde{Q}$ its double. Then write $f = \chi_{\tilde{Q}}f + \chi_{c\tilde{Q}}f = f_I + f_0$. We must show that

$$\frac{1}{|Q|}\int_Q M(f_I)^\alpha dx \leq CM(f)^\alpha(\bar{x})$$

and

$$\frac{1}{|Q|}\int_Q M(f_0)^\alpha dx \leq CM(f)^\alpha(\bar{x}) .$$

As for the first inequality,

$$\left(\frac{1}{|Q|}\int_Q M(f_I)^\alpha dx\right)^{1/\alpha} \leq \left(\frac{1}{|\tilde{Q}|}\int_{\tilde{Q}} |f_I| dx\right) = \frac{1}{|\tilde{Q}|}\int_{\tilde{Q}} |f| .$$

(This is an immediate consequence of the weak type estimate for M on L^1.) This shows that

$$\left(\frac{1}{|Q|}\int_Q M(f_I)^\alpha dx\right) \le \left(\frac{1}{|\tilde{Q}|}\int_{\tilde{Q}} |f|\right)^\alpha \le M(f)^\alpha(\overline{x}) .$$

As for the second inequality, choose a cube C centered at $\overline{x}$ such that

$$\frac{1}{|C|}\int_C |f_0|\,dx \ge \frac{1}{2} M(f_0)(\overline{x}) .$$

Now, if $x \in Q$ then any cube C' centered at x which intersects ${}^c\tilde{Q}$ is contained inside a cube of comparable volume centered at $\overline{x}$ so that

$$\frac{1}{|C'|}\int_{C'} |f_0| \le A \frac{1}{|C|}\int_C |f_0|\,dx .$$

We have proven that for all $x \in Q$, $M(f_0)(x) \le A \frac{1}{|C|} \int_C |f_0|\,dx$ so that

$$\frac{1}{|Q|}\int_Q M(f_0)^\alpha dx \le A^\alpha \left(\frac{1}{|C|}\int_C |f_0|\,dx\right)^\alpha \le A^\alpha M(f)^\alpha(\overline{x}) .$$

Let us begin to discuss the weight theory by showing that the Hardy-Littlewood maximal operator is bounded on $L^p(w)$ if and only if $w \in A^p$, $1 < p < \infty$ [12]. In the first place if $f = w^{-1/(p-1)}\chi_Q$ and if M is bounded on $L^p(w)$ one sees right away that

$$w(Q)\left(\frac{1}{|Q|}\int_Q w^{-1/(p-1)}dx\right)^p \le C \int_Q w^{-p/(p-1)}w\,dx$$

or

$$\frac{w(Q)}{|Q|}\left(\frac{1}{|Q|}\int_Q w^{-1/(p-1)}\right)^{p-1} \leq C\,.$$

Conversely, assume $w \in A^p$. Calderón-Zygmund decompose $f \in L^p(w)$ at heights C^k where $k \in Z$, and C is large (to be described later) and get Calderon-Zygmund cubes $\{Q_j^k\}_{j=1}^{\infty}$ so that

$$\frac{1}{|Q_j^k|}\int_{Q_j^k} f \sim C^k \quad \text{and} \quad \{Mf > \gamma C^k\} \subseteq \cup \tilde{Q}_j^k$$

(γ is a large constant dependent only on n). Then

$$\int_{R^n} (Mf)^p w\,dx \leq C' \sum_{k,j} w(\tilde{Q}_j^k) C^k \leq C' \sum_{k,j} w(\tilde{Q}_j^k)\left(\frac{1}{|Q_j^k|}\int_{Q_j^k} f\right)^p .$$

Now $w \in A^p \Longrightarrow w \in A^\infty$ therefore $w(\tilde{Q}_j^k) \leq C'' w(Q_j^k)$ and so the above expression is

$$\leq C''' \sum_{k,j} w(Q_j^k)\left(\frac{\sigma(Q_j^k)}{|Q_j^k|}\right)\left(\frac{1}{\sigma(Q_j^k)}\int_{Q_j^k} f\sigma^{-1}\sigma dx\right)^p \leq \text{by the } A^p \text{ condition on}$$

$$(*) \quad \leq C''' \sum_{k,j} \sigma(Q_j^k)\left(\frac{1}{\sigma(Q_j^k)}\int_{Q_j^k} (f\sigma^{-1})\,\sigma dx\right)^p \quad \text{where} \quad \sigma = w^{-1/(p-1)}\,.$$

So far we have used only arithmetic. Now we come to the main point. If $E_j^k = Q_j^k - \bigcup_{\ell > k} Q_j^\ell$ then choosing C large enough insures that $|E_j^k| > \frac{1}{2}|Q_j^k|$, and since $\sigma \in A^{p'} \Longrightarrow \sigma \in A^\infty$ we have $\sigma(E_j^k) > \eta\sigma(Q_j^k)$ and so

$$(*) \qquad \leq C \sum \sigma(E_j^k) \left(\frac{1}{\sigma(Q_j^k)} \int_{Q_j^k} (f\sigma^{-1})\, d\sigma \right) \leq C \int_{R^n} M_\sigma^p(f\sigma)\, d\sigma$$

where $M_\sigma(f)(x) = \sup_{x \in Q} \frac{1}{\sigma(Q)} \int_Q |f|\, d\sigma$. Now the same proof that works to show that the Hardy-Littlewood maximal operator is bounded on $L^p(R^n)$ proves that if σ satisfies a doubling condition, then for $p > 1$, M_σ is bounded on $L^p(d\sigma)$.

It follows that the second inequality is

$$\leq C \int f^p \sigma^{1-p}\, dx = \int f^p w\, dx .$$

Note that the operator $M_\mu f(x) = \sup_{x \in Q} \frac{1}{\mu(Q)} \int_Q |f|\, d\mu$ and its boundedness on $L^p(d\mu)$ enter in a crucial way the proof of the weight norm inequalities for the Hardy-Littlewood operator. This operator M_μ is interesting in its own right, since it is natural to ask what happens if we replace the God-given Lebesgue measure by another measure $d\mu$.

In fact, if μ is any measure finite on compact sets and if, in the definition of M_μ we insist that the balls be *centered* at x then M_μ is bounded on $L^p(\mu)$, $p > 1$. The proof of this remarkable theorem relies on a refinement of the usual Vitali covering lemma due to Besocovitch. We should also remark that the original proof of the weight norm inequalities for M on R^n made use of M_μ as well. In fact, if $w \in A^p$ it is not hard to see that

$$M(f)(x) \leq C M_w(f^p)^{1/p}(x) .$$

(Indeed,

$$\frac{1}{|Q|}\int_Q f\,dx = \frac{1}{|Q|}\int_Q f w^{1/p} w^{-1/p} dx \le \frac{1}{|Q|}\left(\int_Q f^p w\,dx\right)^{1/p}\left(\int_Q w^{-p'/p} dx\right)^{1/p'}$$

$$= \frac{w(Q)}{|Q|}^{1/p}\left(\frac{1}{w(Q)}\int_Q f^p w\right)^{1/p}\left(\frac{1}{|Q|}\int_Q w^{-1/(p-1)} dx\right)^{(p-1)/p} \le CM_w(f^p)^{1/p}(x)$$

The proof would be complete if $M_w(f^p)^{1/p}$ were bounded on $L^p(w)$. Unfortunately, $M_w(f^p)^{1/p}$ is bounded on $L^q(w)$ only for $q > p$. But if we notice that $w \in A_p \Longrightarrow w \in A_{p-\varepsilon}$ then the proof is complete.

Anyway, it is clear that the operator M_μ arises naturally. Next we shall study its n-parameter variant $M_\mu^{(n)}$ just the strong maximal operator, only taken relative to the measure μ:

$$M_\mu^{(n)}(f)(x) = \sup_{x \in R} \frac{1}{\mu(R)} \int_R |f|\,d\mu\,.$$

R a rectangle with sides parallel to the axes. Unlike the case where $n = 1$, restrictions must be placed on μ whether or not R is centered at x.

The following result gives conditions on μ which are rather unrestrictive, and which guarantee the boundedness on $L^p(\mu)$ on $M_\mu^{(n)}$ [13]:

THEOREM. *Suppose* μ *is absolutely continuous and non-negative on* R^n, *and that the Radon-Nikodym derivative of* μ, $w(x)$ *satisfies an* A^∞ *condition in each variable separately, uniformly in the other variables. Then* $M_\mu^{(n)}$ *is bounded on* $L^p(d\mu)$ *for all* $p > 1$.

We require the following lemma.

LEMMA. *If* $w(x_1,\cdots,x_n)$ *is uniformly in* A^∞ *in each of the variables separately, then* w *satisfies the following: If* R *is any rectangle with its sides parallel to the axes and* $E \subseteq R$ *is such that* $\frac{|E|}{|R|} > \frac{1}{2}$, *then* $\frac{\int_E w}{\int_R w} > \eta$, *for some* $\eta > 0$.

Proof. The proof is by induction on n. Assume this for $n-1$. Consider a rectangle R as above, $R = I \times J$ where I is n-1 dimensional and J one-dimensional, and a subset E of R such that

$$\frac{|E|}{|R|} > \frac{1}{2}.$$

For each $x \in I$ let $J_x = \{(x,y) | y \in J\}$ be a vertical segment. Since $\frac{|E|}{|R|} > \frac{1}{2}$ it is easy to see that for x in a set I′ of measure $\geq \frac{1}{100}|I|$ we have $|E \cap J_x| > \frac{1}{100}|J_x|$. Now since w is uniformly A^∞ in the x_n variable,

$$\int_{J_{x'} \cap E} w\, dx_n \geq \eta \int_{J_{x'}} w\, dx_n, \quad x \in I'. \tag{3.1}$$

But also if we fix any $x_n \in J$ then

$$\int_{I'} w\, dx' \geq \eta' \int_I w\, dx' \tag{3.2}$$

by induction. It follows by integrating (3.1) in $x' \in I'$ that

$$\int_E w\, dx \geq \eta \int_{I' \times J} w\, dx$$

and integrating (3.2) in $x_n \epsilon J$ gives,

$$\int_{I'\times J} w\, dx \geq \eta' \int_{I\times J} w\, dx\ ,$$

which shows that $\int_E w > \eta\eta' \int_R w$ and proves the lemma.

Proof of the Theorem. We prove a covering lemma, namely, if $\{R_k\}$ are rectangles with sides parallel to the axes, there exists $\{\widetilde{R}_k\}$ so that $w(\cup R_k L \leq Cw(\cup\widetilde{R}_k)$ and

$$\|\sum \chi_{\widetilde{R}_k}\|_{L^{p'}(w)} \leq Cw(\cup\widetilde{R}_k)^{1/p'}\ .$$

To prove this order R_k by decreasing x_n side length, and then select a rectangle R when

$$|R \cap [\cup(\widetilde{R}_k)_d]| < \frac{1}{2}\ |R|$$

where the union is taken over all those k such that $\widetilde{R}_k$ precedes R and $\widetilde{R}_k \cap R = \emptyset$.

Now if we slice R, an unselected rectangle, with a hyperplane perpendicular to the x_n direction we have

$$|S \cap [\cup(\widetilde{S}_k)_d]| > \frac{1}{2}\ |S| \Longrightarrow w(S \cap [\cup(\widetilde{S}_k)_d]) > \eta w(S)$$

so that $M_w^{(n-1)}(\chi_{\cup(\widetilde{R}_k)_d}) > \eta$ on $\cup R_k$. By induction, $w(\cup R_k) \leq Cw(\cup(\widetilde{R}_k)_d)$ Now the $\widetilde{R}_k$ have disjoint parts property with respect to dx and so they have this property with respect to w dx also. It follows that

$$w(\cup(\widetilde{R}_k)_d) \leq \sum w(\widetilde{R}_k)_d \leq C \sum w(\widetilde{R}_k) \leq C'w(\cup\widetilde{R}_k)\ .$$

Therefore

$$w(\cup R_k) \leq Cw(\cup\widetilde{R}_k)\ .$$

Now to estimate $\|\Sigma \chi_{\tilde{R}_k}\|_{L^{p'}(w)}$, let us slice the $\tilde{R}_k$ with a hyperplane perpendicular to x_n, calling the slices $\tilde{S}_k$. Then $|\tilde{S}_k - \bigcup_{j<k} \tilde{S}_j| > \frac{1}{2} |\tilde{S}_k|$ so since $w \in A^\infty$ in $x_1,\cdots,x_{n-1}$ we have $w(\tilde{S}_k - \bigcup_{j<k} \tilde{S}_j) > \eta w(\tilde{S}_k)$ (if $w(E) = \int_E w(x_1,\cdots,x_{n-1},x_n)dx_1,\cdots,dx_{n-1}$) and we test

$$\int_{R^{n-1}} \sum \chi_{\tilde{S}_k} \phi w = \sum \int_{\tilde{S}_k} \phi w dx \leq C \sum w(\tilde{E}_k) \frac{1}{w(\tilde{S}_k)} \int_{\tilde{S}_k} \phi w\, dx \leq$$

$$\leq C \int_{\tilde{S}_k} M_w^{(n-1)}(\phi) w\, dx .$$

$(\tilde{E}_k = \tilde{S}_k - \bigcup_{j<k} \tilde{S}_j)$. By induction $M_w^{(n-1)}$ is bounded on $L^p(w)$ so this last integral is

$$\leq \|\phi\|_{L^p(w)} \|\chi_{\cup \tilde{S}_k}\|_{L^{p'}(w)} \leq \|\phi\|_{L^p(w)} w(\cup \tilde{S}_k)^{1/p} .$$

This shows that $\|\Sigma \chi_{\tilde{S}_k}\|_{L^{p'}(w)} \leq C w(\cup \tilde{S}_k)^{1/p'}$. Raising this to the pth power and integrating in x_n we have

$$\|\sum \chi_{\tilde{R}_k}\|_{L^p(w)} \leq C w(\cup \tilde{R}_k) .$$

REMARK. Given this covering lemma, cover the set $\{M_w^{(n)}(f) > \alpha\}$ by R_k such that $\frac{1}{w(R_k)} \int_{R_k} fw > \alpha$. Then we need only estimate $w(\cup \tilde{R}_k)$ of the covering lemma. But

$$w(\cup \tilde{R}_k) \leq \sum w(\tilde{R}_k) \leq \frac{1}{\alpha} \int f \sum \chi_{\tilde{R}_k} w \leq \frac{1}{\alpha} \|f\|_{L^p(w)} \|\sum \chi_{\tilde{R}_k}\|_{L^{p'}(w)}$$

$$\leq \frac{1}{2} \|f\|_{L^p(w)} w(\cup \tilde{R}_k)^{1/p'} .$$

The maximal operator is weak type (p,p), $p > 1$ and we are finished by interpolation.

One application of this theorem is that with it, one can obtain weighted norm inequalities for multi-parameter maximal operators which cannot be handled directly through iteration. We give the following example.

Suppose $\mathfrak{R}$ denotes the family of rectangles with side lengths of the form s,t, and $s\cdot t$ in R^3, where s and $t > 0$ are arbitrary. (Suppose the sides are also parallel to the axes.) Define the corresponding maximal operator M by

$$Mf(x) = \sup_{x \in R \in \mathfrak{R}} \frac{1}{|R|} \int_R |f|\, dt\,.$$

Then it is natural to ask for which weights w do we have

$$\int Mf^p w \le C \int f^p w\,.$$

The answer is $A^p(\mathfrak{R})$ where this class is defined in the obvious way [14]:

$$w \in A^p(\mathfrak{R}) \text{ if and only if } \left(\frac{1}{|R|}\int_R w\right)\left(\frac{1}{|R|}\int_R w^{-1/(p-1)}\right)^{p-1} \le C$$

for all $R \in \mathfrak{R}$. To prove this result we need the following.

LEMMA. *If* $w \in A^p(\mathfrak{R})$ *then* w *satisfies a reverse Hölder inequality.*

Proof. Since w is uniformly A^p in the x_1 variable, w will satisfy a reverse Hölder inequality uniformly in that variable:

$$\left(\frac{1}{|I|}\int_I w(x_1,p)^{1+\delta}\, dx_1\right)^{1/(1+\delta)} \le C\left(\frac{1}{|I|}\int_I w(x_1,p)\, dx_1\right).$$

(C independent of p). Fix I, an interval of the x_1-axis, and define a measure in the x_2, x_3 plane whose Radon-Nikodym derivative $W(p)$ is defined by

$$W(p) = \left(\frac{1}{|I|}\int_I w^{1+\delta}(x_1,p)\,dx_1\right)^{1/(1+\delta)} .$$

We claim that W satisfies an A^∞ condition uniformly (in I) relative to the class of rectangles whose side lengths are t, $|I|t$ in the x_2,x_3 plane.

Then let S be such a rectangle in the x_2,x_3 plane and $E \subseteq S$ such that $|E|/|S| \le 1/2$. Then

$$\int_E W(p)\,dp \le C \int_E \left(\frac{1}{|I|}\int_I w(x_1,p)\,dx_1\right)dp = C\,\frac{1}{|I|}\iint_{E\times I} w(x_1,p)\,dx_1 dp .$$

Since $w \in A^\infty(\mathfrak{R})$, $\iint_{E\times I} w \le (1-\eta) \iint_R w$ and so $\int_E W(p)\,dp \le C(1-\eta) \int_S W(p)\,dp$, and by taking δ small enough, C will be so close to 1 that $C(1-\eta) < 1$ and W is uniformly A^∞ on the collection of all such S. Therefore since S is just 1-parameter (just a linear change of variable in one of the x_2 or x_3 variable away from squares) we have that W satisfies a reverse Hölder inequality: (For δ' some value $< \delta$)

$$\left(\frac{1}{|S|}\int_W W^{1+\delta}\,dp\right)^{1/(1+\delta')} \le C\,\frac{1}{|S|}\int_S W .$$

This shows that

$$\left(\frac{1}{|S|}\int_S \left(\frac{1}{|I|}\int_I w^{1+\delta}dx_1\right)^{(1+\delta)^{-1}(1+\delta')} dp\right)^{(1+\delta')^{-1}} \le C'\,\frac{1}{|S|}\int_S \left(\frac{1}{|I|}\int_I w\,dx_1\right) dp$$

and so

$$\left(\frac{1}{|R|}\int_R w^{1+\delta'}\right)^{1/(1+\delta')} \le C'\frac{1}{|R|}\int_R w\,, \quad R \in \mathfrak{R}\,.$$

Now on to the theorem. Because $w \in A^p(\mathfrak{R})$ and $w^{-1/(p-1)} \in A^{p'}(\mathfrak{R})$ satisfies a reverse Holder inequality, $w \in A^{p-\varepsilon}(\mathfrak{R})$ for some $\varepsilon > 0$. It follows that

$$Mf(x) \le CM_w(f^{p-\varepsilon})^{1/(p-\varepsilon)}\,,$$

and it just remains to show that M_w is bounded on $L^p(w)$.

A quick review of the proof that $M_w^{(n)}$, $n = 3$, is bounded on $L^p(w)$ reveals that all we really used was that w satisfy an A^∞ condition in the x_1 and x_2 variables as well as a doubling condition in the x_3 variable: $w((R)_d) \le Cw(R)$. All of these are satisfied by our w here, and this concludes the proof since $M_w(f) \le M_w^{(3)}(f)$.

Now we wish to relate some of our results on multi-parameter maximal functions to the theory of multiplier operators.

We shall work in R^2, and consider the following basic question: For which sets $S \subseteq R^2$ is $\chi_S(\xi)$ a multiplier on $L^p(R^2)$ for some $p \ne 2$? For χ_S to be a multiplier of course means that, if for $f \in C_c^\infty(R^2)$ we set $Tf\,\hat{}\,(\xi) = \chi_S(\xi)\hat{f}(\xi)$ then we have the a priori estimate

$$\|Tf\|_{L^p(R^2)} \le C\|f\|_{L^p(R^2)}\,.$$

In his celebrated theorem, Charles Fefferman showed that if S is a nice open set in R^2 whose boundary has some curvature then $\chi_S(\xi)$ will only be an L^2 multiplier [15]. The other nice regions left are those whose boundaries are comprised of polygonal segments. If S is a convex polygon then χ_S will obviously be an L^p multiplier for all p, $1 < p < \infty$, just because of the boundedness of the Hilbert transform on the L^p spaces in R^1. The case that remains is the one we consider here. Let $\theta_1 > \theta_2 > \theta_3 \cdots > \theta_n > \theta_{n-1} \to 0$ be a given sequence of angles θ and let

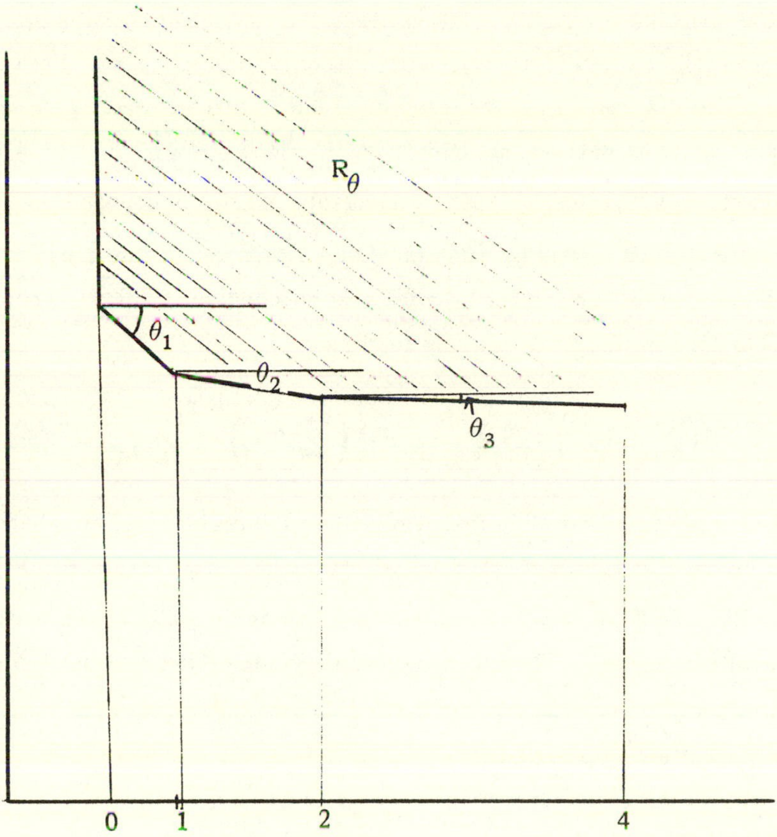

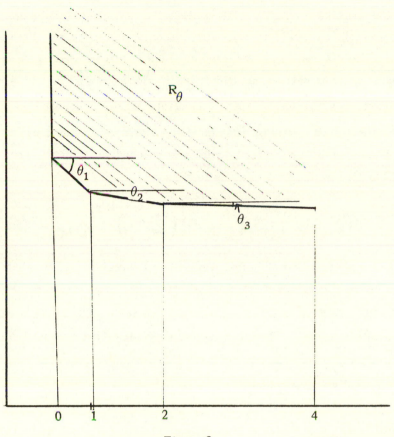

Figure 2

R_θ be the polygonal region pictured above. Then we shall define T_θ by

$$T_\theta f\,\hat{}\,(\xi) = \chi_{R_\theta}(\xi)\hat{f}(\xi)\,.$$

Consider as well the maximal operator M_θ defined by

$$M_\theta f(x) = \sup_{x \in R \in B_\theta} \frac{1}{|R|} \int_R |f|\,dt$$

where B_θ is the family of all rectangles in R^2 which are oriented in one of the directions θ_k, but whose side lengths are arbitrary. We claim that the behavior of T_θ on $L^p(R^2)$ for $p > 2$ is linked with the behavior of M_θ on $L^{(p/2)'}((p/2)'$ is the exponent dual to $p/2$). More precisely, suppose that T_θ is bounded on $L^p(R^2)$ and we assume the weakest possible estimate on M_θ, namely $|\{M_\theta \chi_E > \frac{1}{2}\}| \leq C|E|$. Then M_θ is of weak type on $L^{(p/2)'}$. Conversely, if M_θ is bounded on $L^{(p/2)'}(R^2)$ then T_θ is bounded on $L^q(R^2)$, for $p' < q < p$ [16].

To prove this assume first that T_θ is bounded on $L^p(R^2)$. Then the first step is to notice that this implies that

$$\left\|\left(\sum |T_k f_k|^2\right)^{1/2}\right\|_{L^p(R^2)} \leq C\left\|\left(\sum |f_k|^2\right)^{1/2}\right\|_{L^p(R^2)}.$$

This is proven by observing that if we dilate the region R_θ by a huge factor ρ to get R_θ^ρ and translate R_θ^ρ properly (by τ_k) then R_θ^{ρ,τ_k} looks like a half plane with boundary line making an angle of θ_k with the positive x axis. Then if $r_k(t)$ are the Rademacher functions, and $T_\theta^{\rho,\tau_k} f\hat{} = \chi_{R_\theta^{\rho,\tau_k}} f\hat{}$ then

$$T_\theta^{\rho,\tau_k} f = e^{-i\tau_k \cdot x} T_\theta^\rho(e^{i\tau_k \cdot x} f)$$

and

$$\left\|\left(\sum |T_\theta^{\rho,\tau_k} f_k|^2\right)^{1/2}\right\|_{L^p}^p = \left\|\left(\sum |T_\theta^\rho(e^{i\tau_k \cdot x} f_k)|^2\right)^{1/2}\right\|_{L^p}^p$$

$$= \int_{R^2}\int_0^1 \left|\sum r_k(t)\, T_\theta^\rho(e^{i\tau_k \cdot x} f_k)(x)\right|^p dt\, dx = \int_0^1 \int_{R^2} \left|T_\theta^\rho\left(\sum r_k(t)\, e^{i\tau_k \cdot x} f_k\right)\right|^p dx\, dt$$

$$\leq C \int_{R^2}\int_0^1 \left|\sum r_k(t)\, e^{i\tau_k \cdot x} f_k(x)\right|^p dt\, dx \leq C' \left\|\left(\sum |f_k(x)|^2\right)^{1/2}\right\|_{L^p}.$$

Taking the limit as $\rho \to \infty$, we have

$$\left\| \left(\sum |T_k f_k|^2 \right)^{1/2} \right\|_{L^p(R^2)} \leq C \left\| \left(\sum |f_k|^2 \right)^{1/2} \right\|_{L^p(R^2)},$$

where T_k is the Hilbert transform in the direction θ_k.

The next step is to use the above inequality to prove a covering lemma for rectangles in $\mathfrak{B}_\theta$. Let $\{R_k\}$ be a sequence of such rectangles. Select R given that $\tilde{R}_1, \tilde{R}_2, \cdots, \tilde{R}_{k-1}$ have been chosen provided $|R \cap [\bigcup_{j<k} \tilde{R}_j]| < \frac{1}{2}|R|$. Then if R is unselected $M_\theta(\chi_{\cup \tilde{R}_j}) \geq \frac{1}{2}$ on R so that

$$|\cup R_k| \leq \left|\left\{M_\theta(\chi_{\cup \tilde{R}_k}) > \frac{1}{2}\right\}\right| \leq C|\cup \tilde{R}_k| .$$

Let $E_k = \tilde{R}_k - \bigcup_{j<k} \tilde{R}_j$ and let $f_k = \chi_{\tilde{E}_k}$. Then looking at the picture below, since $|\tilde{E}_k| \geq \frac{1}{2}|\tilde{R}_k|$ on at least a set of measure $> \frac{1}{100}|\tilde{I}_k|$ the

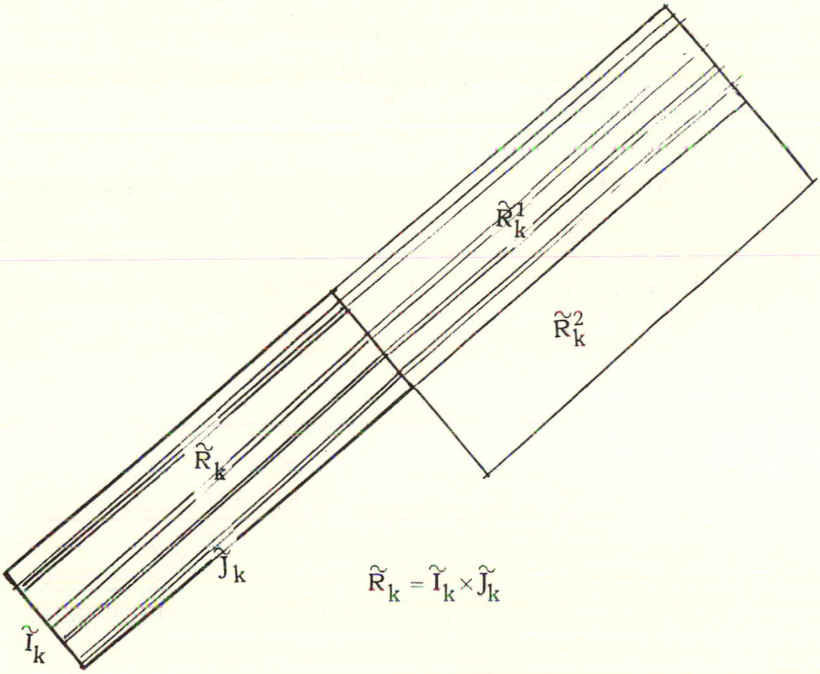

Figure 3

segments pictured contain at least 1/100 of their measure in $\tilde{E}_k$. If we duplicate the rectangle $\tilde{R}_k$ as shown, on these segments $T_k f_k > 1/100$. Applying $T_k' =$ Hilbert transform in the direction perpendicular to θ_k to $T_k f_k$ we see that $T_k'(T_k f_k) > 1/100$ on all of $\tilde{R}_k^2$. Repeating twice more we get

$$\left\| \left(\sum \chi_{\tilde{R}_k} \right)^{1/2} \right\|_{L^p} \le C \left\| \left(\sum |\chi_{\tilde{E}_k}|^2 \right)^{1/2} \right\|_{L^p} \le C \|\chi_{\cup \tilde{R}_k}\|_{L^p} .$$

This shows that M_θ is of weak type $(p/2)'$.

Conversely, assume that M_θ is bounded on $L^{(p/2)'}$. Define $S_k = \{\xi = (\xi_1, \xi_2) | 2^k \le \xi_1 < 2^{k+1}\}$. Then if S_k also stands for the multiplier operator corresponding to S_k, we see that

$$\|T_\theta f\|_q \sim \left\| \left(\sum |S_k T_\theta f|^2 \right)^{1/2} \right\|_q = \left\| \left(\sum |S_k T_k f|^2 \right)^{1/2} \right\|_q .$$

To estimate $\|(\sum |S_k T_k f|^2)^{1/2}\|_q^2$, let $\|\phi\|_{(q/2)'} = 1$ and let us estimate

$$\int \sum |T_k S_k f|^2 \phi \le \sum \int |T_k(S_k f)|^2 \phi .$$

But in R^1 we have the classical weight norm inequality for the Hilbert transform:

$$\int |Hf|^2 \phi \le C \int |f|^2 M(\phi^{1+\varepsilon})^{1/(1+\varepsilon)} \text{ for all } \varepsilon > 0 .$$

It follows that (3.3) is

$$\le \sum \int |S_k(f)|^2 M_\theta(\phi^{1+\varepsilon})^{1/(1+\varepsilon)} . \tag{3.4}$$

M_θ is bounded on $L^{(q/2)'/(1+\varepsilon)}$ if ε is sufficiently small. It follows that (3.4) is

$$\leq C\left\|\left(\sum |S_k f|^2\right)\right\|_{L^{q/2}} \leq C'\|f\|_{L^q}^2$$

proving that T_θ is bounded on L^q.

4. H^p *spaces – one and several parameters*

In this lecture we wish to discuss another chapter of harmonic analysis relating to differentiation theory and singular integrals, namely Hardy Space theory. In this lecture, we shall discuss the one-parameter theory, and, in the next, the theory in several parameters. In the beginning when H^p spaces were first considered, they were spaces of complex analytic functions in $R_+^2 = \{z = x + iy \mid x \in R^1, y \in R^1, y > 0\}$ which satisfies the size restriction

$$\left(\int_{-\infty}^{+\infty} |F(x+iy)|^p dx\right)^{1/p} \leq C \quad \text{for all} \quad y \in R_+^1 .$$

One of the main reasons for introducing these spaces was the connection with the Hilbert transform. If $F(z) = u(z) + iv(z)$ is analytic with u and v real, and if F is sufficiently nice then F will have boundary value $u(x) + iv(x)$ where $v(x)$ is the Hilbert transform of $u(x)$. It turns out, since

$$\int_{-\infty}^{+\infty} |F(x+it)|\, dx$$

increases as $t \to 0$, we have

$$\|F\|_{H^1} \overset{\text{def}}{=} \sup_{t>0} \int_{-\infty}^{+\infty} |F(x+it)| dx \sim \|u\|_{L^1(R^1)} + \|v\|_{L^1(R^1)} .$$

So we may view the space H^1 through its boundary values as the space of all real valued functions $f \in L^1(R^1)$ whose Hilbert transforms are L^1 as well.

If we want a theory of $H^p(R^n)$ then, following Stein and Weiss we may consider the functions $F(x,t)$ in $R_+^{n+1} = \{(x,t) | x \in R^n, t>0\}$ whose values lie in $R^{n+1} : F(x,t) = (u_0(x,t),\cdots,u_n(x,t))$ where the $u_i(x,t)$ satisfy the "Generalized Cauchy-Riemann equations,"

$$\sum_{i=0}^{n} \frac{\partial u_i}{\partial x_i}(x,t) \equiv 0 \quad (t = x_0)$$

and

$$\frac{\partial u_i}{\partial x_j} \equiv \frac{\partial u_j}{\partial x_i} \quad \text{for all } i,j \; .$$

These Stein-Weiss analytic functions are then said to be $H^p(R_+^{n+1})$ if and only if

$$\sup_{t>0} \left(\int_{R^n} |F(x,t)|^p dx \right)^{1/p} = \|F\|_{H^p(R^n)} \qquad [17] .$$

Again, these functions have an interpretation in terms of singular integrals, since if a Stein-Weiss analytic function $F(x,t)$ is sufficiently "nice" on $\bar{R}_+^{n+1}$, then the boundary values $u_i(x)$ satisfy $u_i(x) = R_i[u_0](x)$ where R_i is the ith Riesz transform given by $R_i(f)(x) = f * \frac{c_n x_i}{|x|^{n+1}}$. In particular we may consider an $H^1(R_+^{n+1})$ function (by identifying functions in R_+^{n+1} with their boundary values) as a function f with real values in $L^1(R^n)$ each of whose Riesz transforms $R_i f$ also belong to $L^1(R^n)$. An interesting feature of H^p spaces is that they are intimately connected to differentiation theory as well as singular integrals. To discuss this, let us make some well-known observations. For a harmonic function $u(x,t)$ which is continuous on $\overline{R_+^{n+1}}$ and bounded there, u is given as an average of its boundary values according to the Poisson integral:

$$u(x,t) = f * P_t(x);\ f(x) = u(x,0) \text{ and } P_t(x) = \frac{c_n t}{(|x|^2+t^2)^{(n+1)/2}} .$$

Let $\Gamma(x) = \{(y,t) \mid |x-y| < t\}$. Then since convolving with P_t at a point x can be dominated by an appropriate linear combination of averages of f over balls centered at x of different radii, it follows that

$$\text{if } u^*(x) = \sup_{(y,t)\in\Gamma(x)} |u(y,t)| , \text{ then } u^*(x) \le cMf(x) .$$

Unfortunately, if $u(x,t)$ is harmonic, for $p \le 1$, $u = P[f]$, and $\int_{R^n} |u(x,t)|^p dx \le C$ then the domination $u^* \le CMf$ is not useful, since M is not bounded on L^p, and it is not true in general that $u^*(x) < \infty$ for a.e. $x \in R^n$. On the other hand, suppose F is Stein-Weiss analytic, say $F \in H^1(R^{n+1}_+)$.

Then a beautiful computation shows that if $1 > \alpha > 0$ is close enough to 1 $\left(\alpha \ge \frac{n-1}{n}\right)$ then $\Delta(|F|^\alpha) \ge 0$ so that $|F|^\alpha$ is subharmonic. If $s(x,t)$ is subharmonic and has boundary values $h(x)$ then s is dominated by the averages of h, i.e.,

$$s(x,t) \le P[h](x,t) .$$

Applying this to $G = |F|^\alpha$ (which has $\int G^{1/\alpha}(x,t)dx \le C$ for all $t > 0$) we see that $G^* \le M(h)$ for some $h \in L^{1/\alpha}$. Now M *is* bounded on $L^{1/\alpha}$ so that $M(h) \in L^{1/\alpha}$ and so $G^* \in L^{1/\alpha}$. It follows that $F^* \in L^1$. Just as for a random $f \in L^1(R^n)$ we do not necessarily have $R_i f \in L^1(R^n)$ (singular integrals do not preserve L^1) it is also not true that for an arbitrary L^1 function f that for $u = P[f]$, $u^* \in L^1$. But if $f \in H^1(R^{n+1}_+)$ then $u^* \in L^1(R^n)$. Thus the nontangential maximal function

$$F^*(x) = \sup_{(y,t)\in\Gamma(x)} |F(y,t)| \in L^1(R^n)$$

if and only if the analytic function $f \in H^1(R^{n+1}_1)$.

We know so far that we can characterize H^p functions in terms of singular integrals and maximal functions. There is another characterization which is of great importance. To discuss it, let us return to H^p functions in R^2_+ as complex analytic functions, $F = u + iv$. It is an interesting question as to whether the maximal function characterization of H^p can be reformulated entirely in terms of u. That is, is it true that $F^* \in L^p$ if and only if $u^* \in L^p$? In fact, this is true, and the best way to see this is by introducing a special singular integral, the Lusin-Littlewood-Paley-Stein area integral,

$$S^2(u)(x) = \iint_{\Gamma(x)} |\nabla u|^2(y,t)\,dt\,dy$$

which we already considered in the first lecture. As we shall see later, for a harmonic function $u(x,t)$, $\|S(u)\|_{L^p} \approx \|u^*\|_{L^p}$ for all $p > 0$ [18]. The importance of S here is that the area integral is invariant under the Hilbert transform, i.e.,

$$S(u) \equiv S(v), \text{ since } |\nabla v| = |\nabla u| .$$

When we combine the last two results, we immediately see that

$$\|u^*\|_{L^p} < \infty \Longleftrightarrow \|F^*\|_{L^p} < \infty \Longleftrightarrow F \in H^p(R^2_+) .$$

It is interesting to note that the first proof of $\|S(u)\|_{L^p} \sim \|u^*\|_{L^p}$, $1 \geq p > 0$ was obtained by Burkholder, Gundy, and Silverstein [19] by using probabilistic arguments involving Brownian motion. Nowadays direct real variable proofs of this exist as we shall see later on.

To summarize, we can view functions f in H^p spaces by looking at their harmonic extensions u to R^{n+1}_+ and requiring that u^* or $S(u)$ belong to $L^p(R^n)$.

It turns out that there is another important idea which is very useful concerning H^p spaces and their real variable theory. So far, we have spoken of H^p functions only in connection with certain differential equations. Thus, if we wanted to know whether or not $f \in H^p$ we could take $u = P[f]$ which of course satisfies $\Delta u = 0$.

This is not necessary. If f is a function and $\phi \in C_c^\infty(R^n)$ with $\int_R \phi_n = 1$, then we may form $f^*(x) = \sup_{(t,y)\in\Gamma(x)} |f * \phi_t(y)|$, $\phi_t(x) = t^{-n}\phi(x/t)$ and if $\psi \in C_c^\infty(R^n)$ is suitably non-trivial (say radial, non-zero) and $\int\psi = 0$ we may form

$$S_\psi^2(f)(x) = \iint_{\Gamma(x)} |f * \psi_t(y)|^2 \frac{dy\,dt}{t^{n+1}} \quad [18].$$

Then C. Fefferman and E. M. Stein have shown that

$$\|f\|_{H^p(R^n)} \sim \|f^*\|_{L^p(R^n)} \sim \|S_\psi(f)\|_{L^p(R^n)} \quad \text{for } 0 < p < \infty .$$

Thus, it is possible to think of H^p spaces without any reference to particular approximate identities like $P_t(x)$ which relate to differential equations.

In addition to understanding the various characterizations of H^p spaces, another important aspect is that of duality of H^1 with BMO, which we shall now discuss.

A function $\phi(x)$, locally integrable on R^n is said to belong to the class BMO of functions of bounded mean oscillation provided

$$\frac{1}{|Q|}\int_Q |\phi(x) - \phi_Q|\,dx \leq M \quad \text{for all cubes } Q \text{ in } R^n,$$

where $\phi_Q = \frac{1}{|Q|}\int_Q \phi$. The BMO functions are really functions defined modulo constants and $\|\ \|_{BMO}$ is defined to be $\sup \frac{1}{|Q|}\int_Q |\phi - \phi_Q|$.

According to a celebrated theorem of C. Fefferman and Stein, BMO is the dual of H^1 [18]. This result's original proof involves knowing that singular integrals map L^∞ to BMO and also a characterization of BMO functions in terms of their Poisson integrals which we now describe.

Suppose $\mu \geq 0$ is a measure in R_+^{n+1} and $Q \subseteq R^n$ is a cube. Let $S(Q) = \{y,t) | y \in Q, 0 < t < \text{side length } (Q)\}$. Then we say that μ is a Carleson measure on R_+^{n+1} iff $\mu(S(Q)) \leq C|Q|$. The basic property that characterizes Carleson measures is

$$\iint_{R_+^{n+1}} |u(x,t)|^p d\mu(x,t) \leq C_p \int_{R^n} u^*(x)^p dx \qquad p > 0$$

for all functions u on R_+^{n+1}. In connection with this type of measure there is the characterization of functions in $BMO(R^n)$ in terms of their Poisson integrals. A function $\phi(x)$ on R^n with Poisson integral $u(x,t)$ is in BMO if and only if the associated measure

$$d\mu(x,t) = |\nabla u|^2(x,t)\, t\, dx\, dt$$

is a Carleson measure. C. Fefferman proved this and used it to prove that every function in BMO acts continuously on H^1:

$$\left| \int_{R^n} f(x)\phi(x)\, dx \right| \leq C \|f\|_{H^1(R^n)} \|\phi\|_{BMO(R^n)} .$$

These are the basic facts of H^p spaces that will concern us here and which we shall later generalize to product spaces.

Let us now prove that for a harmonic function $u(x,t)$ in R_+^{n+1},

$$\|S(u)\|_{L^p} \sim \|u^*\|_{L^p} \quad \text{for } p > 0 .$$

We begin with the estimate $\|u^*\|_p \leq C_p \|S(u)\|_p$, and to do this we shall show that

$$|\{u^* > C\alpha\}| \leq C \left[\frac{1}{\alpha^2} \int_{S(u) \leq \alpha} S^2(u)(x)\,dx + |\{S(u) > \alpha\}| \right].$$

From this our claim follows. This is because for $\lambda_g(\alpha) = |\{|g| > \alpha\}|$ we have

$$\|u^*\|^p_{L^p} \sim \int_0^\infty \alpha^{p-1} \lambda_{u^*}(\alpha) d\alpha \lesssim \int_0^\infty \alpha^{p-1} \left\{ \frac{1}{\alpha^2} \int_0^\alpha \beta \lambda_{S(u)}(\beta) d\beta + \lambda_{S(u)}(\alpha) \right\} d\alpha$$

$$\leq \int_0^\infty \beta \lambda_{S(u)}(\beta) \int_\infty^\infty \alpha^{p-3} d\alpha d\beta + \int_0^\infty \alpha^{p-1} \lambda_{S(u)}(\alpha) d\alpha .$$

Assuming $p < 2$ as we clearly may, this is

$$\lesssim \int_0^\infty \beta^{p-1} \lambda_{S(u)}(\beta) d\beta \sim \|S(u)\|^p_p .$$

To prove the estimate on $|\{u^* > C\alpha\}|$, we set the notation that $\tilde{E} = \left\{ M(\chi_E) > \frac{1}{2} \right\}$, and then claim

I. $$\int_{S(u) \leq \alpha} S^2(u)(x)\,dx \geq c \iint_R |\nabla u(y,t)|^2 t\,dt\,dy \quad \text{where } \Re = \bigcup_{x \notin \{S(u) > \alpha\}} \Gamma(x) .$$

In fact, if $(y,t) \in \Re$ then

$$|B(y,t) \cap \{S(u) > \alpha\}| < \frac{1}{2} |B(y,t)| .$$

Then

$$\int_{S(u)\leq\alpha} S^2(u)(x)\,dx = \int_{x\in\{S(u)\leq\alpha\}} \left(\int_{T(x)} |\nabla u|^2(y,t)t^{1-n}dy\,dt \right) dx$$

(4.1)

$$= \iint_{R_+^{n+1}} |\nabla u|^2(y,t)t^{1-n}|\{x|(y,t)\in\Gamma(x), x\notin\{S(u)>\alpha\}\}|\,dy\,dt\ .$$

But for $(y,t)\in R$,

$$|\{x|(y,t)\in\Gamma(x), x\notin\{S(u)>\alpha\}\}| = |B(y,t)\cap {}^c\{S(u)>\alpha\}| \geq \frac{1}{2}\ |B(y,t)|$$

and (4.1) is

$$(4.1) \geq c \iint |\nabla u|^2(y,t)\,t\,dy\,dt$$

as claimed.

II. The next step is to write $|\nabla u|^2 = \Delta(u^2)$, and apply Green's theorem to $\mathfrak{R}$:

$$\iint_R \Delta(u^2)(y,t)\,t\,dy\,dt = \int_{\partial R} \frac{\partial(u^2)}{\partial n}\,t - u^2\,\frac{\partial(t)}{\partial n}\,d\sigma\,.$$

Now $\frac{-\partial t}{\partial n} \geq c > 0$ for some c so the above gives

$$\int_{\partial R} u^2 d\sigma \leq C \left\{ \int_{S(u)\leq\alpha} S^2(u)(x)\,dx + \int_{\partial R} u(\nabla u)\,t\,d\sigma \right\}.$$

Since, for purposes of all estimates we may assume that u is rather nice, we may assume $u(\nabla u)t$ vanishes at $t = 0$, so

$$\int_{\partial R} u(\nabla u)t\,d\sigma = \int_{\partial R} u(\nabla u)t\,d\sigma$$

where ∂R is the part of ∂R above $\{S(u) > \alpha\}$. It is not hard to see that $|\nabla u|t \le \alpha$ on ∂R so that

$$\int_{\partial R} |u|\,|\nabla u|t\,d\sigma \le \alpha|\{S(u) > \alpha\}|^{1/2}\left(\int_{\partial R} u^2 d\sigma\right)^{1/2}.$$

Putting all of our estimates together, we see that

$$\int_{\partial R} u^2 d\sigma \le C\left\{\int_{S(u)\le\alpha} S^2(u)(x)\,dx + \alpha^2\,|S(u) > \alpha\}|\right\}.$$

III. Next we wish to define a function $\tilde{f}$ by $\tilde{f}(x) = u(x, \tau(x))$ where $(x, \tau(x)) \in \partial R$ defines the function τ. We claim that in the region R

$$|u| \le P[\tilde{f}] + C\alpha\,. \tag{4.2}$$

This is done by harmonic majorization. It is enough to show this on ∂R, and this in turn is just saying that for any point $p \in \partial R$, $|U(p)|$ is dominated by the average over a relative ball on ∂R of $u + C\alpha$. This follows from the estimate $|\nabla u|t \le C\alpha$ on ∂R. Anyway, from (4.2) we have, for $x \notin \{S(u) > \alpha\}$, $u^*(x) \le CP[f]^*(x) + C\alpha$, so that finally

$$|\{u^* > C'\alpha\}| \le \{M\tilde{f}(x) > \alpha\}| \le \frac{C}{\alpha^2}\|\tilde{f}\|_2^2$$

$$\le \frac{C}{\alpha^2}\int_{S(u)\le\alpha} S^2(u)(x)\,dx + C|\{S(u) > \alpha\}|\,.$$

This completes the proof.

The proof that $\|u^*\|_p \leq C_p\|S(u)\|_p$ which we just gave has been lifted from Charles Fefferman and E. M. Stein's Acta paper [18]. To prove the reverse inequality we want to go via a different route, and we shall follow Merryfield here [20]. We prove the following lemma. In the next lecture we show how this lemma proves $\|u^*\|_p \geq C_p\|S(u)\|_p$.

LEMMA. *Let* $f(x)$ *and* $g(x) \in L^2(R^n)$, *and suppose* $\phi \in C_c^\infty(R^n)$ *radial and* $u = P[f]$. *Then*

$$\iint_{R_+^{n+1}} |\nabla u|^2(x,t)|g * \phi_t(x)|^2 t \, dt\, dx$$

$$\leq \int_{R^n} f^2(x)\phi^2(x)\, dx + \iint_{R_+^{n+1}} u^2(x,t)|g * \psi_t(x)|^2 \frac{dt}{t}\, dx$$

for some $\psi \in C_c^\infty(R^n)$ *with* $\int\psi = 0$ (ψ *real-valued*).

Proof.

$$\iint_{R_+^{n+1}} \Delta(u^2)(x,t)|g * \phi_t(x)|^2 t\, dt\, dx = \int_{R_+^{n+1}} \nabla \cdot \nabla (u^2)|g*\phi_t(x,t)|^2 t\, dt\, dx$$

$$= -\iint_{R_+^{n+1}} \nabla(u^2)(x,t) \cdot \nabla[(g * \phi_t(x,t)^2)t]dt\, dx = -2\iint_{R_+^{n+1}} u(x,t)\nabla u(x,t) \cdot \nabla[(g * \phi_t)^2](x,t)\, t\, dt\, dx$$

$$-2\iint_{R_+^{n+1}} u(x,t)\frac{\partial u}{\partial t} u(x,t)(g * \phi_t)^2(x,t)\, dt\, dx$$

$$= -2\iint_{R_+^{n+1}} u\nabla u \cdot 2(g*\phi_t)(x)\nabla[g*\phi_t(x)]t\, dt\, dx - 2\iint_{R_2^{n+1}} u\frac{\partial u}{\partial t}(g*\phi_t)^2 dt\, dx = I + II,$$

where

$$I = \iint_{R_+^{n+1}} u(t\nabla(g*\phi_t))\, t^{-1/2} \cdot \nabla u(g*\phi_t)\, t^{1/2} dt\, dx$$

$$\leq \left(\iint_{R_+^{n+1}} u^2 |g*\psi_t|^2 \frac{dt}{t}\, dx \right)^{1/2} \left(\iint_{R_+^{n+1}} |\nabla u|^2 |g*\phi_t(x)|^2\, t\, dt\, dx \right)^{1/2},$$

and

$$II = \iint_{R_+^{n+1}} u \frac{\partial u}{\partial t} (g*\phi_t)^2\, dt\, dx = - \iint_{R_+^{n+1}} u \frac{\partial}{\partial t} (u \cdot (g*\phi_t)^2)\, dt\, dx + \int_{R^n} u^2(x,0) g^2(x)\, dx$$

$$= - \iint_{R_+^{n+1}} u \frac{\partial u}{\partial t} (g*\phi_t)^2\, dt\, dx - \iint u^2 2(g*\phi_t) \frac{\partial}{\partial t} (g*\phi_t)\, dt\, dx + \int_{R^n} f^2 g^2 dx\,.$$

We see that

$$2 \iint_{R_+^{n+1}} u \frac{\partial u}{\partial t} (g*\phi_t)^2 dt\, dx \leq \iint_{R_+^{n+1}} u^2 (g*\phi_t) \frac{\partial}{\partial t} (g*\phi_t)\, dt\, dx + \int_{R^n} f^2 g^2\, dx$$

but

$$\iint_{R_+^{n+1}} u^2 (g*\phi_t) \frac{\partial}{\partial t} (g*\phi_t)\, dt\, dx = \iint_{R_+^{n+1}} u^2 (g*\phi_t)\, t \frac{\partial}{\partial t} (g*\phi_t) \frac{dt}{t}\, dx\,.$$

But

$$t\frac{\partial}{\partial t}(g*\phi_t) = \left[\sum_i \frac{\partial}{\partial x_i}(x_i\phi)\right]_t * g = \sum_i t\frac{\partial}{\partial x_i}[(x_i\phi)_t * g]$$

so

$$S = \sum \iint -2u\frac{\partial u}{\partial x_i}(g*\phi_t)[(x_i\phi)_t * g]\,dt\,dx + \int -u^2 t\frac{\partial}{\partial x_i}(g*\phi_t)(g*(x_i\phi))\frac{dt\,dx}{t}$$

$$\leq \sum\left\{\left(\iint u^2(g*(x_i\phi)_t)^2\frac{dx\,dt}{t}\right)^{1/2}\left(\iint |\nabla u|^2(g*\phi_t)^2\,t\,dt\,dx\right)^{1/2}\right.$$

$$\left. + \left(\iint u^2\left(g*\left(\frac{\partial\phi}{\partial x_i}\right)_t\right)^2\frac{dt\,dx}{t}\right)^{1/2}\left(\iint u^2(g*(x_i\phi))^2\frac{dt\,dx}{t}\right)^{1/2}\right\}.$$

Putting this together gives

$$\iint |\nabla u|^2(g*\phi_t)^2\,t\,dx\,dt$$

$$\leq C\left[\sum\left\{\iint u^2\left(g*\left(\frac{\partial\phi}{\partial x_i}\right)_t\right)^2\frac{dt\,dx}{t} + \iint u^2(g*(x_i\phi)_t)^2\frac{dt\,dx}{t}\right\} + \int_{R^n} f^2g^2\right]$$

$$+ \iint u^2\left(g*\sum\frac{\partial}{\partial x_i}(\alpha_i\phi)\right)^2\frac{dt\,dx}{t}.$$

5. *More on* H^p *spaces*

At this point we wish to discuss the theory of multi-parameter H^p spaces and BMO. We saw, in the last lecture, that $H^p(R^n)$ could be defined either by maximal functions or by Littlewood-Paley-Stein theory. All of these spaces, H^p and BMO were invariant under the usual dilations on R^n, $x \to \delta x$, and this is hardly a surprise, since they can be defined by the maximal functions and singular integrals which are invariant under these dilations. Here we shall define H^p and BMO spaces which are invariant under the dilations (in R^2) $(x_1,x_2) \to (\delta_1 x_1, \delta_2 x_2)$, $\delta_1, \delta_2 > 0$. For convenience we shall work in R^2 but all of this could just as well be carried out in $R^n \times R^m$, $n, m > 1$. We shall call our H^p and BMO spaces "product H^p and BMO" and denote them by $H^p(R^2_+ \times R^2_+)$ and $BMO(R^2_+ \times R^2_+)$ so that we reserve $H^p(R^2)$ for the one-parameter space of functions on R^2.

Let $(x_1,x_2) \in R^2$ and denote by $\Gamma(x)$ the set

$$\{(y_1,t_1,y_2,t_2) |\, |y_i - x_i| < t_i,\ t_i > 0\} \subseteq R^2_+ \times R^2_+ .$$

Let $u(x,t)$ be a function in $R^2_+ \times R^2_+$, $x \in R^2$, $t \in R_+ \times R_+$, which is biharmonic, i.e., harmonic in each half plane separately. Then the non-tangential maximal function and area integral of u are defined by

$$u^*(x_1,x_2) = \sup_{(y,t) \in \Gamma(x_1,x_2)} |u(y_1,t_1,y_2,t_2)|$$

and

$$S^2(u)(x) = \iint_{\Gamma(x)} |\nabla_1 \nabla_2 u(y,t)|^2 dy_1 dy_2 dt_1 dt_2 .$$

More generally, if $\phi \in C_c^\infty(R^2)$ and if

$$\int \phi = 1,\ \phi_{t_1,t_2}(x_1,x_2) = t_1^{-1} t_2^{-1} \phi\left(\frac{x_1}{t_1}, \frac{x_2}{t_2}\right)$$

then for a function f on R^2

$$f^*(x) = \sup_{(y,t) \in \Gamma(x)} |f * \phi_{t_1,t_2}(x)|$$

and if $\psi \in C_c^\infty(R^2)$ and

$$\begin{cases} \int \psi(x_1,x_2)dx_1 = 0 \quad \text{for all } x_2 \in R^1 \\ \\ \int \psi(x_1,x_2)dx_2 = 0 \quad \text{for all } x_1 \in R^1 , \end{cases}$$

then

$$S_\psi^2(f)(x) = \iint_{\Gamma(x)} |f * \psi_{t_1,t_2}(y_1,y_2)|^2 \, dy_1 \, dy_2 \frac{dt_1}{t_1^{\;2}} \frac{dt_2}{t_2^{\;2}} .$$

Given $f(x_1, x_2)$, we define its bi-Poisson integral by $u(x_1,t_1,x_2,t_2) = P[f](x,t) = f * P_{t_1,t_2}(x_1,x_2)$, where the bi-Poisson (or just Poisson for short) kernel is defined by $P_{t_1,t_2}(x_1,x_2) = t_1^{-1}t_2^{-1} P\left(\frac{x_1}{t_1}\right) \dot{P}\left(\frac{x_2}{t_2}\right)$ and where P is the 1-dimensional Poisson kernel. Then, of course, $P[f]$ is bi-harmonic in $R_+^2 \times R_+^2$.

In analogy with the 1-parameter case, we define $f \in H^p(R_+^2 \times R_+^2)$ if and only if $u^* \in L^p(R^2)$ where $u = P[f]$. It is not hard to see, just as in the single parameter case that for any $\phi \in C_c^\infty(R^2)$, $\int \phi = 1$,

$$\|f^*\|_{L^p} \sim \|u^*\|_{L^p} \quad \text{for } p > 0 ,$$

so that we may use any approximate identity which is sufficiently nice to define product H^p spaces. In terms of area integrals, we also have, for $\psi(x_1,x_2)$ suitably nontrivial, say ψ even in x_1, x_2 and not $\equiv 0$,

$$\|S_\psi f\|_{L^p} \sim \|S(u)\|_{L^p} \quad p > 0, \; u = P[f] .$$

To complete the chain of equivalences, we would like to know that $\|S(u)\|_{L^p} \sim \|u^*\|_{L^p}$.

In fact, this is true, but is not obvious, and so we intend to present the proof here. The proofs are by iteration, but they are not of the same totally straightforward nature as the iteration in the Jessen-Marcinkiewicz-Zygmund theorem. Often, this is the case in the analysis of product domains, namely, the proof is by iteration, but this requires a different way of looking at the one-parameter case than one is used to.

Proof that of $S(u) \in L^p$, *then,* $u^* \in L^p$ (Gundy-Stein) [21]. We can assume that $u = P[f]$. Then since the 2-parameter area integral is invariant under taking Hilbert transforms in each variable separately, we see that we may write

$$f = f_{++} + f_{+-} + f_{-+} + f_{--}$$

where $\hat{f}_{++}$ is supported in $\xi_1, \xi_2 > 0$ and $\hat{f}_{+-}$ is supported in $\xi > 0$, $\xi_2 < 0$, etc., and we have $S(f_{\pm\pm}) \in L^p$. By reflection we may assume f is analytic (i.e., P[f] is bi-analytic) and then show that $f^* \in L^p$. But for $u = [f]$ a bi-analytic function, we know that for $\alpha > 0$, $|u(x_1,t_1,x_2,t_2)|^\alpha$ is subharmonic in each half plane $(x_i,t_i) \in R^2_+$ separately; this implies that

$$|u(x,t)|^\alpha \leq P[|f(x)|^\alpha] .$$

If $\alpha < p$,

$$u^*(x)^\alpha \leq M(|f|^\alpha) \in L^{p/\alpha}(R^2), \ \text{if } f \in L^p .$$

So what we must show is that

$$\|f\|_{L^p} \leq C_p \|S(f)\|_{L^p}, \quad p > 0 .$$

To show this let us define some notation. In R^1, if $f(x)$ has Poisson integral u, we let Q_t be the operator (or kernel) which takes f to $t\nabla u(x,t)$, so that

$$S^2(f)(x) = \iint_{\Gamma(x)} |f * Q_t(y)|^2 \frac{dy\,dt}{t^2} .$$

Then going back to our present situation where f is given on R^2, we define

$$Q_t^1 f(x_1,x_2) = \int_{-\infty}^{\infty} f(x_1 - y, x_2) Q_t(y)\,dy$$

and Q_t^2 similarly. Then let us define a Hilbert space valued function

$$F(x_1,x_2) \in L^2\left(\Gamma(0); \frac{dy\,dt}{t^2}\right)$$

by

$$F(x_1,x_2)(y_2,t_2) = Q_{t_2}^2 f(x_1,x_2+y_2) .$$

Now we know that in the one-parameter case $\|S^1 f\|_p \geq c_p \|f\|_p$ and a glance at the proof of this fact reveals that it remains valid for Hilbert space valued functions. Fix x_2. Then

$$\int_{-\infty}^{\infty} S^1(F)(x_1,x_2)^p\,dx_1 \geq c_p \int_{-\infty}^{\infty} |F(x_1,x_2)|^p_{L^2(\Gamma)}\,dx_1$$

and integrating this in x_2,

$$\iint_{R^2} S^1(F)(x_1,x_2)^p\,dx_1dx_2 \geq c_p \iint_{R^2} |F(x_1,x_2)|^p_{L^2(\Gamma)}\,dx_1dx_2 . \tag{5.1}$$

But fixing x_1, since $|F(x_1,x_2)|_{L^2(\Gamma)}$ is the value of the one-parameter integral of $f(x_1,\cdot)$ at x_2, we have

$$\int_{R^1} |F(x_1,x_2)|^p_{L^2(\Gamma)} dx_2 \geq c_p \int_{R^1} |f(x_1,x_2)|^p dx_2$$

and so (5.1) is greater than or equal to

$$c'_p \iint_{R^2} |f|^p dx_1 dx_2 .$$

On the other hand when the S^1 operator acts on the first variable we have

$$S^1(F)(x_1,x_2) = S(f)(x_1,x_2) ,$$

the two-parameter area integral of f.

Next, let us prove that $\|S(u)\|_{L^p} \leq c_p \|u^*\|_{L^p}$ [21]. The proof is a simple iteration of the one-parameter case given previously. To begin with, we recall Merryfield's lemma: Let $\phi \in C_c^\infty(R^1)$ be supported in $[-1,+1]$ and have $\int\phi = 1$. Then there exists $\psi \in C_c^\infty(R^1)$ whose support is also contained in $[-1,+1]$ with $\int\psi = 0$ and such that if $u = P[f]$,

$$\iint_{R^2_+} |\nabla u|^2(x,t)\,(g*\phi_t(x))^2\, t\, dt\, dx \leq C \left\{ \int_{R^1} f^2(x)\, g^2(x)\, dx + \right.$$

$$\left. + \iint_{R^2_+} u^2(x,t)\,(g*\psi_t(x))^2 \frac{dt}{t}\, dx \right\} .$$

Introduce the notation $t\nabla u(x,t) = Q_t f(x)$, $u(x,t) = P_t f(x)$, $g*\phi_t(x) = \widetilde{P}_t(g)(x)$ and $g*\psi_t(x) = \widetilde{Q}_t(g)(x)$, Q_t^i, $i = 1,2$, will denote the operator acting in the ith variable. Then we estimate

$$(5.2)\qquad \iint_{R^2_+\times R^2_+} [Q^1_{t_1} Q^2_{t_2} f(x_1,x_2)]^2 [\widetilde{P}^1_{t_1} \widetilde{P}^2_{t_2} g(x_1,x_2)]^2 \frac{dt_1 dt_2}{t_1 t_2}\, dx_1 dx_2 \,.$$

Fix x_2, t_2. Then (5.2) is

$$\leq \iint_{x_2,t_2} \iint_{x_1,t_2} (P^1_{t_1} Q^2_{t_2} f(x))^2 (\widetilde{Q}^1_{t_1} \widetilde{P}^2_{t_2} g(x))^2 \frac{dt\,dx}{t}$$

$$+ \iint_{x_2 t_2} \int_{x_1} [Q^2_{t_2} f(x)]^2 [\widetilde{P}^2_{t_2} g(x)]^2\, dx_1 \frac{dt_2 dx_1}{t_2} = \mathrm{I} + \mathrm{II}\,.$$

Now

$$\mathrm{I} = \iint_{x_1 t_1} \iint_{x_2 t_2} P^2_{t_2} P^1_{t_1} f(x))^2 (\widetilde{Q}^2_{t_2} \widetilde{Q}^1_{t_1} g(x))^2 \frac{dt\,dx}{t}$$

$$+ \iint_{x_1,t_1} \int_{x_2} (P^1_{t_1} f(x))^2 (\widetilde{Q}_{t_1} g(x))^2\, dx_2 \frac{dt_1}{t_1}\, dx_1 \,.$$

Then

$$\mathrm{II} = \int_{x_1} \iint (P^2_{t_2} f(x))^2 (\widetilde{Q}_{t_2} g(x))^2\, dx_2\, dt_2/t_2\, dx_1 + \iint_{x_1 x_2} f^2(x)\, g^2(x)\, dx_1 dx_2 \,.$$

So the inequality we seek in the product case is

$$(5.3)\quad \int_{(R^2_+)^2} (Q_t f)^2(x)(\tilde{P}_t g)^2(x)\,dxdt/t \le C\left\{\int_{(R^2_+)^2} (P_t f(x))^2(\tilde{Q}_t g(x))^2 dxdt/t\right.$$

$$+\int_{x_1\epsilon R^1}\left(\int_{(x_2,t_2)\epsilon R^2_+} (P^2_{t_2} f(x))^2(\tilde{Q}^2_{t_2})^2(x)\,dx_2 dt_2/t_2\right)$$

$$\left.+\int_{x_2\epsilon R^1}\left(\int_{(x_1,t_1)\epsilon R^2_+} P^{(1)}_{t_1} f(x)^2\cdot\tilde{Q}_{t_1} g(x)^2\,dx_1 dt_1/t_1\right)dx_2+\int_{R^2} f^2(x)\,g^2(x)\,dx\right\}.$$

It is now easy to see that $\|S(u)\|_{L^p} \le C_p\|u^*\|_{L^p}$, $p > 0$.

In order to simplify things a little, we shall take a modified definition of u^* in what follows, namely

$$u^*(x) = \sup_{(y,t)\epsilon\Gamma_{10^{10}}(x)} |u(y,t)|$$

where

$$\Gamma_{10^{10}}(x) = \{(y,t)\mid |y_i - x_i| < 10^{10} t_i,\ i = 1,2\}.$$

This is an irrelevant change, since a trivial computation shows that $\|u^*\|_p$ is, for a larger aperture, no more than a constant times $\|u^*\|_p$ for a smaller aperture, the constant depending only on the apertures involved.

In (5.3), take $\phi(x) = 1$ for all $|x| < 1/3$, and $g(x) = \chi_{u^*(x)\le\alpha}(x)$. Let us estimate

$$\int_{M(\chi_{u^*>\alpha})<1/200} S^2(u)(x)\,dx \quad \text{when} \quad u = P[f]$$

$$\leq \iint |\nabla_1\nabla_2 u|^2(y,t)|R(y,t)\cap\{M(\chi_{u^*>\alpha})<1/200\}|\,dy\,dt$$

$$\leq \iint_{R^*} |\nabla_1\nabla_2 u|^2(y,t)\,t_1t_2\,dy\,dt$$

where $R^* = \{(y,t)|\ |R(y,t)\cap\{u^*>\alpha\}| < \frac{1}{200}|R(y,t)|\}$ and where $R(y;t)$ is the rectangle in R^2 with sides parallel to the axes and with side lengths $2t_1, 2t_2$ centered at y. Notice that if $|R(y;t)\cap\{u^*>\alpha\}| < \frac{1}{100}|R(y;t)|$ then $g*\phi_t(y) = \tilde{P}_t g(y) > c$ for some $c > 0$. It follows that

$$\int_{M(\chi_{\{u^*>\alpha\}})<1/200} S^2(u)(x)\,dx \leq \iint_{(R^2_+)^2} |\nabla_1\nabla_2 u|^2(y,t)\tilde{P}_t(g)^2(y)\,dy\,t_1t_2\,dt$$

$$\leq C\left\{\iint_{(R^2_+)^2} u^2(y,t)\tilde{Q}_t(g)^2(y)dy\,\frac{dt}{t_1t_2} + \int_{y_1}\iint_{(y_2,t_2)\epsilon R^2_+} P^2_{t_2}f(y)^2\cdot\tilde{Q}^2_{t_2}(g)^2(y)dy\,\frac{dt_2}{t_2}\right.$$

$$\left. + \int_{y_2}\iint_{(y_1,t_1)\epsilon R^2_+} P^{(1)}_{t_1}f(y)^2\cdot\tilde{Q}_{t_1}(g)^2(y)\,dy\,\frac{dt_1}{t_1} + \iint_{R^2} f^2g^2\,dy\right\}$$

$= i + ii + iii + iv$.

Consider i: If $\tilde{Q}_t(g)(y) \neq 0$ then $u^*(x) \leq \alpha$ for some $x \epsilon R(y;t)$. But then $|u(y,t)| \leq \alpha$ so i is less than or equal to

$$\alpha^2 \iint_{(R_+^2)^2} \tilde{Q}_t(g)^2(y)dy \frac{dt}{t_1 t_2} = \alpha^2 \iint_{(R_+^2)^2} \tilde{Q}_t(1-g)^2(y)dy \frac{dt}{t} \leq \alpha^2 \|1-g\|_{L^2}^2 \leq \alpha^2 |\{u^* > \alpha\}| .$$

Consider ii: If $\tilde{Q}_{t_2}^{(2)}(g)(y_1,y_2) \neq 0$, then there exists x_2 such that $|x_2 - y_2| < t_2$ and $u^*(y_1,x_2) \leq \alpha$. This implies that $|P_{t_2}^{(2)} f(y_1,y_2)| \leq \alpha$ so ii is less than or equal to

$$\alpha^2 \int_{y_1} \left(\iint_{(y_2,t_2)} \tilde{Q}_{t_2}^{(2)}(g)^2(y) \frac{dt_2}{t_2} dy_2 \right) dy_1 .$$

Again

$$\int_{y_1} \left(\iint_{y_2,t_2} \tilde{Q}_{t_2}^{(2)}(g)^2(y) \frac{dt_2}{t_2} dy_2 \right) dy_1 = \int_{y_1} \left(\iint_{y_2,t_2} \tilde{Q}_{t_2}^{(2)}(1-g)^2(y)dy_2 \frac{dt_2}{t_2} \right) dy_1$$

$$\leq \int_{y_1} \left(\int_{y_2} (1-g)^2(y_1,y_2)\, dy_2 \right) dy_1 \leq |\{u^* > \alpha\}| .$$

(iii) is similar to (ii).

Finally, (iv) is less than or equal to

$$\int_{u^*(x) \leq \alpha} f^2(x)\, dx \leq \int_{u^*(x) \leq \alpha} (u^*)^2(x)\, dx .$$

So we have

$$\int_{\{u^* \leq \alpha\}} S^2(u)(x)\, dx \leq C \quad \alpha^2 |\{u^* > \alpha\}| + \int_{u^*(x) \leq \alpha} u^{*2}(x)\, dx$$

and we have seen before that this implies that

$$\|S(u)\|_{L^p} \le C_p \|u^*\|_{L^p}, \quad 2 > p > 0 .$$

The next topic that we shall consider is that of duality of H^1 and BMO in the product setting. In the classical case there were four results which expressed this duality.

1) The characterization of Carleson measures μ for which the Poisson transform $f \to P[f]$ is bounded from $L^p(dx)$ to $L^p(d\mu)$, $p > 1$.

2) The characterization of functions in $BMO(R^1)$ by a condition on their Poisson integrals in terms of Carleson measures.

3) The characterization of functions in the dual of H^1 by the BMO condition.

4) The atomic decomposition of H^1.

Let us try to guess what the analogous theory should look like in product spaces. For simplicity we consider the dual of $H^1(R^2_+ \times R^2_+)$. Then what should an element of $BMO(R^2_+ \times R^2_+)$ look like? We might look at tensor products of functions in $BMO(R^1)$ to get a feel for the answer. So, for example if ϕ_1 and ϕ_2 are in $BMO(R^1)$ then $\phi_1(x_1)\phi_2(x_2)$ might be our model. Of course, this function $\phi(x_1,x_2)$ satisfies

$$(5.4) \qquad \frac{1}{|R|} \int_R |\phi(x_1,x_2) - c_1(x_1) - c_2(x_2)|^2 \, dx_2 dx_2 \le C$$

for the appropriate choice of functions c_1 and c_2 of the x_1, x_2 variable.

A Carleson measure in $R^2_+ \times R^2_+$ would be a non-negative measure μ for which

$$(5.5) \qquad \iint_{(R^2_+)^2} P[f]^p \, d\mu \le C_p \iint_{R^2} |f(x)|^p \, dx , \quad p > 1 ,$$

where P is the bi-Poisson integral. The obvious guess is that μ

satisfies (5.5) if and only if $\mu(S(R)) \leq C|R|$ for all rectangles $R \subseteq R^2$ with sides parallel to the axes, where the Carleson region $S(R)$ is defined by $S(I \times J) = S(I) \times S(J)$ for $R = I \times J$. In terms of these Carleson measures, it is not hard to show that ϕ satisfies (5.4) if and only if its bi-Poisson integral u satisfies

$$d\mu = |\nabla_1 \nabla_2 u|^2(y,t)\, t_1 t_2 dt \quad \text{is a Carleson measure.}$$

And finally, all of this in some sense is equivalent to asserting that every $f \in H^1(R^2_+ \times R^2_+)$ can be written as $\Sigma \lambda_k a_k$ where $\Sigma |\lambda_k| \leq C\|f\|_{H^1}$ and $a_k(x_1,x_2)$ are "atoms," i.e., a_k is supported in a rectangle $R_k = I_k \times J_k$ such that

$$\begin{cases} \displaystyle\int_{I_k} a_k(x_1,x_2)\,dx_1 = 0 & \text{for all } x_2 \\ \displaystyle\int_{J_k} a_k(x_1,x_2)\,dx_2 = 0 & \text{for all } x_1 \end{cases}$$

and

$$\|a_k\|_2 \leq \frac{1}{|R_k|^{1/2}}\,.$$

In 1974 [22], L. Carleson showed that $\mu(S(R)) \leq C|R|$ was not sufficient to guarantee the inequality

$$\iint_{(R^2_+)^2} |P[f]|^p\, d\mu \leq C_p \int_{R^2} |f|^p\, dx\,.$$

From here it is not difficult to produce examples of functions $\phi(x_1,x_2)$ which satisfy

$$\frac{1}{|R|}\int_R |\phi(x_1,x_2)-C_1(x_1)-C_2(x_2)|^2\,dx_1dx_2 \le C$$

where C_1, C_2 depend on R, yet $\phi \notin L^p(R^2)$ for any $p>2$. Therefore, this condition is not strong enough to force ϕ to belong to the dual of H^1.

In other words the simple picture of the structure of $H^1(R^2_+ \times R^2_+)$ and $BMO(R^2_+ \times R^2_+)$ suggested above as the obvious guess is completely wrong.

Rather one considers the role of rectangles to be played instead by arbitrary open sets. Although this may seem a bit frightening at first glance, it turns out, and this is of course the final test of the theory, that nearly all the classical theory of H^p and BMO can easily be carried out using the approach suggested here.

By way of introduction, we shall prove that for any function $\phi \in H^1(R^2_+ \times R^2_+)^*$, if $u = P[\phi]$ we have a Carleson condition with respect to open sets satisfied by the appropriate measure. To describe this result, we make the following definition ([23], [24], and [25]).

Let $\Omega \subseteq R^2$ be an arbitrary open set, and let $R(y;t)$ be the rectangle in R^2 centered at $(y_1,y_2) = y$ and with side lengths $2t_1$ and $2t_2$. Then $S(\Omega)$ the Carleson region above Ω is defined as

$$S(\Omega) = \bigcup_{R \subseteq \Omega} S(R) = \{(y,t) \in (R^2_+)^2 \,|\, R(y;t) \subseteq \Omega\}\,.$$

Then we say that $\mu \ge 0$ in $(R^2_+)^2$ is a Carleson measure if and only if $\mu(S(\Omega)) \le C|\Omega|$ for every open set $\Omega \subseteq R^2 \cdot f \in H^1(R^2_+ \times R^2_+)^*$ if and only if for $u = P[f]$,

$$d\mu = |\nabla_1 \nabla_2 u|^2 t_1 t_2 dt_1 dt_2 dy_1 dy_2 \text{ is a Carleson measure.}$$

In fact, this follows immediately from the inequality (5.3). To see this, notice that $|\nabla_1 \nabla_2 u|$ is invariant under the Hilbert transform $H_{x_i} (i = 1,2)$ so that if we prove this when $f \in L^\infty(R^2)$, we will have proven it also when f is of the form

$$g_1 + H_{x_1}g_2 + H_{x_2}g_3 + H_{x_1}H_{x_2}g_4 \quad g_i \in L^\infty .$$

A function $a(x)$ on R^2 will be in $H^1(R_+^2 \times R_+^2)$ if and only if $a \in L^1(R^2)$, $H_{x_1}a, H_{x_2}a$, and $H_{x_1}H_{x_2}a \in L^1$.

In fact, if $a \in H^1(R_+^2 \times R_+^2)$ then $S(a) \in L^1(R^2)$ hence so are $S(H_{x_1}a)$, $S(H_{x_2}a)$ and $S(H_{x_1}H_{x_2}a)$; therefore $H_{x_i}a, H_{x_1}H_{x_2}a \in L^1$. Conversely, if $a, H_{x_i}a$, and $H_{x_1}H_{x_2}a \in L^1(R^2)$ then we can form F_{++}, F_{+-}, F_{-+}, and $F_{--} \in L^1(R^2)$ such that $a = \Sigma F_{\pm\pm}$ and reflections of the $F_{\pm\pm}$ are boundary values of bi-analytic functions. A bianalytic function F with (distinguished) boundary values in $L^1(R^2)$ has $F^* \in L^1$ by a subharmonicity argument applied to $|F|^\alpha$, $\alpha < 1$. So $a^* \in L^1$ and $a \in H^1$. Let $\Phi \in H^1(R_+^2 \times R_+^2)^*$. Define a map from $H^1(R_+^2 \times R_+^2) \xrightarrow{\vartheta} \bigoplus_{i=1}^{4} L^1(R^2)_i$ by

$$\vartheta(f) = (f, H_{x_1}f, H_{x_2}f, H_{x_1}H_{x_2}f) .$$

Then $\|\vartheta f\|_{\oplus L^1} \sim \|f\|_{H^1}$. ϑ is obviously one to one, so $\vartheta^{-1} = \mathfrak{J}$ exists and is bounded on $\mathrm{Im}(\vartheta)$. The map $\Phi \circ \mathfrak{J}$ extends, by Hahn-Banach to an element of the dual $\oplus L^1 = \oplus L^\infty$. Then

$$\Phi(f) = \Phi \circ \mathfrak{J}(f, H_{x_1}f, H_{x_2}f, H_{x_1}H_{x_2}f)$$

$$= \int fg_1 + H_{x_1}f \cdot g_2 + H_{x_2}f \cdot g_3 + H_{x_1}H_{x_2}f \cdot g_4$$

$$= \int f \cdot (g_1 H_{x_1}g_2 + H_{x_2}g_3 + H_{x_1}H_{x_2}g_4) \quad f \in H^1 .$$

Thus every element of $(H^1)^*$ is of the form

$$g_1 + H_{x_1}g_2 + H_{x_2}g_3 + H_{x_1}H_{x_2}g_4 \quad g_i \in L^\infty .$$

So it suffices to show that if $f \in L^\infty(R^2)$ with $u = P[f]$ then

$$d\mu = |\nabla_1\nabla_2|^2(y,t)\,t_1t_2\,dy\,dt \text{ is a Carleson measure.}$$

But in (5.3), take $g = \chi_\Omega(x_1,x_2)$, and notice that if $\int\phi = 1$, $\operatorname{supp}\phi(x) \subseteq [-1,1]$ then $\tilde{P}_t g(x) = 1$ if $(x,t) \in S(\Omega)$. This is because for such (x,t), $R(x;t) \subseteq \Omega$ and $g*\phi_t(x) = \int_{R^2}\phi_t(x-u)\,du = 1$. It follows from (5.3) that

$$\iint_{S(\Omega)} |\nabla_1\nabla_2 u|^2 t_1t_2\,dx\,dt \le C\|f\|_\infty^2\Bigg(\iint |\tilde{Q}_t g|^2 \frac{dt}{t}\,dx$$

$$+\int_{x_1}\Bigg(\iint_{(x_2t_2)} |\tilde{Q}_{t_2} g|^2 \frac{dt_2}{t_2} dx_2\Bigg) dx_1 + \int_{x_2}\Bigg(\iint_{(x_1,t_1)} |\tilde{Q}_{t_1} g|^2 \frac{dt_1}{t_1}\,dx_1\Bigg)\,dx_2 + \int g^2 dx\Bigg)$$

$$\le C\|f\|_\infty^2\|g\|_2^2 \le C\|f\|_\infty^2|\Omega| .$$

6. *Duality of* H^1 *and* BMO *and the atomic decomposition*

In this lecture we shall consider in greater detail the spaces $H^p(R_+^2\times R_+^2)$ and $BMO(R_+^2\times R_+^2)$, which we discussed briefly in section 5. There we saw that in product spaces, the most obvious guesses at characterizations of H^p atoms of BMO failed. In order to circumvent these difficulties we must take a slightly different approach than we are used to in the classical 1-parameter case.

In what follows we shall be working with functions in $H^p(R_+^2\times R_+^2)$ or $BMO(R_+^2\times R_+^2)$ only. The theory for $R_+^2\times R_+^2\times\cdots\times R_+^2$ or for $R_+^{n+1}\times R_+^{m+1}$ is quite similar and only requires minor changes. Now let $\mathfrak{R}$ be the family of all rectangles with sides parallel to the axes and $\mathfrak{R}_d$ be the subfamily of $\mathfrak{R}$ whose sides are dyadic intervals.

If $f(x_1,x_2)$ is a sufficiently nice function on R^2, and $\psi \in C^\infty(R^1)$, ψ is even, $\psi \not\equiv 0$ real valued and $\operatorname{supp}(\psi) \subseteq [-1,1]$, and ψ has a large number of moments vanishing, then for

$$\psi_{t_1t_2}(x_1,x_2) = \psi\left(\frac{x_1}{t_1}\right)\psi\left(\frac{x_2}{t_2}\right)t_1^{-1}t_2^{-1}, \quad \int_0^\infty |\hat{\psi}(\xi)|^2 d\xi/\xi = 1$$

we have

$$f(x_1,x_2) = \iint_{R_+^2\times R_+^2} f*\psi_{t_1,t_2}(y_1,y_2)\psi_{t_1,t_2}(x_1-y_1,x_2-y_2)dy_1dy_2\frac{dt_1dt_2}{t_1t_2}$$

In fact, taking Fourier transforms of both sides, for the right-hand side we have

$$\iint_{R_+^2\times R_+^2} \hat{f}(\xi)|\hat{\psi}(t_1\xi_1,t_2\xi_2)|^2\frac{dt_1dt_2}{t_1t_2} = \hat{f}(\xi)\int_0^\infty\int_0^\infty |\hat{\psi}(t_1\xi_1,t_2\xi_2)|^2\frac{dt_1dt_2}{t_1t_2} = \hat{f}(\xi)$$

We can use this representation to decompose the function f as follows: $R \in \mathfrak{R}_d$. Set $\mathfrak{A}(R) = \{(y,t) \in R_+^2\times R_+^2 | y \in R,\ \ell_1 < t_1 \le 2\ell_i$ where ℓ_i, $i = 1,2$ is the side length of R in the x_i direction$\}$. Since $R_+^2\times R_+^2 = \bigcup_{R\in\mathfrak{R}_d} \mathfrak{A}(R)$, if we define

$$f_R(x_1,x_2) = \iint_{\mathfrak{A}(R)} f(y,t)\psi_{t_1t_2}(x_1-y_1,x_2-y_2)dy\frac{dt}{t_1t_2}$$

where $f(y,t) = f*\psi_t(y)$, then $f = \sum_{R\in\mathfrak{R}_d} f_R$, and each f_R is supported in $\tilde{R}$ the double of R and has the property that

$$\begin{cases} \displaystyle\int_{\tilde{I}} f_R(x_2,x_2)\,dx_1 = 0 & \text{for all } x_2 \\[2ex] \displaystyle\int_{\tilde{J}} f_R(x_1,x_2)\,dx_2 = 0 & \text{for all } x_1 \end{cases}$$

where $\tilde{R} = \tilde{I} \times \tilde{J}$.

It will be convenient to define a norm $|\ |_R$ on functions supported on a rectangle R, as follows.

$$|f|_R = \sum_{|\alpha|=0}^{N} \left\| \frac{\partial^\alpha f}{\partial x_1^{\alpha_1} \partial x_2^{\alpha_2}} \right\|_\infty |I|^{\alpha_1} |J|^{\alpha_2}$$

where N is a large integer. With these preliminaries we can pass to a theorem characterizing $(H^1)^*$ in a number of useful ways.

THEOREM [25]. *For a function on* R^2 *the following are equivalent*:

(1) $\phi \in H^1(R^2_+ \times R^2_+)^*$.

(2) $\phi = g_1 + H_{x_1}(g_2) + H_{x_2}(g_3) + H_{x_1}H_{x_2}(g_4)$ *for some* g_1, g_2, g_3 *and* g_4 *in* $L^\infty(R^2)$.

(3) *If* $u = P[\phi]$ *in* $R^2_+ \times R^2_+$, *then*

$$\iint_{S(\Omega)} |\nabla_1 \nabla_2 u|^2(y,t)\, t_1 t_2\, dt \le C|\Omega|, \quad \textit{for all open sets } \Omega \subseteq R^2 .$$

(4) *If* $\phi(y,t) = \phi * \psi_t(y)$, *then*

$$\iint_{S(\Omega)} |\phi(y,t)|^2 dy\, \frac{dt}{t} \le C|\Omega|, \quad \textit{for all open sets } \Omega \subseteq R^2 .$$

(5) ϕ *can be written in the form* $\sum_{R \in R_d} c_R b_R$ *where* $b_R(x_1,x_2)$ *are supported in* $\tilde{R}$, $|b_R|_R \le 1$ *and* $\sum_{R \subseteq \Omega} c_R^2 \le C|\Omega|$ *for all* Ω *open in* R^2.

Proof. To begin with, we proved in section 5 that (1) $\Longrightarrow$ (2). It is also trivial that (2) $\Longrightarrow$ (1), since if $f \in H^1(R_+^2 \times R_+^2)$,

$$\int f(x) H_{x_1} H_{x_2}(g)(x)\,dx = \int H_{x_1} H_{x_2}(f)(x)\, g(x)\,dx$$

and since $f \in H^1$, $H_{x_1} H_{x_2}(f) \in L^1$.

Next, we recall that (2) $\Longrightarrow$ (3) was also proven in the preceding section.

Now we claim that (3) or (4) implies (1). We show that (4) implies (1), the other proof being similar. We do this via the atomic decomposition of H^1 which we shall describe here only enough to derive our implication. We shall present the decomposition of H^1 in greater detail later. Let $f \in H^1(R_+^2 \times R_+^2)$. Then $S_\psi(f) \in L^1(R^2)$ (this follows by vector iteration, just as in the argument that $S(f) \in L^1$ implies $f \in L^1$).

Consider the sets $\Omega_k = \{S_\psi(f) > 2^k\}$, $k \in Z$. Set

$$a_k(x_1,x_2) = \sum_{\substack{R \in \mathfrak{R}_d \\ |R \cap \Omega_k| > 1/2|R| \\ |R \cap \Omega_{k+1}| < 1/2|R|}} f_R(x)\,.$$

Then, as we shall elaborate later on $\tilde{a}_k(x_1,x_2) = \dfrac{a_k(x_1,x_2)}{2^k|\tilde{\Omega}_k|}$ is an $H^1(R_+^2 \times R_+^2)$ atom where

$$\tilde{\Omega}_k = \{(x_1,x_2) | M^{(2)}(\chi_{\Omega_k})(x_1,x_2) > 1/10\}\,.$$

Then $f = \Sigma \lambda_k a_k$ where $\lambda_k = 2^k|\tilde{\Omega}_k|$, and by the strong maximal theorem $|\tilde{\Omega}_k| \le C|\Omega_k|$ so that

$$\sum \lambda_k \le C\|S_\psi(f)\|_{L^1} \le C'\|f\|_{H^1} .$$

Now consider $\phi(x_1,x_2)$ satisfying (4). Then it will be enough to show that

$$\left| \int_{R^2} \tilde{a}_k(x_1,x_2)\,\phi(x_1,x_2)\,dx \right| \le C$$

and then simply sum over k. But

$$(6.1) \qquad \int_{R^2} a_k(x_1,x_2)\,\phi(x_1,x_2)\,dx_1dx_2 = \frac{1}{2^k|\tilde{\Omega}_k|}\int \sum f_R(x)\,\phi(x)\,dx$$

where the sum is taken over

$$\mathfrak{R}_k = \left\{ R \in \mathfrak{R}_d \,\middle|\, |R \cap \Omega_k| > \tfrac{1}{2}\,|R| \ \text{ but } \ |R \cap \Omega_{k+1}| \le \tfrac{1}{2}\,|R| \right\}.$$

Then (6.1) becomes

$$\sum_{R \in \mathfrak{R}_k} \phi(x) \iint_{\mathfrak{A}(R)} f(y,t)\,\psi_t(x-y)\,\frac{dy\,dt}{t_1t_2}\,dx \cdot \frac{1}{2^k|\tilde{\Omega}_k|}$$

$$= \frac{1}{2^k|\tilde{\Omega}_k|} \bigcup_{R \in \mathfrak{R}_k} \iint_{\mathfrak{A}(R)} f(y,t)\,\phi(y,t)\,dy\,\frac{dt}{t_1t_2}$$

$$\le \frac{1}{2^k|\tilde{\Omega}_k|}\left(\bigcup_{R \in \mathfrak{R}_k} \iint_{\mathfrak{A}(R)} |f(y,t)|^2 dy\,\frac{dt}{t_1t_2}\right)^{1/2}\left(\bigcup_{R \in \mathfrak{R}_k} \iint_{\mathfrak{A}(R)} |\phi(y,t)|^2 dy\,\frac{dt}{t_1t_2}\right)^{1/2} .$$

$$(6.2)\qquad \int_{\tilde{\Omega}_k/\Omega_{k+1}} |S_\psi(f)^2(x)\,dx \geq \bigcup_{R\in\mathfrak{R}_k} \iint_{\mathfrak{A}(R)} |f(y,t)|^2 dy\,\frac{dt}{t_1 t_2}\,.$$

To show this

$$\int_{\tilde{\Omega}_k/\Omega_{k+1}} S_\psi^2(f)(x)\,dx = \int_{x\in\tilde{\Omega}_k/\Omega_{k+1}} \left(\iint_{\Gamma(x)} |f(y,t)|^2 dy\,\frac{dt}{t_1^2 t_2^2} \right) dx$$

$$\geq \iint |f(y,t)|^2 |\{x \in \tilde{\Omega}_k/\Omega_{k+1} | (y,t) \in \Gamma(x)\}|\, dy\,\frac{dt}{t_1^2 t_2^2}\,.$$

Suppose that $(y,t) \in \mathfrak{A}(R)$, $R \in \mathfrak{R}_k$. Then if the aperture of Γ is large enough, then $(y,t) \in \Gamma(x)$ for all $x \in R$. Since

$$|\{x \in (\tilde{\Omega}_k/\Omega_{k+1}) \cap R\}| \geq \frac{1}{2}\,|R|\,,\quad R \in \mathfrak{R}_k$$

we see that

$$|\{x \in \tilde{\Omega}_k/\Omega_{k+1}, (y,t) \in \Gamma(x)\}| \geq \frac{1}{2}\, t_1 t_2 \quad \text{(observe that } t_1 t_2 \sim |R|\,)$$

for $(y,t) \in \bigcup_{R\in R_k} \mathfrak{A}(R)$, and this proves (6.1).

But then

$$\int_{\tilde{\Omega}_k/\Omega_{k+1}} S_\psi(f)^2(x)\,dx \leq (2^{k+1})^2 |\tilde{\Omega}_k|$$

and combining this with (6.1) yields

$$\bigcup_{R\in\mathfrak{R}_k} \iint_{\mathfrak{A}(R)} |f(y,t)|^2 dy\,\frac{dt}{t_1^2 t_2^2} \leq C 2^k\, |\tilde{\Omega}_k|\,.$$

As for

$$\bigcup_{R\in\mathfrak{R}_k} \iint\limits_{\mathfrak{A}(R)} |\phi(y,t)|^2\, dy \frac{dt}{t_1 t_2},$$

if we observe that for any $R \in \mathfrak{R}_k$, $\mathfrak{A}(R)$ is contained in $S(\tilde{\Omega}_k)$, we get

$$\bigcup_{R\in\mathfrak{R}_k} \iint\limits_{\mathfrak{A}(R)} |\phi(y,t)|^2\, dy \frac{dt}{t_1 t_2} \le \iint\limits_{S(\tilde{\Omega}_k)} |\phi(y,t)|^2\, dy \frac{dt}{t_1 t_2} \le C|\tilde{\Omega}_k|$$

by (4). This shows that $|\int \tilde{a}_k \cdot \phi\, dx| \le C$ and completes the proof that (4) implies (1).

We shall show next that (2) implies (4). Let $g \in L^\infty(R^2)$. We claim that if $\Omega \subset R^2$, then

$$\iint\limits_{S(\Omega)} |g(y,t)|^2\, dy \frac{dt}{t_1 t_2} C\|g\|_\infty^2 |\Omega| ,$$

where $g(y,t) = g * \psi_{t_1 t_2}(y)$. To show this, observe that since $\mathrm{supp}(\psi) \subseteq [-1,1]$, $\mathrm{supp}(\psi_{t_1,t_2}(\cdot - y)) \subseteq R(y,t)$. Hence, if $(y,t) \in S(\Omega)$, $g * \psi_t(y) = (g\chi_\Omega) * \psi_t(y)$ and so

$$\iint\limits_{S(\Omega)} |g(y,t)|^2\, dy \frac{dt}{t_1 t_2} \le \iint\limits_{(R_+^2)^2} |(g\chi_\Omega)(y,t)|^2\, dy \frac{dt}{t_1 t_2} .$$

An easy application of Plancherel's formula says that this is, in turn,

$$\|g\chi_\Omega\|_2^2 \int_0^\infty |\hat{\psi}(\xi)|^2 \frac{d\xi}{\xi} .$$

Since, $\int_{-1}^{+1} \psi = 0$,

$$\begin{cases} \hat{\psi}(0) = 0 \text{ and } \hat{\psi} \in C^\infty \\ \hat{\psi}(\xi) = 0(|\xi|^{-N}) \text{ as } |\xi| \to \infty, \text{ for each } N > 0 . \end{cases}$$

so

$$\int_0^\infty |\hat{\psi}(\xi)|^2 \frac{d\xi}{\xi} < \infty .$$

It follows that

$$\iint_{S(\Omega)} |g(y,t)|^2 dy \frac{dt}{t_1 t_2} \leq C \|g\|_\infty^2 |\Omega|$$

as claimed.

If we wish to prove that the same Carleson condition holds for g replaced by $H_{x_1} H_{x_2}(g)$, then we proceed as follows. Observe that

$$\left(\iint_{S(\Omega)} |[H_{x_1} H_{x_2}(g)] * \psi_t(y)|^2 dy \frac{dt}{t_1 t_2} \right)^{1/2} = \left(\iint_{S(\Omega)} |g * \psi_t(y)|^2 dy \frac{dt}{t_1 t_2} \right)^{1/2}$$

where $\Psi = H_{x_1} H_{x_2}(\psi)$. The function Ψ splits into a product of $\Psi^{(1)}(x_1) \cdot \Psi^{(2)}(x_2)$ where $\Psi^{(i)}$ is odd, C^∞ and decreasing at ∞ like $|x_i|^{-N}$ (depending on how many moments of ψ vanish). Now suppose we choose $\eta(x)$ on R^1 so that $\text{supp}(\eta) \subseteq (1/4, 4)$, $\eta \in C^\infty(R^1)$, η even and $\sum_{k=-\infty}^{\infty} \eta\left(\frac{x}{2^k}\right) \equiv 1$. Let $\eta_0(x) = \sum_{k \leq 0} \eta\left(\frac{x}{2^k}\right)$ and for $k > 0$, let $\eta_k(x) = \eta\left(\frac{x}{2^k}\right)$. Set $\psi_{k,j}(x_1, x_2) = \Psi^{(1)}(x_1)\, \eta_k(x_1) \cdot \Psi^{(2)}(x_2)\, \eta_k(x_2)$. Then

(a) $\text{supp}(\Psi_{k,j}) \subseteq 4R(0; 2^k, 2^j)$

(b) $\Psi_{k,j}$ is odd in each variable separately

(c) Ψ_{kj} is $C_c^\infty(R^2)$

and

(d) $\left\|\left(\frac{\partial}{\partial x}\right)^{\alpha}\Psi_{k,j}\right\|_\infty = 0(2^{-kN-jN})$ as $k,j \to \infty$ if $|\alpha| \le 2$

By Minkowski's inequality, we have

$$(6.3)\quad \left(\iint_{S(\Omega)} |g * \Psi_t|^2 dy \frac{dt}{t_1 t_2}\right)^{1/2} \le \sum_{k,j}\left(\iint_{S(\Omega)} |g*(\Psi_{k,j})_t|^2 dy \frac{dt}{t_1 t_2}\right)^{1/2} .$$

Now, to estimate

$$\sum_{k,j} |g*(\Psi_{k,j})_t|^2 dy \frac{dt}{t_1 t_2}$$

we use the same argument as that given above, except that now $\text{supp}(\Psi_{k,j}) \le 4R(0; 2^k, 2^j)$ and not the unit square. If $(y,t) \in S(\Omega)$, then $R(y;t) \subseteq \Omega$ and the support of $(\Psi_{k,j})_t(\cdot - y)$ will be contained in

$$R(y, 2^k t_1, 2^j t_2) \subseteq \{M^{(2)}(\chi_\Omega) > 2^{-(k+j)}\} = \tilde{\Omega}_{k,j} .$$

Thus

$$\iint_{S(\Omega)} |g*(\Psi_{k,j})_t(y)|^2 dy \frac{dt}{t_1 t_2} = \iint_{S(\Omega)} |(g\chi_{\tilde{\Omega}_{kj}}) * \Psi_{kj}|^2 dy \frac{dt}{t_1 t_2}$$

$$\le \|g\chi_{\tilde{\Omega}_{kj}}\|_2^2 \int_0^\infty\int_0^\infty |\Psi_{kj}(\xi_1,\xi_2)|^2 \frac{d\xi_1}{\xi_1}\frac{d\xi_2}{\xi_2} \le \|g\|_\infty^2 |\tilde{\Omega}_{kj}| \left(\int_0^\infty\int_0^\infty |\Psi_{kj}(\xi_1,\xi_2)|^2 \frac{d\xi_1}{\xi_1}\frac{d\xi_2}{\xi_2}\right) .$$

By the strong maximal theorem $|\widetilde{\Omega}_{kj}| \leq C(k+j)2^{k+j}$, and it is easy to see that $\int_0^\infty \int_0^\infty |\hat{\Psi}_{kj}|^2 \, d\xi$ decreases like a large power of $2^{-(k+j)}$ as $k,j \to \infty$. So our desired estimate follows from (6.2).

So far we have proven the equivalence of (1), (2), (3) and (4). We shall not go into the details of the equivalence of (5) except to say the proof is given in the Annals paper of Chang-Fefferman [25]. Rather, let us point out a beautiful application of the equivalence of (5) with the other definitions of BMO which occurs already in the one-parameter setting. This is the theorem of A. Uchiyama [26], which tells us which families of multipliers homogeneous on R^n of degree 0 determine $H^1(R^n)$. He showed that for multiplier operators $I, K_1, K_2, \cdots, K_m$ with multipliers $1, \theta_i(\xi)$ that $f, K_i f \in L^1(R^n)$ implies $f \in H^1(R^n)$ if and only if the θ_i separate antipodal points of S^{n-1}, i.e., if and only if for every $\xi \in S^{n-1}$, there exists i such that $\theta_i(\xi) \neq \theta_i(-\xi)$.

The way Uchiyama proves this is to show that the dual statement is true, namely, every $\phi \in BMO(R^n)$ can be written as

$$(6.4) \qquad g_0 + \sum K_i g_i \quad \text{for some } g_0, g_1, \cdots, g_m \in L^\infty .$$

This depends on a simple lemma.

LEMMA. *If* θ_i *are as above, then given* $f \in L^2$, *and a vector* $\nu \in C^{m+1}$, *there exist functions* $g_0, \cdots, g_m \in L^2$ *so that*

$$g_0 + \sum K_i g_i = f \quad \textit{and} \quad g(x) = (g_0(x), g_1(x), \cdots, g_m(x)) \quad \nu \quad \textit{for all} \ x \in R^n .$$

To prove the formula (6.4), Uchiyama decomposes $\phi = \Sigma C_I \phi_I$ as in our (5), and applies the lemma to get functions $g_I(x)$ such that $Kg_I(x) = C_I \phi_I(x)$ for which $g(x)$ is perpendicular to the correct ν, and the result, when modified only slightly to $\tilde{g}_I$, has the property that $\Sigma \tilde{g}_I \in L^\infty$. For the details see Uchiyama's recent paper in Acta [26].

Now, finally we wish to discuss the atomic decomposition of $H^1(R^2_+ \times R^2_+)$ in greater detail. There are interesting applications of this decomposition besides duality with $BMO(R_+^2 \times R_+^2)$ which was presented above. We shall be content with one more application here which sheds a good deal of light on the nature of these atoms. Namely, we intend to give a second proof, directly by real variables, that on $R_+^2 \times R_+^2$ if $S_\psi(f) \in L^1(R^2)$ then $f^* \in L^1(R^2)$ [27]. Suppose $S_\psi(f) \in L^1(R^2)$. Then in our discussion of duality we defined atoms

$$\tilde{a}_k(x) = \frac{1}{2^k|\tilde{\Omega}_k|} \sum_{R \in R_k} f_R(x_1,x_2) .$$

To simplify this notation we define $\omega = \tilde{\Omega}_k$ and $A(x) = 2^k|\tilde{\Omega}_k|\tilde{a}_k(x)$. Let ϕ^1 and $\phi^2 \in C_c^\infty(R^1)$ with $\phi^i(x) \geq 0$, $\int \phi^i = 1$, and $\mathrm{supp}(\phi^i) \subset [-1,+1]$. Set

$$\phi_{t_1,t_2}(x_1,x_2) = t_1^{-1}t_2^{-1}\phi^1\left(\frac{x_1}{t_1}\right)\phi^2\left(\frac{x_2}{t_2}\right) .$$

Define $\tilde{\omega} = M^{(2)}(\chi_\omega) > \frac{1}{10^{10}}$. We need to estimate $A * \phi_{t_1,t_2}(x)$ for $x \notin \tilde{\omega}$. To do this let us make the following definitions. If R is a rectangle then R_1, R_2 will be its sides, so that $R = R_1 \times R_2$. Let

$$A_j^1(x) = \sum_{\substack{R \in \mathfrak{R}_k \\ |R_1| = 2^j}} f_R(x), \quad A_j^2(x) = \sum_{\substack{R \in \mathfrak{R}_k \\ |R_2| = 2^j}} f_R(x) .$$

Then to estimate $A * \phi_{t_1,t_2}(x)$, since $\mathrm{supp}(\phi_t(\cdot - x)) \subseteq R(x;t) = S$, in the definition of A we need only consider those f_R for which $\tilde{R} \cap S \neq 0$. For any such rectangle $R \in \mathfrak{R}_k$, since $R \subset \omega$,

$$\text{minimum}\left(\frac{|R_1|}{|S_1|}, \frac{|R_2|}{|S_2|}\right) < \left(\frac{|S \cap \tilde{\omega}|}{|S|}\right)^{1/2} = \rho$$

where $\tilde{\omega}$ denotes again $M^{(2)}(\chi_\omega) > \frac{1}{10^{10}}$. Then

$$A * \phi_{t_1 t_2}(x) = \sum_{2^j/|S_1|<\rho} A_j^1 * \phi_{t_1 t_2}(x) + \sum_{2^j/|S_2|<\rho} A_j^2 * \phi_{t_1 t_2}(x)$$

$$- \sum_{R \in \mathfrak{B}} f_R * \phi_{t_1 t_2}(x)$$

where $\mathfrak{B} \subset \mathfrak{R}_k$ consists of rectangles R so that $\frac{|R_1|}{|S_1|} < \rho$, $\frac{|R_2|}{|S_2|} < \rho$ and $\widetilde{R} \cap S \neq \emptyset$. Thus $R \subseteq \widetilde{S}$ for all $R \in \mathfrak{B}$, and the reason this subtracted term occurs is that we have double counted these f_R whose R sides are both very small.

In order to estimate $A_j^1 * \phi_{t_1 t_2}(x)$ we use the following trivial lemma.

LEMMA. *On* R^1 *suppose that* $\phi(x) \in C^\infty$ *and is supported in an interval* ϑ. *Suppose* $\alpha(x)$ *is supported on disjoint subintervals of* ϑ, I_k *whose lengths are all* $\leq \gamma|\vartheta|$. *Assume that* $\alpha(x)$ *has* N *vanishing moments over each* I_k. *Then*

$$\left| \int_\vartheta \alpha(x)\phi(x)dx \right| \leq C \|\phi^{(N+1)}\|_\infty (\gamma|\vartheta|)^{N+1} \int |\alpha(x)|dx \, .$$

We estimate $A_j^1 * \phi_{t_1 t_2}(x)$ using the fact that for each fixed x_2, $A_j^1(\cdot, x_2)$ has N vanishing moments over disjoint x_1 intervals over length $2 \cdot 2^j$. (Actually, we sould have to break up A_j^1 into 3 pieces to insure this, but we spare the reader this trivial complication.) It follows from the lemma that

$$|A_j^1 *_1 \phi_{t_1}^1(x)| \leq_{t_1} \left(\frac{2^j}{|S_1|}\right)^{N+1} (|A_j^1| *_1 \chi_{[-t_1 t_1]}(x)) \, .$$

Convolving in the x_2 variable, we have

$$|A_j^1 * \phi_{t_1, t_2}(x)| \leq C \; \frac{2^j}{|S_1|}^{N+1} \; \frac{1}{|\widetilde{S}|} \int_{\widetilde{S}} |A_j^1| dx' \, .$$

For this we get

$$\left|\sum_{2^j/|S_1|<\rho} A_j^1 * \phi_{t_1,t_2}(x)\right| \le C\rho^{N/2}\frac{1}{|\widetilde{S}|}\int_{\widetilde{S}}\Big(\sum |A_j^1|^2\Big)^{1/2}\left(\sum_{2^j/|S_1|<\rho}\Big(\frac{2^j}{|S_1|}\Big)^N\right)^{1/2}$$

$$\le C\rho^{N/2}\frac{1}{|\widetilde{S}|}\int_{\widetilde{S}}\Big(\sum |A_j^1|^2\Big)^{1/2}.$$

By symmetry

$$\left|\sum_{2^j/|S_2|<\rho} A_j^2 * \phi_{t_1t_2}(x)\right| \le C\rho^{N/2}\frac{1}{|\widetilde{S}|}\int_{\widetilde{S}}\Big(\sum |A_j^2|^2\Big)^{1/2}.$$

Now let $R \subset \widetilde{S}$, with $\dfrac{|R_1|}{|S_1|} \le \dfrac{|R_2|}{|S_2|}$. Then

$$|f_R * \phi_{t_1}(x)| \le C\Big(\frac{|R|}{|S|}\Big)^{N/2}\frac{1}{t_1}\chi_{[-t_1,t_2]} * |f_R|(x)$$

and also

$$|f_R * \phi_{t_1,t_2}(x)| \le C\Big(\frac{|R|}{|S|}\Big)^{N/2}\frac{1}{|\widetilde{S}|}\int_{\widetilde{S}}|f_R|.$$

Thus

$$\sum_{R\epsilon\mathfrak{R}_k}|f_R * \phi_{t_1t_2}(x)| \le C\frac{1}{|\widetilde{S}|}\int_{\widetilde{S}}\Big(\sum_{R\epsilon\mathfrak{R}_k} f_R^2(x)\Big)^{1/2}\left(\sum_{R\epsilon\mathfrak{R}_k}\Big(\frac{|R|}{|S|}\Big)^{N/2}\right)^{1/2}\cdot R^{N/4}$$

$$\le C\rho^{N/4}\frac{1}{|\widetilde{S}|}\int_{\widetilde{S}}\sum_{R\epsilon\mathfrak{R}_k}|f_R|^2{}^{\;1/2}.$$

To sum up our findings, we have seen that if $x \notin \widetilde{\omega}$ then

$$|A * \phi_{t_1t_2}(x)| \leq C\left(\frac{|S \cap \omega|}{|\tilde{S}|}\right)^{N/4}\left[\frac{1}{|\tilde{S}|}\int_{\tilde{S}} J(y)\,dy\right] \tag{6.5}$$

where

$$J(x) = \left(\sum (A_j^1)^2\right)^{1/2} + \left(\sum (A_j^2)^2\right)^{1/2} + \left(\sum_{R\in\mathfrak{R}_k} f_R\right)^{1/2}.$$

To finish the proof, we need another lemma:

LEMMA. *Let* $g(x) = \sum_{R\in\mathfrak{B}} f_R(x)$ *where* B *is a collection of dyadic rectangles. Then*

$$\|g\|_2^2 \leq \bigcup_{R\in\mathfrak{B}} \int_{\mathfrak{A}(R)} |f(y,t)|^2\,dy\,\frac{dt}{t_1t_2}.$$

Proof. Let $\|h\|_{L^2(R^2)} = 1$. Then

$$\int g(x)\,h(x)\,dx = \int \sum_{R\in\mathfrak{B}} \iint_{\mathfrak{A}(R)} f(y,t)\,\psi_t(x-y)\,dy\,\frac{dt}{t_1t_2} \cdot h(x)\,dx$$

$$= \sum_{R\in\mathfrak{B}} \iint_{\mathfrak{A}(R)} f(y,t)\,h(y,t)\,dy\,\frac{dt}{t_1t_2}$$

$$\leq \left(\bigcup_{R\in\mathfrak{B}} \iint_{\mathfrak{A}(R)} f(y,t)|^2 dy\,\frac{dt}{t_1t_2}\right)^{1/2}\left(\iint_{(R_+^2)^2} |h(y,t)|^2 dy\,\frac{dt}{t_1t_2}\right)^{1/2}$$

$$\leq C\left(\iint |f(y,t)|^2 dy\,\frac{dt}{t_1t_2}\right)^{1/2} \|S_\psi(h)\|_2 \leq C'\left(\iint |f(y,t)|^2 dy\,\frac{dt}{t_1t_2}\right)^{1/2}.$$

Now, notice that, by the lemma,

$$\|J\|_2^2 \le \iint_{R\epsilon_k^{(R)}} |f(y,t)|^2 dy \frac{dt}{t_1 t_2} \le \int_{\tilde{\Omega}_k/\Omega_{k+1}} S_\psi^2(f)(x)\,dx \le C \cdot 2^{2k} \cdot |\omega| \, .$$

The same estimate holds for $\|A\|_2^2$. Then

$$\int_{\tilde{\omega}} A^* \le |\omega|^{1/2} \left(\int_{R^2} A^{*2} \right)^{1/2} \le C|\omega|^{1/2} \|A\|_2 \le C2^k |\omega| \, .$$

Also away from $\tilde{\omega}$,

$$A^*(x) \le M^{(2)}(\chi_{\tilde{\omega}})^{10}(x) \cdot M^{(2)}(J)(x) \, ,$$

so

$$\int_{c\,\tilde{\omega}} A^* dx \le \left(\int_{R^2} M^{(2)}(\chi_{\tilde{\omega}})^{20}(x)\,dx \right)^{1/2} \left(\int_{R^2} M^{(2)}(J)^2(x)\,dx \right)^{1/2}$$

$$\le C|\omega|^{1/2} |\omega|^{1/2} 2^k = C2^k |\omega| \, .$$

It follows that $\|A^*\|_1 \le C2^k|\omega|$ and also $\|\tilde{a}_k\|_1 \le C$.

ROBERT FEFFERMAN
DEPARTMENT OF MATHEMATICS
UNIVERSITY OF CHICAGO
CHICAGO, ILLINOIS 60637

BIBLIOGRAPHY

[1] B. Jessen, J. Marcinkiewicz and A. Zygmund, Notes on the Differentiability of Multiple Integrals, Fund. Math. 24, 1935.

[2] E. M. Stein and S. Wainger, Problems in Harmonic Analysis Related to Curvature, Bull. AMS. 84, 1978.

[3] A. Cordoba and R. Fefferman, A Geometric Proof of the Strong Maximal Theorem, Annals of Math., 102, 1975.

[4] J. O. Stromberg, Weak Estimates on Maximal Functions with Rectangles in Certain Directions, Arkiv fur Math., 15, 1977.

[5] A. Cordoba and R. Fefferman, On Differentiation of Integrals, Proc. Nat. Acad. of Sci., 74, 1977.

[6] A. Nagel, E. M. Stein, and S. Wainger, Differentiation in Lacunary Directions, Proc. Nat. Acad. Sci., 75, 1978.

[7] A. Cordoba, Maximal Functions, Covering Lemmas, and Fourier Multipliers, Proc. Symp. in Pure Math., 35, Part I, 1979.

[8] F. Soria, Examples and Counterexamples to a Conjecture in the Theory of Differentiation of Integrals, to appear in Annals of Math.

[9] B. Muckenhoupt, Weighted Norm Inequalities for the Hardy Maximal Function, Trans. of the AMS, 165, 1972.

[10] R. Hunt, B. Muckenhoupt, and R. Wheeden, Weighted Norm Inequalities for the Conjugate Function and Hilbert Transform, Trans. AMS, 176, 1973.

[11] R. R. Coifman and C. Fefferman, Weighted Norm Inequalities for Maximal Functions and Singular Integrals, Studia Math., 51, 1974.

[12] M. Christ and R. Fefferman, A Note on Weighted Norm Inequalities for the Hardy-Littlewood Maximal Operator, Proceedings of the AMS, 84, 1983.

[13] R. Fefferman, Strong Differentiation with Respect to Measures, Amer. Jour. of Math., 103, 1981.

[14] ________, Some Weighted Norm Inequalities for Cordoba's Maximal Function, to appear in Amer. Jour. of Math.

[15] C. Fefferman, The Multiplier Problem for the Ball, Annals of Math., 94, 1971.

[16] A. Cordoba and R. Fefferman, On the Equivalence between the Boundedness of Certain Classes of Maximal and Multiplier Operators in Fourier Analysis, Proc. Nat. Acad. Sci., 74, No. 2, 1977.

[17] E. M. Stein and G. Weiss, On the Theory of H^p Spaces, Acta. Math., 103, 1960.

[18] C. Fefferman and E. M. Stein, H^p Spaces of Several Variables, Acta Math., 129, 1972.

[19] D. Burkholder, R. Gundy, and M. Silverstein, A Maximal Function Characterization of the Class H^p, Trans. AMS, 157, 1971.

[20] K. Merryfield, Ph.D. Thesis: H^p Spaces in Poly-Half Spaces, University of Chicago, 1980.

[21] R. Gundy and E. M. Stein, H^p Theory for the Polydisk, Proc. Nat. Acad. Sci., 76, 1979.

[22] L. Carleson, A Counterexample for Measures Bounded on H^p for the Bi-Disc, Mittag-Leffler Report No. 7, 1974.

[23] S. Y. Chang, Carleson Measure on the Bi-Disc, Annals of Math., 109, 1979.

[24] R. Fefferman, Functions on Bounded Mean Oscillation on the Bi-Disc, Annals of Math., 10, 1979.

[25] S. Y. Chang and R. Fefferman, A Continuous Version of the Duality of H^1 and BMO on the Bi-Disc, Annals of Math., 1980.

[26] A. Uchiyama, A Constructive Proof of the Fefferman-Stein Decomposition of $BMO(R^n)$, Acta. Math., 148, 1982.

ELLIPTIC BOUNDARY VALUE PROBLEMS ON LIPSCHITZ DOMAINS

Carlos E. Kenig*

Dedicated to the memory of Jack P. Burke

PREFACE

This paper is an outgrowth of a series of lectures I presented at the Summer Symposium of Analysis in China (SSAC), held at Peking University in September, 1984. The material in the introduction and parts (a) and (b) of Section 1 comes from the expository article 'Boundary value problems on Lipschitz domains' ([19]), which I wrote jointly with D. S. Jerison in 1980. The rest of the paper can be considered as a sequel to that article. Some of the material in part (b) of Section 2, and all of Section 3 comes from the recent expository article "Recent progress on boundary value problems on Lipschitz domains" ([23]). The results explained in Section 2, (b) and Section 3 are unpublished. Full details will appear elsewhere in several joint papers.

Acknowledgements. I would like to thank Peking University, and the organizing committee of the SSAC, Professors M. T. Cheng, S. L. Wang, S. Kung, D. G. Deng and R. Long for their invitation to participate in the SSAC, and for their warm hospitality during my visit to China. I would also like to thank Professor E. M. Stein for his many efforts to make the SSAC a success. Thanks are also due to Mr. You Zhong and Mr. Wang Wengshen for taking careful notes of my lectures.

Finally, I would like to thank B. Dahlberg, E. Fabes, D. Jerison and G. Verchota for the many discussions and fruitful collaborations that we

*Supported in part by the NSF.

have had throughout the years, which resulted in the work explained in this paper.

Introduction

A harmonic function u is a twice continuously differentiable function on an open subset of R^n, $n \geq 2$, satisfying the *Laplace equation* $\Delta u = \sum_{j=1}^{n} \frac{\partial^2 u}{\partial X_j^2} = 0$. Harmonic function arise in many problems in mathematical physics. For example, the function measuring gravitational or electrical potential in free space is harmonic. A steady state temperature distribution in a homogeneous medium also satisfies the Laplace equation. Moreover, the Laplace equation is the simplest, and thus the prototype, of the elliptic equations, or systems of equations. These in turn also have many applications to mathematical physics and geometry. A first step in the understanding of this more general situation is the study of the Laplacian. This will be illustrated very clearly later on.

Initially we will be concerned with the two basic boundary value problems for the Laplace equation, the Dirichlet and Neumann problems. Let D be a bounded, smooth domain in R^n and let f be a smooth (i.e. C^∞) function on ∂D, the boundary of D. The classical *Dirichlet problem* is to find and describe a function u that is harmonic in D, continuous in $\bar{D}$, and equals f on ∂D. This corresponds to the problem of finding the temperature inside a body D when one knows the temperature f on ∂D. The classical *Neumann problem* is to find and describe a function u that is harmonic in D, belongs to $C^1(\bar{D})$, and satisfies $\frac{\partial u}{\partial N} = f$ on ∂D, where $\frac{\partial u}{\partial N}$ represents the normal derivative of u on ∂D. This corresponds to the problem of finding the temperature inside D when one knows the heat flow f through the boundary surface ∂D.

Our main purpose here is to describe results on the boundary behavior of u in the case of smooth domains, and to study in detail the extension of these results to the case of minimally smooth domains, where we allow corners and edges, i.e. *Lipschitz domains*. This class of domains is important from the point of view of applications, and also from the mathematical point of view. Their importance resides in the fact that this is a dilation invariant class of domains with some smoothness. They have the borderline amount of regularity necessary for the validity of the results we are going to expound on.

In a smooth domain, *the method of layer potentials*, (which we are going to develop soon) yields the existence of a solution u to the Dirichlet problem with boundary data $f \in C^{k,\alpha}(\partial D)$, and the bound

$$\|u\|_{C^{k,\alpha}(\bar{D})} \le C_{k,\alpha}\|f\|_{C^{k,\alpha}(\partial D)} \qquad \begin{array}{l} k = 0,1,2,\cdots . \\ 0 < \alpha < 1 \end{array}$$

What happens if the size of f is measured in some other norm, like the L^2 norm? This is of interest as a measure of the variation in data even if we are only concerned with continuous functions: if $f_1 - f_2$ has small L^2 norm we want to know that the corresponding solutions u_1 and u_2 are near each other. The wisdom of hindsight tells us that as long as we are going to examine all continuous functions in L^2 norm, it is no harder to consider arbitrary functions in L^2. Another reason to consider the L^2 norm is that it is better suited for the Neumann problem, even on smooth domains. We will also consider how our results change if we consider L^p norms, $p \neq 2$, as the smoothness of the domain decreases.

We will then show the flexibility of our methods by considering extensions of our results to *systems of elliptic equations* in Lipschitz domains. The specific systems of equations that we will study are the Navier systems which arise in the linear and infinitesimal theory of elasticity, when the displacements or the surface forces are given on the boundary of a homogeneous and isotropic elastic body D. These systems are the prototype of the second order elliptic systems of equations. We will also study the so-called Stokes problem; this is the linearized stationary problem of the mathematical theory of viscous incompressible flow.

Before going on to study the general situation, we will formulate appropriate theorems, by examining a model case, namely the Laplacian in the unit ball B. In this case we have a lot of symmetry at our disposal and everything can be done explicitly. Let $d\sigma$ denote surface measure of ∂B.

THEOREM. *Suppose that* $1 < p \leq \infty$ *and* $f \in L^p(\partial B, d\sigma)$. *Then, there exists a unique harmonic function* u *in* B *such that* $\lim_{r\to 1} u(rQ) = f(Q)$ *for almost every* $Q \in \partial B$, *and*

$$(*) \qquad \int_{\partial B} u^*(Q)^p \, d\sigma(Q) \leq C_p \int_{\partial B} |f(Q)|^p \, d\sigma(Q) ,$$

where $u^*(Q) = \sup_{0 \leq r < 1} |u(rQ)|$.

The theorem asserts that $f_r(Q) = u(rQ)$ converges to $f(Q)$ not only in L^p norm, but also in the sense of Lebesgue's dominated convergence.

In the analogous estimates to (*) in the Neumann problem, u is replaced by the gradient of u. In that case the estimate fails for $p = \infty$, even if $\frac{\partial u}{\partial N}$ is continuous.

In both the case of the Dirichlet problem and the Neumann problem, the radial limit can be replaced by a non-tangential limit: if X tends to Q with $|X-Q| < (1+\alpha)\operatorname{dist}(X, \partial B)$, for some fixed $\alpha > 0$, then $u(X)$ has the limit $f(Q)$ for almost every Q.

The theorem is most easily proved by writing down a formula for the solution, $u(X) = \int_{\partial B} P(X,Q) f(Q) \, d\sigma(Q)$, where $P(X,Q) = \frac{1}{\omega_n} \frac{1-|X|^2}{|X-Q|^n}$.
The estimate now follows as an easy consequence of the Hardy-Littlewood maximal theorem. An analogous formula holds for the Neumann problem.

This time, it is more difficult to obtain the estimates. One needs to use the Calderon-Zygmund theory of singular integrals, and the Hardy-Littlewood maximal theorem.

The case of the Laplacian in the ball is relatively easy because of the existence of explicit formulas for the solution. What should we do in the case of a general domain, where explicit formulas are not available? What should we do to study systems of equations? What happens to our solutions as the domains become less smooth? We hope to give a systematic answer to these questions in the rest of this paper.

§1. *Historical comments and preliminaries*

(a) *The method of layer potentials for Laplace's equation on smooth domains.*

DEFINITION. A bounded domain D is called a Lipschitz domain with Lipschitz constant less than or equal to M if for any $Q \in \partial D$ there is a ball B with center at Q, a coordinate system (isometric to the usual coordinate system) $x' = (x_1, \cdots, x_{n-1})$, x_n, with origin at Q and a function $\phi : R^{n-1} \to R$ such that

$$\phi(0) = 0,\ |\phi(x') - \phi(y')| \leq M|x' - y'| \text{ and } D \cap B = \{X = (x', x_n) : x_n > \phi(x')\} \cap B .$$

If for each Q the function ϕ can be chosen in $C^1(R^{n-1})$, then D is called a C^1 domain. If in addition, $\nabla\phi$ satisfies a Holder condition of order α,

$$|\nabla\phi(x') - \nabla\phi(y')| \leq C|x' - y'|^{\alpha} ,$$

we call D a $C^{1,\alpha}$ domain.

Notice that the cone $\Gamma_e = \{(x', x_n) : x_n < -M|x'|\}$ satisfies $\Gamma_e \cap B \subset {}^cD$. Similarly, $\Gamma_i = \{(x', x_n) : x_n > M|x'|\}$ satisfies $\Gamma_i \cap B \subset D$. Thus, Lipschitz domains satisfy the interior and exterior cone condition.

The function ϕ satisfying the Lipschitz condition $|\phi(x') - \phi(y')| \leq M|x' - y'|$ is differentiable almost everywhere and $\nabla\phi \in L^{\infty}(R^{n-1})$, $\|\nabla\phi\|_{\infty} \leq M$.

Surface measure σ is defined for each Borel subset $E \subset \partial D \cap B$ by

$$\sigma(E) = \int_{E^*} (1 + |\nabla\phi(x')|^2)^{1/2} dx' ,$$

where $E^* = \{x' : (x', \phi(x')) \in E\}$.

The unit outer normal to ∂D given in the coordinate system by $(\nabla\phi(x'), -1)/(1 + |\nabla\phi(x')|^2)^{1/2}$ exists for almost every x'. The unit normal at Q will be denoted by N_Q. It exists for almost every $Q \in \partial D$, with respect to $d\sigma$.

In order to motivate the use of the method of layer potentials, we need to recall some formulas from advanced calculus, and some definitions. We will start with the derivation corresponding to the Dirichlet problem.

We first recall the fundamental solution $F(X)$ to Laplace's equation in $R^n : \Delta F = \delta$, where

$$F(X) = \begin{cases} -\dfrac{1}{(n-2)\,\omega_n |X|^{n-2}} & n > 2 \\[2ex] \dfrac{1}{2\pi} \log |X| & n = 2 \end{cases}$$

where ω_n is the surface area of the unit sphere in R^n. $F(X)$ is the electrical potential in free space induced by a unit charge at the origin. It provides a formula for a solution ω to the equation $\Delta\omega = \psi$, with $\psi \in C_0^\infty(R^n)$,

$$\omega(X) = F * \psi(X) = \int_{R^n} F(X-Y)\psi(Y)dY .$$

It will be convenient to put $F(X,Y) = F(X-Y)$. Notice that $\Delta_Y F(X,Y) = \delta(X-Y)$. The fundamental solution in a bounded domain is known as the Green function $G(X,Y)$. It is the function on $\overline{D} \times \overline{D}$ continuous for $X \neq Y$ and satisfying $\Delta_Y G(X,Y) = \delta(X-Y)$, $X \in D$; $G(X,Y) = 0$, $X \in D$, $Y \in \partial D$.

$G(X,Y)$ as a function of Y is the potential induced by a unit charge at X that is grounded to zero potential on ∂D. The Green function can be obtained if one knows how to solve the Dirichlet problem. In fact, let $u_X(Y)$ be the harmonic function with boundary values $u_X(Y)|_{\partial D} = F(X,Y)|_{\partial D}$. Then,

$$G(X,Y) = F(X,Y) - u_X(Y) . \tag{1}$$

On the other hand, if we know $G(X,Y)$, we can formally write down the solution to the Dirichlet problem.

In fact,

$$u(X) = \int_D u(Y)\delta(X,Y)\,dY = \int_D u(Y)\Delta_Y G(X,Y)\,dY =$$

$$= \int_D [u(Y)\Delta_Y G(X,Y) - \Delta u(Y) \cdot G(X,Y)]\,dY =$$

$$= \int_{\partial D} \left[u(Q)\,\frac{\partial G}{\partial N_Q}(X,Q) - \frac{\partial u}{\partial N_Q}(Q)\,G(X,Q)\right] d\sigma(Q) =$$

$$= \int_{\partial D} u(Q)\,\frac{\partial}{\partial N_Q} G(X,Q)\,d\sigma(Q) ,$$

where the fourth equality follows from Green's formula. Thus, we have derived the formula

$$u(X) = \int_{\partial D} f(Q)\,\frac{\partial G}{\partial N_Q}(X,Q)\,d\sigma(Q) \tag{2}$$

for the harmonic function u with boundary values f. The problem with formula (2) is of course that we don't know $G(X,Q)$. Because of formulas (1) and (2), C. Neumann proposed the formula

$$w(X) = \int_{\partial D} f(Q) \frac{\partial F}{\partial N_Q}(X,Q)\, d\sigma(Q) =$$

$$= \frac{1}{\omega_n} \int_{\partial D} f(Q) \frac{\langle X-Q, N_Q \rangle}{|X-Q|^n}\, d\sigma(Q)$$

as a first approximation to the solution of the Dirichlet problem, $\Delta u = 0$ in D, $u|_{\partial D} = f$.

$w(X)$ is known as the double layer potential of f. First of all w is harmonic in D. Also, one can show that as $X \to Q \in \partial D$, $w(X) \to \frac{1}{2} f(Q) + Kf(Q)$, where K is the operator on ∂D given by

$$Kf(Q) = \frac{1}{\omega_n} \int_{\partial D} \frac{\langle Q-P, N_P \rangle}{|P-Q|^n} f(P)\, d\sigma(P)\ .$$

If Kf were zero, we would be done, and in some sense it is true that Kf is small compared to $\frac{1}{2} f$, when the domain D is smooth. In fact, ∂D has dimension $n-1$, while it is easy to see that if ∂D is C^∞,

$$\left| \frac{\langle Q-P, N_P \rangle}{|P-Q|^n} \right| \le \frac{C}{|P-Q|^{n-2}}\ .$$

Thus, the operator $K: C(\partial D) \to C(\partial D)$ is compact. Therefore, by the general theory of Fredholm, the operator $T = \frac{1}{2} I + K$ is invertible modulo a finite dimensional subspace of $C(\partial D)$. If D and ${}^cD$ are connected, T is actually invertible on $C(\partial D)$. Therefore, the solution to the Dirichlet problem may be written

$$u(X) = \frac{1}{\omega_n} \int_{\partial D} g(Q) \frac{<X-Q, N_Q>}{|X-Q|^n} \, d\sigma(Q) ,$$

where $g = T^{-1}(f)$. The operator K is compact on $C(\partial D)$ even in $C^{1,\alpha}$ domains $\left(\text{because in this case } \left|\frac{<Q-P,N_P>}{|P-Q|^n}\right| \leq \frac{C}{|P-Q|^{n-1+\alpha}}\right)$ and so this procedure solves the classical Dirichlet problem in that case too. If D is a C^∞ domain, K is compact from $C^{k,\alpha}(\partial D)$ to $C^{k,\alpha}(\partial D)$, $k = 1, 2, \cdots, 0 < \alpha < 1$. Hence, if $f \in C^\infty(\partial D)$, $u \in C^\infty(\bar{D})$. This approach can also be used to obtain results for $f \in L^p(\partial D)$ in C^∞ domains, and even on $C^{1,\alpha}$ domains. We will now sketch the extension of the theorem for the ball stated in the introduction to $C^{1,\alpha}$ domains. We first define non-tangential approach regions as follows:

$$\Gamma_\beta(Q) = \{X \in D : |X-Q| < (1+\beta) \operatorname{dist}(X, \partial D)\} .$$

The non-tangential maximal function, with opening β, of a function w defined in D is

$$N_\beta(w)(Q) = \sup\{|w(x)| : X \in \Gamma_\beta(Q)\} .$$

Because of the estimate $\left|\frac{<Q-P,N_P>}{|P-Q|^n}\right| \leq C/|P-Q|^{n-1+\alpha}$, on $C^{1,\alpha}$ domains, it is easy to see that K is compact as a mapping on $L^p(\partial D)$. Also, standard arguments show that

$$N_\beta(w)(Q) \leq C_\beta\{M(f)(Q) + M(Kf)(Q)\} ,$$

where M is the Hardy-Littlewood maximal operator on ∂D, and w is the double layer potential of f. Finally, from L^p bounds for M, K and $T^{-1} = \left(\frac{1}{2} I + K\right)^{-1}$, one obtains:

THEOREM. *Let* D *be a* $C^{1,\alpha}$ *domain,* $1 < p < \infty$. *If* $f \in L^p(\partial D, d\sigma)$, *and* $u(x) = \frac{1}{\omega_n} \int_{\partial D} \frac{\langle X-Q, N_Q \rangle}{|X-Q|^n} T^{-1} f(Q)\, d\sigma(Q)$, *then* $\|N_\beta u\|_{L^p(d\sigma)} \leq C_p \|f\|_{L^p(d\sigma)}$ *and the harmonic function* u *tends to* f *non-tangentially.*

The difficulty in the case of C^1 and Lipschitz domains, is that, in this case the size estimate on the kernel of K is

$$\left| \frac{\langle Q-P, N_P \rangle}{|P-Q|^n} \right| \leq \frac{C}{|P-Q|^{n-1}},$$

and so, even the L^p boundedness, much less the compactness of K, is far from obvious.

Let us now turn to the Neumann problem. Let D be a smooth domain. We seek to solve $\Delta u = 0$ in D, $\frac{\partial u}{\partial N}\big|_{\partial D} = f$. By Green's formula, we must have $\int_{\partial D} f\, d\sigma = 0$. When D and ${}^cD$ are connected this is the only compatibility condition needed. We will only consider that case for simplicity. A good first guess at the solution u is the so-called single layer potential of f given by $v(x) = \int_{\partial D} f(Q) F(X,Q)\, d\sigma(Q) = C_n \int_{\partial D} f(Q) \frac{1}{|X-Q|^{n-2}}\, d\sigma(Q)$. Once again v is harmonic in D, and $\frac{\partial v}{\partial N_Q}(Q) = \frac{1}{2} f(Q) - K^* f(Q)$, where K^* is the adjoint of K above, i.e. $K^* f(P) = \frac{1}{\omega_n} \int_{\partial D} \frac{\langle P-Q, N_P \rangle}{|P-Q|^n} f(Q)\, d\sigma(Q)$. Thus, K^* is compact from $C^{k,\alpha}(\partial D)$ to $C^{k,\alpha}(\partial D)$, and Fredholm's theory says that $\tilde{T} = \frac{1}{2} I - K^*$ is invertible on the subspace of $C^{k,\alpha}(\partial D)$ of functions of mean value 0. Therefore the solution to the Neumann problem can be written

$$u(X) = \int_{\partial D} (T^{-1} f)(Q) F(X,Q)\, d\sigma(Q),$$

and if $f \in C^\infty(\partial D)$, $u \in C^\infty(\bar{D})$. If D is $C^{1,\alpha}$, T is also invertible on the subspace of $L^p(\partial D)$ of functions with mean value, $1 < p < \infty$. Hence:

THEOREM. *Let* D *be a* $C^{1,\alpha}$ *domain,* (D *and* ${}^cD$ *connected*). *Let* $1 < p < \infty$. *Assume that* $f \in L^p(\partial D, d\sigma)$, $\int_{\partial D} f\, d\sigma = 0$. *Then,* $u(X) = \int_{\partial D} (T^{-1}f)(Q) F(X,Q)\, d\sigma(Q)$ *is harmonic in* D, $\nabla u(X)$ *has non-tangential limits* $\nabla u(Q)$ *for a.e.* $Q \in \partial D$, $f(Q) = \langle N_Q, \nabla u(Q)\rangle$, *and* $\|N_\beta(\nabla u)\|_{L^p(d\sigma)} \le C_p \|f\|_{L^p(d\sigma)}$.

What do we do when ∂D is merely C^1, or even merely Lipschitz? As I mentioned before, the L^p boundedness of K is even in doubt. In 1977, A. P. Calderon ([1]) showed that for any C^1 domain, $K: L^p(\partial D) \to L^p(\partial D)$, $1 < p < \infty$ is a bounded operator. Shortly afterwards, Fabes, Jodeit and Riviere ([11]) showed that K is in fact compact in this case. They were thus able to extend the theorems above to the case of C^1 domains.

Before going on to the main subject matter of this paper, i.e. the method of layer potentials on Lipschitz domains, I want to discuss another important method for the Dirichlet problem for Laplace's equation.

(b) *The method of harmonic measure*

Another way of studying the Dirichlet problem for Laplace's equation is in terms of the notion of *harmonic measure*. Let D be a bounded Lipschitz domain in R^n. As we mentioned before, then D satisfies the exterior cone condition, and so, by a classical result of Zaremba and Lebesgue, we can solve the classical Dirichlet problem for Δ in D, for any $f \in C(\partial D)$. Given $X \in D$, the maximum principle shows that the mapping $f \mapsto u(X)$ defines a positive continuous linear functional on $C(\partial D)$. Therefore, by the Riesz representation theorem, there is a unique positive Borel probability measure ω^X on ∂D such that

$$u(X) = \int_{\partial D} f(Q)\, d\omega^X(Q)$$

ω^X is called the harmonic measure for D, evaluated at X. For example, harmonic measure for the unit ball B, evaluated at the origin is a constant multiple of surface measure: $\omega^0 = \sigma/\sigma(\partial B)$. (This follows from the mean value property of harmonic functions.) For different X, the measures ω^X are mutually absolutely continuous (a simple consequence of Harnack's principle). We fix $X_0 \in D$, and denote $\omega = \omega^{X_0}$. The importance of harmonic measure to the boundary behavior of harmonic functions on Lipschitz domains can be illustrated by the following theorem of Hunt and Wheeden (1967): If u is a positive harmonic function in a Lipschitz domain D, then u has finite non-tangential limits almost everywhere with respect to ω. Conversely, given any set $E \subset \partial D$, with $\omega(E) = 0$, there is a positive harmonic function u in D with $\lim u(X) = +\infty$ as $X \to Q$, for every $Q \in E$. Despite its advantages, harmonic measure has some inherent difficulties. First, it is hard to calculate it explicitly. Second, it is tied up to the maximum principle, positivity, and the Harnack principle, and so it is not useful for the Neumann problem, or for the Dirichlet problem for systems of equations.

In general, harmonic measure may be very different from surface measure. If D is a $C^{1,\alpha}$ domain, then harmonic measure and surface measure are essentially identical in that each is a bounded multiple of the other. This can be proved by the classical method of layer potentials. Along the same lines, as we saw before, one can use layer potentials to solve the Dirichlet and Neumann problems with boundary data in L^p.

On C^1 domains, it is no longer true that harmonic measure is a bounded multiple of surface measure, or vice versa. Moreover, as was explained before, the applicability of the method of layer potentials is not obvious. The situation for general Lipschitz domains is even less obvious.

In 1977, B. E. J. Dahlberg ([4]) proved that on a C^1 or even a Lipschitz domain, harmonic measure and surface measure are mutually absolutely continuous. Using a quantitative version of mutual absolute continuity, and the theory of weighted norm inequalities, he proved ([5])

that in a Lipschitz domain D one can solve the Dirichlet problem as in the theorem above with $f \in L^2(\partial D, d\sigma)$. In fact, he showed that given a Lipschitz domain D, there exists $\varepsilon = \varepsilon(D)$ such that this can be done for $f \in L^p(\partial D, d\sigma)$, $2-\varepsilon \le p \le \infty$. Also, simple examples to be presented later show that given $p < 2$, we can find a Lipschitz domain D for which this cannot be done in L^p. By establishing further properties for harmonic measure on C^1 domains, he was able to show the results above in the range $1 < p < \infty$ for C^1 domains. (The best possible regularity result for harmonic measure on C^1 domains is due to D. Jerison and C. Kenig (1981): if $k = \frac{d\omega}{d\sigma}$, then $\log k \in VMO(\partial D)$.)

A shortcoming of Dahlberg's method of proof, as was explained before, is that, by studying harmonic measure, it relied on positivity and the Harnack principle. This made the method inapplicable to the Neumann problem, or to systems of equations. Also the method does not provide useful representation formulas for the solution.

(c) *The method of layer potentials revisited*

In 1979, D. Jerison and C. Kenig [16], [17] were able to give a simplified proof of Dahlberg's results, using an integral identity that goes back to Rellich ([30]). However, the method still relied on positivity. Shortly afterwards, D. Jerison and C. Kenig ([18]) were also able to treat the Neumann problem on Lipschitz domains, with $L^2(\partial D, d\sigma)$ data and optimal estimates. To do so, they combined the *Rellich type formulas* with Dahlberg's results on the Dirichlet problem. Thus, it still relied on positivity, and dealt only with the L^2 case, leaving the corresponding L^p theory open.

In 1981, R. Coifman, A. McIntosh and Y. Meyer [2] established the boundedness of the *Cauchy integral on any Lipschitz curve*, opening the door to the applicability of the method of layer potentials to Lipschitz domains. This method is very flexible, does not relie on positivity, and does not in principle differentiate between a single equation or a system of equations. The difficulty then becomes the solvability of the integral equations, since unlike in the C^1 case, the Fredholm theory is not applicable, because on a Lipschitz domain operators like the operator K in part (a) are not compact, as simple examples show.

For the case of the Laplace equation, with $L^2(\partial D, d\sigma)$ data, this difficulty was overcome by G.C. Verchota ([33]) in 1982, in his doctoral dissertation. He made the key observation that the Rellich identities mentioned before are the appropriate substitutes to compactness, in the case of Lipschitz domains. Thus, Verchota was able to recover the L^2 results of Dahlberg [5] and of Jerison and Kenig [18], for Laplace's equation on a Lipschitz domain, but using the method of layer potentials. These results of Verchota's will be explained in the first part of Section 2.

In 1984, B. Dahlberg and C. Kenig ([16]) were able to show that given a Lipschitz domain $D \subset \mathbb{R}^n$, there exists $\varepsilon = \varepsilon(D) > 0$ such that one can solve the Neumann problem for Laplace's equation with data in $L^p(\partial D, d\sigma)$, $1 < p \leq 2+\varepsilon$. Easy examples (to be presented later) show that this range of p's is optimal. Moreover, they showed that the solution can be obtained by the method of layer potentials, and that Dahlberg's solution of the L^p Dirichlet problem can also be obtained by the method of layer potentials. They also obtained end point estimates for the *Hardy space* $H^1(\partial D, d\sigma)$, which generalize the results for $n = 2$ in [20] and [21], and for C^1 domains in [12]. The key idea in this work is that one can estimate the regularity of the so-called Neumann function for D, by using the *De Giorgi-Nash regularity theory* for elliptic equations with bounded measurable coefficients. This, combined with the use of the so-called '*atoms*' yields the desired results. These results will be explained in the second part of Section 2.

Also in 1984, B. Dahlberg, C. Kenig and G. Verchota ([8]) and E. Fabes, C. Kenig and G. Verchota ([13]) were able to extend the ideas of Verchota to be able to obtain results for L^2 boundary value problems for some systems of equations on Lipschitz domains. The systems treated are those that arise in *linear elastostatics* and in *linear hydro-*

statics. The results obtained had not been previously available for general Lipschitz domains, although a lost of work had been devoted to the case of piecewise linear domains. (See [24], [25] and their bibliographies.) For the case of C^1 domains, these results for the systems of elastostatics had been previously obtained by A. Gutierrez ([15]), using compactness and the Fredholm theory. This is of course, not available for the case of Lipschitz domains. The authors use once more the method of layer potentials. Invertibility is shown again by means of Rellich type formulas. This works very well in the Dirichlet problem for the Stokes system (see part (b) of Section 3), but serious difficulties occur for the systems of elastostatics (see part (a) of Section 3). These difficulties are overcome by proving a *Korn type inequality at the boundary*. The proof of this inequality proceeds in three steps. One first establishes it for the case of small Lipschitz constant. One then proves an analogous inequality for non-tangential maximal functions on any Lipschitz domain, by using the ideas of G. David ([10]), on increasing the Lipschitz constant. Finally, one can remove the non-tangential maximal function, using the results on the Dirichlet problem for the Stokes system, which are established in part (b) of Section 3.

As a final comment, I would like to point out that even though throughout this paper we have emphasized *non-tangential maximal function estimates*, also optimal *Sobolev space estimates* hold. All the Sobolev estimates can be proved in a unified fashion, using square functions and a variant of some of the real variable arguments used in part (b) of Section 3. The details will appear in a forthcoming paper of B. Dahlberg and C. Kenig, [7].

§2. *Laplace's equation on Lipschitz domains*

(a) *The* L^2 *theory*

A bounded Lipschitz domain $D \subset R^n$ is one which is locally given by the domain above the graph of a Lipschitz function. Such domains satisfy both the interior and exterior cone condition. For such a domain D, the non-tangential region of opening β at a point $Q \in \partial D$ is $\Gamma_\beta(Q) = \{X \in D : |X-Q| < (1+\beta)\operatorname{dist}(X, \partial D)\}$. All the results in this paper are valid, when suitably interpreted for all bounded Lipschitz domains in R^n, $n \geq 2$, with the non-tangential approach regions defined above. For simplicity, in this exposition we will restrict ourselves to the case $n \geq 3$ (and sometimes even to the case $n = 3$), and to domains $D \subset R^n$, $D = \{(x,y) : y > \phi(x)\}$, where $\phi : R^{n-1} \to R$ is a Lipschitz function with Lipschitz constant M, i.e. $|\phi(x) - \phi(x')| \leq M|x-x'|$. $D^- = \{(x,y) : y < \phi(x)\}$. For fixed $M' < M$, $\Gamma_e(x) = \{(z,y) : (y-\phi(x)) < -M'|z-x|\} \subset D^-$ and $\Gamma_i(x) = \{(z,y) : (y-\phi(x)) > M'|z-x|\} \subset D$. Points in D will usually be denoted by X, while points on ∂D by $Q = (x, \phi(x))$ or simply by x. N_x or N_Q will denote the unit normal to $\partial D = \Lambda$ at $Q = (x, \phi(x))$. If u is a function defined on $R^n \backslash \Lambda$ and $Q \in \partial D$, $u^{\pm}(Q)$ will denote $\lim_{\substack{X \to Q \\ X \in \Gamma_i(Q)}} u(X)$ or $\lim_{\substack{X \to Q \\ X \in \Gamma_e(Q)}} u(X)$, respectively. If u is a function defined on D, $N(u)(Q) = \sup_{X \in \Gamma_i(Q)} |u(X)|$.

We wish to solve the problems

$$(D) \begin{cases} \Delta u = 0 \text{ in } D \\ u|_{\partial D} = f \in L^2(\partial D, d\sigma) \end{cases}, \qquad (N) \begin{cases} \Delta u = 0 \text{ in } D \\ \frac{\partial u}{\partial N}\Big|_{\partial D} = f \in L^2(\partial D, d\sigma) \end{cases}.$$

The results here are

THEOREM 2.1.1. *There exists a unique* u *such that* $N(u) \in L^2(\partial D, d\sigma)$, *solving (D), where the boundary values are taken non-tangentially a.e.. Moreover, the solution* u *has the form*

$$u(X) = \frac{1}{\omega_n} \int_{\partial D} \frac{\langle X-Q, N_Q \rangle}{|Q-X|^n}\, g(Q)\, d\sigma(Q),$$

for some $g \in L^2(\partial D, d\sigma)$.

THEOREM 2.1.2. *There exists a unique* u *tending to* 0 *at* ∞, *such that* $N(\nabla u) \in L^2(\partial D, d\sigma)$, *solving* (N) *in the sense that* $N_Q \cdot \nabla u(X) \to f(Q)$ *as* $X \to Q$ *non-tangentially a.e.. Moreover, the solution* u *has the form*

$$u(X) = \frac{1}{\omega_n(n-2)} \int_{\partial D} \frac{1}{|X-Q|^{n-2}} g(Q)\, d\sigma(Q) ,$$

for some $g \in L^2(\partial D, d\sigma)$.

In order to prove the above theorems, we introduce the double and single layer potentials

$$\mathcal{K}g(X) = \frac{1}{\omega_n} \int_{\partial D} \frac{\langle X-Q, N_Q \rangle}{|X-Q|^n} g(Q)\, d\sigma(Q)$$

and

$$Sg(X) = -\frac{1}{\omega_n(n-2)} \int_{\partial D} \frac{1}{|X-Q|^{n-2}} g(Q)\, d\sigma(Q) .$$

If $Q = (x, \phi(x))$, $X = (z, y)$, then

$$\mathcal{K}g(z,y) = \frac{1}{\omega_n} \int_{R^{n-1}} \frac{y - \phi(x) - (z-x) \cdot \nabla\phi(x)}{[|x-z|^2 + [\phi(x) - \phi(z)]^2]^{n/2}} g(x)\, dx$$

$$Sg(z,y) = -\frac{1}{\omega_n(n-2)} \int_{R^{n-1}} \frac{\sqrt{1 + |\nabla\phi(x)|^2}}{[|x-z|^2 + |\phi(x) - y]^2]^{\frac{n-2}{2}}} g(x)\, dx .$$

THEOREM 2.1.3. a) *If* $g \in L^p(\partial D, d\sigma)$, $1 < p < \infty$, *then* $N(\nabla Sg)$, $N(\mathcal{K}g)$ *also belong to* $L^p(\partial D, d\sigma)$ *and their norms are bounded by* $C\|g\|_{L^p(\partial D, d\sigma)}$.

(b) $$\lim_{\varepsilon\to 0}\frac{1}{\omega_n}\int_{|x-z|>\varepsilon}\frac{\phi(z)-\phi(x)-(z-x)\cdot\nabla\phi(x)}{[|x-z|^2+[\phi(x)-\phi(z)]^2]^{n-2}}\,g(x)\,dx = Kg(z)\quad \textit{exists}$$

a.e. and $\|Kg\|_{L^p(\partial D,d\sigma)} \le C\|g\|_{L^p(\partial D,d\sigma)}$ $1 < p < \infty$.

$$\lim_{\varepsilon\to 0}\frac{1}{\omega_n}\int_{|z-x|>\varepsilon}\frac{(z-x,\phi(z)-\phi(x))\sqrt{1+|\nabla\phi(x)|^2}}{[|z-x|^2+[\phi(z)-\phi(x)]^2]^{n/2}}\,g(x)\,dx\quad \textit{exists a.e. and in}$$

$L^p(\partial D, d\sigma)$, *and its* L^p *norm is bounded by* $C\|g\|_{L^p(\partial D,d\sigma)}$, $1 < p < \infty$.

(c) $(\mathcal{K}g)^{\pm}(Q) = \pm\frac{1}{2}\,g(Q) + Kg(Q)$

$$(\nabla Sg)^{\pm}(z) = \pm\frac{1}{2}\,g(z)\,N_z + \frac{1}{\omega_n}\lim_{\varepsilon\to 0}\int_{|z-x|>\varepsilon}\frac{(z-x,\phi(z)-\phi(x))\sqrt{(1+|\nabla\phi(x)|}}{[|z-x|^2+[\phi(z)-\phi(x)]^2]^{n/2}}\,g(x)\,dx\,.$$

COROLLARY 2.1.4. $(N_z\nabla Sg)^{\pm}(z) = \mp\frac{1}{2}\,g(z) - K^*g(z)$, *where* K^* *is the* $L^2(\partial D, d\sigma)$ *adjoint of* K.

The proof of Theorem 2.1.3 a) follows by well-known techniques from the deep theorem of Coifman, McIntosh and Meyer ([2]).

THEOREM ([2]). *Let* $\theta : R \to R$ *be even, and* C^∞. *Let* $A,B : R^{n-1} \to R$ *be Lipschitz. Let* $K(z,x) = \frac{A(z)-A(x)}{|z-x|^n}\,\theta\left[\frac{B(z)-B(x)}{|z-x|}\right]$. *Then, the maximal operator*

$$M^*g(z) = \sup_{\varepsilon\to 0}\left|\int_{|t-x|>\varepsilon} K(z,x)\,g(x)\,dx\right|$$

is bounded on $L^p(R^{n-1})$, $1 < p < \infty$, *with*

$$\|M^*g\|_p \leq C\|g\|_p \,,$$

where $C = C(M,\theta,n,p)$, *and* $\|\nabla A\|_\infty \leq M$, $\|\nabla B\|_\infty \leq M$.

The proof of (b), (c) follow from the theorem above, together with the following simple lemmas.

LEMMA. *If* $f \in C_0^\infty(R^{n-1})$, *then*

$$\lim_{\varepsilon\to 0} \frac{1}{\omega_n} \int_{|z-x|>\varepsilon} \frac{\phi(z)-\phi(x)-(z-x)\cdot\nabla\phi(x)}{[|x-z|^2+[\phi(x)-\phi(z)]^2]^{n/2}} f(x)\,dx =$$

$$= -\sum_1^{n-1} \frac{1}{\omega_n} \int \frac{z_k-x_k}{|z-x|^{n-1}} \lambda\left[\frac{\phi(z)-\phi(x)}{|x-z|}\right] \frac{\partial f}{\partial x_k}(x)\,dx \,,$$

where $\lambda(0) = 0$, $\lambda'(t) = (1+t^2)^{-n/2}$, *and where the equality holds at every* z *at which* ϕ *is differentiable, i.e. for a.e.z.*

LEMMA. *If* $\alpha \in R^{n-1}$, $a \neq \phi(\alpha)$, $f \in C_0^\infty(R^{n-1})$ *and* λ *is as in the previous lemma, then*

$$\frac{1}{\omega_n}\int \frac{a-\phi(\alpha)-(\alpha-x)\cdot\nabla\phi(x)}{[|\alpha-x|^2+[\phi(x)-a]^2]^{n/2}} f(x)\,dx =$$

$$= \frac{1}{2}\,\mathrm{sign}\,(a-\phi(\alpha))\,f(\alpha) - \frac{1}{\omega_n}\int\sum_1^{n-1} \frac{x_k-\alpha_k}{|x-\alpha|^{n-1}} \lambda\left(\frac{\phi(x)-a}{|x-\alpha|}\right) \frac{\partial f}{\partial x_k}(x)\,dx \,.$$

Moreover, the integral on the right-hand side of the equality is a continuous function of $(\alpha, a) \in R^n$.

It is easy to see that (at least the existence part) of Theorems 2.1.1 and 2.1.2 will follow immediately if we can show that $\left(\frac{1}{2}I+K\right)$ and

$\frac{1}{2}\,I+K^*$ are invertible on $L^2(\partial D, d\sigma)$. This is the result of G. Verchota ([33]).

THEOREM 2.1.5. $\left(\pm\frac{1}{2}\,I+K\right)$, $\left(\pm\frac{1}{2}\,I+K^*\right)$ *are invertible on* $L^2(\partial D, d\sigma)$.

In order to prove this theorem, it suffices to show that $\left(\pm\frac{1}{2}\,I+K^*\right)$ are invertible. In order to do so, we show that if $f \in L^2(\partial D, d\sigma)$, $\|\left(\frac{1}{2}\,I+K^*\right)f\|_{L^2(\partial D, d\sigma)} \approx \|\left(\frac{1}{2}\,I-K^*\right)f\|_{L^2(\partial D, d\sigma)}$, where the constants of equivalence depend only on the Lipschitz constant M. Let us take this for granted, and show, for example, that $\frac{1}{2}\,I+K^*$ is invertible. To do this, note first that if $T=\frac{1}{2}\,I+K^*$, $\|Tf\|_{L^2} \geq C\|f\|_{L^2}$, where C depends only on the Lipschitz constant M. For $0 \leq t \leq 1$, consider the operator $T_t = \frac{1}{2}\,I+K_t^*$, where K_t^* is the operator corresponding to the domain defined by $t\phi$. Then, $T_0 = \frac{1}{2}\,I$, $T_1 = T$, and $\frac{\partial T_t}{\partial t} : L^p(R^{n-1}) \to L^p(R^{n-1})$, $1 < p < \infty$ with bound independent of t, by the theorem of Coifman-McIntosh-Meyer. Moreover, for each t, $\|T_t f\|_{L^2} \geq C\|f\|_{L^2}$, C independent of t. The invertibility of T now follows from the continuity method:

LEMMA 2.1.6. *Suppose that* $T_t : L^2(R^{n-1}) \to L^2(R^{n-1})$ *satisfy*

(a) $\|T_t f\|_{L^2} \geq C_1\|f\|_{L^2}$

(b) $\|T_t f - T_s f\|_{L^2} \leq C_2|t-s|\,\|f\|_{L^2}$, $0 \leq t, s \leq 1$.

(c) $T_0 : L^2(R^{n-1}) \to L^2(R^{n-1})$ *is invertible.*

Then, T_1 is invertible. The proof of 2.1.6 is very simple. We are thus reduced to proving (2.1.7) $\|\left(\frac{1}{2}\,I+K^*\right)f\|_{L^2(\partial D, d\sigma)} \approx \|\left(\frac{1}{2}\,I-K^*\right)f\|_{L^2(\partial D, d\sigma)}$.

In order to prove (2.1.7), we will use the following formula, which goes back to Rellich [30] (also see [28], [29], [27]).

LEMMA 2.1.8. *Assume that* $u \in \mathrm{Lip}(\overline{D})$, $\Delta u = 0$ *in* D, *and* u *and its derivatives are suitably small at* ∞. *Then, if* e_n *is the unit vector in the direction of the* y*-axis,*

$$\int_{\partial D} \langle N, e_n \rangle |\nabla u|^2 \, d\sigma = 2 \int_{\partial D} \frac{\partial u}{\partial y} \cdot \frac{\partial u}{\partial N} \, d\sigma .$$

Proof. Observe that $\operatorname{div}(e_n |\nabla u|^2) = \frac{\partial}{\partial y} |\nabla u|^2 = 2 \frac{\partial}{\partial y} \nabla u \cdot \nabla u$, while $\operatorname{div} \frac{\partial u}{\partial y} \nabla u = \frac{\partial}{\partial y} \nabla u \cdot \nabla u + \frac{\partial u}{\partial y} \cdot \operatorname{div} \nabla u = \frac{\partial}{\partial y} \nabla u \cdot \nabla u$. Stokes' theorem now gives the lemma.

We will now deduce a few consequences of the Rellich identity. Recall that $N_x = (-\nabla\phi(x), 1)/\sqrt{1 + |\nabla\phi(x)|^2}$, so that $\frac{1}{(1+M^2)^{1/2}} \leq \langle N_x, e_n \rangle \leq 1$.

COROLLARY 2.1.9. *Let* u *be as in 2.1.8, and let* $T_1(x)$, $T_2(x)$, $T_{n-1}(x)$ *be an orthogonal basis for the tangent plane to* ∂D *at* $(X, \phi(X))$. *Let* $|\nabla_t u(x)|^2 = \sum_{j=1}^{n-1} |\langle \nabla u(x), T_j(x) \rangle|^2$. *Then,*

$$\int_{\partial D} \left(\frac{\partial u}{\partial N}\right)^2 d\sigma \leq C \int_{\partial D} |\nabla_t u|^2 \, d\sigma .$$

Proof. Let $\alpha = e_n - \langle N_x, e_n \rangle N_x$, so that α is a linear combination of $T_1(x), T_2(x), \cdots, T_{n-1}(x)$. Then, $\frac{\partial u}{\partial y} = \langle N_x, e_n \rangle \frac{\partial u}{\partial N} + \langle \alpha, \nabla u \rangle$. Also, $|\nabla u|^2 = \left(\frac{\partial u}{\partial N}\right)^2 + |\nabla_t u|^2$, and so $\int_{\partial D} \langle N_x, e_n \rangle \left(\frac{\partial u}{\partial N}\right)^2 d\sigma + \int_{\partial D} \langle N_x, e_n \rangle |\nabla_t u|^2 \, d\sigma = 2 \int_{\partial D} \langle N_x, e_n \rangle \left(\frac{\partial u}{\partial N}\right)^2 d\sigma + 2 \int_{\partial D} \langle \alpha, \nabla u \rangle \left(\frac{\partial u}{\partial N}\right) d\sigma$. Hence $\int_{\partial D} \langle N_x, e_n \rangle \left(\frac{\partial u}{\partial N}\right)^2 d\sigma = \int_{\partial D} \langle N_x, e_n \rangle |\nabla_t u|^2 \, d\sigma - 2 \int \langle \alpha, \nabla u \rangle \frac{\partial u}{\partial N} \, d\sigma$. So, $\int \left(\frac{\partial u}{\partial N}\right)^2 d\sigma \leq C \int_{\partial D} |\nabla_t u|^2 \, d\sigma + C\left(\int_{\partial D} |\nabla_t u|^2 \, d\sigma\right)^{1/2} \left(\int_{\partial D} \frac{\partial u}{\partial N}^{\,2} d\sigma\right)^{1/2}$, and the corollary follows.

COROLLARY 2.1.10. *Let* u *be as in 2.1.8. Then,*

$$\int_{\partial D} |\nabla_t u|^2 \, d\sigma \leq C \int_{\partial D} \left(\frac{\partial u}{\partial N}\right)^2 d\sigma \, .$$

Proof. $\int_{\partial D} |\nabla u|^2 d\sigma \leq 2(\int_{\partial D} |\nabla u|^2 d\sigma)^{1/2} \left(\int_{\partial D} |\frac{\partial u}{\partial N}|^2 d\sigma\right)^{1/2}$, by 2.18, and the corollary follows.

COROLLARY 2.1.11. *Let* u *be as in 2.1.8. Then,* $\int_{\partial D} |\nabla_t u|^2 d\sigma \approx \int_{\partial D} |\frac{\partial u}{\partial N}$

In order to prove 2.1.7, let $u = Sg$. Because of 2.1.3c, $\nabla_t u$ is continuous across the boundary, while by 2.1.4, $\left(\frac{\partial u}{\partial N}\right)^{\pm} = \left(\mp \frac{1}{2} I - K^*\right) g$. We now apply 2.1.11 in D and $\bar{D}$, to obtain 1.1.7. This finishes the proof of 2.1.1 and 2.1.2.

We now turn our attention to L^2 regularity in the Dirichlet problem.

DEFINITION 2.1.12. $f \in L_1^p(\Lambda)$, $1 < p < \infty$, if $f(x, \phi(x))$ has a distributional gradient in $L^p(R^{n-1})$. It is easy to check that if F is any extension to R^n of f, then $\nabla_x F(x,\phi(x))$ is well defined, and belongs to $L^p(\Lambda)$. We call this $\nabla_t f$. The norm in $L_1^p(\Lambda)$ will be $\|\nabla_t f\|_{L^p(\Lambda)}$.

THEOREM 2.1.13. *The single layer potential* S *maps* $L^2(\Lambda)$ *into* $L_1^2(\Lambda)$ *boundedly, and has a bounded inverse.*

Proof. The boundedness follows from 2.1.3a). Because of the L^2-Neumann theory, and 2.1.11, $\|\nabla_t S(f)\|_{L^2(\Lambda)} \geq C \left\|\frac{\partial S}{\partial N}(f)\right\|_{L^2(\Lambda)} \geq C\|f\|_{L^2(\Lambda)}$. The argument used in the proof of 2.1.5 now proves 2.1.13.

THEOREM 2.1.14. *Given* $f \in L_1^2(\Lambda)$, *there exists a harmonic function* u, *with* $\|N(\nabla u)\|_{L^2(\Lambda)} \leq C\|\nabla_t f\|_{L^2(\Lambda)}$, *and such that* $\nabla_t u = \nabla_t f$ (a.e.) *nontangentially on* Λ. u *is unique (modulo constants), and we can chose* $u = S(g)$, *where* $g \in L^2(\Lambda)$.

The existence part of 2.1.14 follows directly from 2.1.13.

(b) *The* L^p *theory*

We will start out our treatment of the L^p theory by discussing some counterexamples.

Let $z = x + iy \in \mathbb{C}$, and for $0 < \beta < 2\pi$, let

$$D_\beta = \{z \in \mathbb{C} : |\arg z| < \beta/2\} .$$

We will consider the holomorphic function $f(z) = z^{\pi/\beta}$, which maps D_β conformally onto $D\pi$, the right-hand plane. We will also consider a bounded domain $\Omega_\beta \subset D_\beta$, with the property that $\partial\Omega_\beta \backslash 0$ is smooth, and such that $\partial\Omega_\beta \cap \{|z| < 1\} = \partial D_\beta \cap \{|z| < 1\}$. Let $u(x,y) = \mathrm{Re} f(z)$, and $v(x,y) = u(x,y)|_{\Omega_\beta}$ $\cdot$ v is harmonic in Ω_β, and v is identically zero near the corner of $\partial\Omega_\beta$, and is smooth everywhere else in $\bar{\Omega}_\beta$. Let s be the arc length parametrization of $\partial\Omega_\beta$, starting at 0. Then, it is clear that $\frac{\partial v}{\partial s} \in L^\infty(\partial\Omega_\beta)$. Let $w(x,y) = \mathrm{Im}\, f(z)|_{\Omega_\beta}$. By the Cauchy-Riemann equations,

$$\left|\frac{\partial w}{\partial N}\Big|_{\partial\Omega_\beta}\right| = \left|\frac{\partial v}{\partial s}\Big|_{\partial\Omega_\beta}\right| \in L^\infty(\partial\Omega_\beta) .$$

However, $N(\nabla v)(s) = N(\nabla w)(s) \approx s^{-1+\pi/\beta}$. This function belongs to $L^p(ds)$ if and only if $p\,\pi/\beta - p > -1$. Fix now a $p > 2$, and choose β so close to 2π that $p\,\pi/\beta - p < -1$. Then, $N(\nabla w) \notin L^p(\partial\Omega_\beta)$. If $\left(\frac{1}{2}\, I - K^*\right)$ were invertible in $L^p(\partial\Omega_\beta)$, then, since $\frac{\partial w}{\partial N} \in L^\infty(\partial\Omega_\beta)$, we would have that $\tilde{w}(z) = S\left(\left(\frac{1}{2}\, I - K^*\right)^{-1}\left(\frac{\partial w}{\partial N}\right)\right)(z)$ has a non-tangential maximal function in $L^p(\partial\Omega_\beta)$. By the L^2-uniqueness in the Neumann problem, $w - \tilde{w}$ is constant in Ω_β, but this is a contradiction. This shows that given $p > 2$, we can find a Lipschitz domain so that $\left(\frac{1}{2}\, I - K^*\right)$ is not invertible in L^p. The example can also be used to show that $\frac{1}{2} I+K$ is not always invertible in L^q, when $q<2$. In fact, fix $q<2$, and let p satisfy $\frac{1}{p}+\frac{1}{q}=1$. Choose β so that $p\frac{\pi}{\beta}-p<-1$. Let $B = \left\{|z|<\frac{1}{2}\right\} \cap \Omega_\beta$, and fix $X_* \in \Omega_\beta \setminus \{|z|<1\}$. Let $\omega = \omega^{X_*}$ be harmonic measure evaluated at X_*, and $k=\frac{d\omega}{ds}$. We first claim that $k \notin L^p(ds)$. In fact, let $G(X)$ be the Green's function of Ω_β with pole at X_*. Then, for s near 0, $k(s) = \frac{\partial G}{\partial N}(s) =$

$$\lim_{\varepsilon\to 0} \frac{G(s+\varepsilon N)-G(s)}{\varepsilon} = \lim_{\varepsilon\to 0} \frac{G(s+\varepsilon N)}{\varepsilon} \geq C \lim_{\varepsilon\to 0} \frac{v(s+\varepsilon N)}{\varepsilon} = C\,\frac{\partial v}{\partial N}(s) \approx s^{-1+\pi/\beta},$$

where the first inequality follows from the fact that both G and v are positive, and harmonic on B, and 0 on $\partial\Omega_\beta \cap B$ (this is Lemma 5.10 in [19]). Assume now that $\frac{1}{2} I+K$ were invertible on $L^q(ds)$. Let $g \geq 0 \in C(\partial\Omega_\beta)$, and $h(X)$ be the solution of the Dirichlet problem with data g. Then,

$$h(X_*) = \int_{\partial\Omega_\beta} g\,d\omega = \int_{\partial\Omega_\beta} gk\,ds\,,$$

also, by the L^2-theory, $h(X_*) = \mathcal{K}\left[\left(\frac{1}{2} I+K\right)^{-1}(g)\right](X_*)$, where $\mathcal{K}$ is the double layer potential. Let U be a ball centered at X_*, contained in Ω_β. By the mean value property of harmonic functions and Harnack's principle, we have

$$h(X_*) \leq \frac{1}{|U|}\int_U h \leq C\left(\int_{\partial\Omega_\beta} N(h)^q\,ds\right)^{1/q} \leq C\left(\int_{\partial\Omega_\beta} g^q\,ds\right)^{1/q},$$

because of the second formula for $h(X_*)$, and the assumed L^q boundedness of $\left(\frac{1}{2} I+K\right)^{-1}$. But this implies that $k \in L^p(ds)$, a contradiction.

We now turn to the positive results. They are:

THEOREM 2.2.1. *There exists* $\varepsilon = \varepsilon(M) > 0$ *such that, given* $f \in L^p(\partial D, d\sigma)$, $2-\varepsilon \le p < \infty$, *there exists a unique* u *harmonic in* D, *with* $N(u) \in L^p(\partial D, d\sigma)$ *such that* u *converges non-tangentially almost everywhere to* f. *Moreover, the solution* u *has the form*

$$u(X) = \frac{1}{\omega_n} \int_{\partial D} \frac{<X-Q, N_Q>}{|X-Q|^n} g(Q)\, d\sigma(Q),$$

for some $g \in L^p(\partial D, d\sigma)$.

THEOREM 2.2.2. *There exists* $\varepsilon = \varepsilon(M) > 0$, *such that, given* $f \in L^p(\partial D, d\sigma)$, $2-\varepsilon \le p < \infty$, *there exists a unique* u *harmonic in* D, *tending to* 0 *at* ∞, *with* $N(\nabla u) \in L^p(\partial D, d\sigma)$, *such that* $N_Q \nabla u(X)$ *converges non-tangentially a.e. to* $f(Q)$. *Moreover,* u *has the form*

$$u(X) = \frac{1}{\omega_n(n-2)} \int_{\partial D} \frac{1}{|X-Q|^{n-2}} g(Q)\, d\sigma(Q),$$

for some $g \in L^p(\partial D, d\sigma)$.

THEOREM 2.2.3. *There exists* $\varepsilon = \varepsilon(M) > 0$ *such that given* $f \in L^p_1(\Lambda)$, $1 < p \le 2+\varepsilon$, *there exists a harmonic function* u, *with* $\|N(\nabla u)\|_{L^p(\Lambda)} \le C\|\nabla_t f\|_{L^p(\Lambda)}$, *and such that* $\nabla_t u = \nabla_t f$ (a.e.) *non-tangentially on* $\Lambda \cdot u$ *is unique (modulo constants). Moreover,* u *has the form*

$$u(X) = -\frac{1}{\omega_n(n-2)} \int_{\partial D} \frac{1}{|X-Q|^{n-2}} g(Q)\, d\sigma(Q),$$

for some $g \in L^p(\partial D, d\sigma)$.

The case $p = 2$ of the above theorems was discussed in part (a). The first part of 2.2.1 (i.e. without the representation formula), is due to B. Dahlberg (1977) ([5]). Theorem 2.2.3 was first proved by G. Verchota (1982) ([33]). The representation formula in 2.2.1, Theorem 2.2.2, and the proof that we are going to present of 2.2.3 are due to B. Dahlberg and C. Kenig (1984) ([6]). Just like in part (a), 2.2.1, 2.2.2, and 2.2.3 follow from.

THEOREM 2.2.4. *There exists* $\varepsilon = \varepsilon(M) > 0$ *such that* $\left(\pm\frac{1}{2}I - K^*\right)$ *is invertible in* $L^p(\partial D, d\sigma)$, $1 < p \le 2+\varepsilon$, $\left(\pm\frac{1}{2}I + K\right)$ *is invertible in* $L^p(\partial D, d\sigma)$, $2-\varepsilon \le p < \infty$, *and* $S : L^p(\partial D, d\sigma) \to L^p_1(\partial D, d\sigma)$ *is invertible* $1 < p \le 2+\varepsilon$.

In order to prove Theorem 2.2.4, just as in part (a), it is enough to show that if $u = Sf$, f nice, then, for $1 < p \le 2+\varepsilon$, $\|\nabla_t u\|_{L^p(\partial D, d\sigma)} \approx \left\|\frac{\partial u}{\partial N}\right\|_{L^p(\partial D, d\sigma)}$. This will be done by proving the following two theorems:

THEOREM 2.2.5. *Let* $\Delta u = 0$ *in* D. *Then* $\|N(\nabla u)\|_{L^p(\partial D, d\sigma)} \le C\left\|\frac{\partial u}{\partial N}\right\|_{L^p(\partial D, d\sigma)}$, $1 < p \le 2+\varepsilon$.

THEOREM 2.2.6. *Let* $\Delta u = 0$ *in* D. *Then* $\|N(\nabla u)\|_{L^p(\partial D, d\sigma)} \le C\|\nabla_t u\|_{L^p(\partial D, d\sigma)}$, $1 < p \le 2+\varepsilon$.

We first turn our attention to the case $1 < p < 2$ of Theorem 2.2.5. In order to do so, we introduce some definitions. A surface ball B in Λ is a set of the form $(x, \phi(x))$, where x belongs to a ball in $\mathbb{R}^{n-1}$.

DEFINITION 2.2.7. An atom a on Λ is a function supported in a surface ball B, with $\|a\|_{L^\infty} \le 1/\sigma(B)$, and with $\int_\Lambda a\,d\sigma = 0$. Notice that atoms are in particular L^2 functions. The following interpolation theorem will be of importance to us.

THEOREM 2.2.8. *Let* T *be a linear operator such that* $\|Tf\|_{L^2(\Lambda)} \leq C\|f\|_{L^2(\Lambda)}$, *and such that for all atoms* a, $\|Ta\|_{L^1(\Lambda)} \leq C$. *Then, for* $1 < p < 2$, $\|Tf\|_{L^p(\Lambda)} \leq C\|f\|_{L^p(\Lambda)}$.

For a proof of this theorem, see [3].

Thus, in order to establish the case $1 < p < 2$ of 2.2.5, it suffices to show that if $a = \frac{\partial u}{\partial N}$ is a atom, then $\|N(\nabla u)\|_{L^1(\Lambda)} \leq C$. By dilation and translation invariance we can assume that $\phi(0) = 0$, $\text{supp}\, a \subset B_1 = \{(x, \phi(x)) : |x| < 1\}$. Let B^* be a large ball centered at $(0,0)$ in R^n, which contains $(x, \phi(x))$, $|x| < 2$. The diameter of B^* depends only on M. Since $\|a\|_{L^2(\Lambda)} \leq \frac{1}{\sigma(B_1)^{1/2}} = C$, by the L^2-Neumann theory, $\int_{\partial D \cap B^*} N(\nabla u) \leq C(\int_{\partial D \cap B^*} N(\nabla u)^2 d\sigma)^{1/2} \leq C$. Thus, we only have to estimate $\int_{C_{B^*} \cap \partial D} N(\nabla u) d\sigma)$. We will do so by appealing to the regularity theory for divergence form elliptic equations. Consider the bi-Lipschitzian mapping $\Phi : D \to D^-$ given by $\Phi(x,y) = (x, \phi(x) - [y - \phi(x)])$. Define u^* on D^- by the formula $u^* = u \circ \Phi^{-1}$, u^* verifies (in the weak sense) the equation $\text{div}(A(x,y)\nabla u^*) = 0$, where $A(x,y) = \frac{1}{J\Phi(X)} \cdot (\Phi')(X)$, where $X = \Phi^{-1}(x,y)$. It is easy to see that $A \in L^\infty(D^-)$, and $\langle A(x,y)\xi, \xi\rangle \geq C|\xi|^2$. Notice also that $\text{supp}\, \frac{\partial u}{\partial N} \subset B_1 < B^* \cap \partial D$. Define now

$$B(x,y) = \begin{cases} I & \text{for } (x,y) \in D \\ A(x,y) & \text{for } (x,y) \in D^- \end{cases}, \text{ and } \tilde{u}(x,y) = \begin{cases} u(x,y) & \text{for } (x,y) \in D \\ u^*(x,y) & \text{for } (x,y) \in D^- \end{cases}.$$

Because $\frac{\partial u}{\partial N} = 0$ in $\partial D \setminus B^*$, it is very easy to see that u is a (weak) solution in $R^n \setminus B^*$ of the divergence form elliptic equation with bounded measurable coefficients, $L\tilde{u} = \text{div}\, B(x,y)\nabla\tilde{u} = 0$. In order to estimate u, (and hence ∇u) at ∞, we use the following theorem of J. Serrin and H. Weinberger ([31]).

THEOREM 2.2.9. *Let* $\tilde{u}$ *solve* $L\tilde{u} = 0$ *in* $R^n \backslash B^*$, *and suppose that* $\|\tilde{u}\|_{L^\infty(R^n \backslash B^*)} < \infty$. *Let* $g(X)$ *solve* $Lg = 0$ *in* $|X| > 1$, *with* $g(X) \approx |X|^{2-n}$. *Then,* $\tilde{u}(X) = u_\infty + a g(X) + v(X)$, *where* $Lv = 0$ *in* $R^n \backslash B^*$, *and* $|v(X)| \le C \|u\|_{L^\infty(R^n \backslash B^*)} \cdot |X|^{2-n-\nu}$, *where* $\nu > 0$, $C > 0$ *depend only on the ellipticity constants of* L. *Moreover,* $a = C \int B(X) \nabla u(X) \cdot \nabla \psi(X$ *where* $\psi \in C^\infty(R^n)$, $\psi \equiv 0$ *for* $|X|$ *in* $2B^*$, *and* $\psi \equiv 1$ *for large* X.

Let us assume for the time being that $\tilde{u}$ is bounded and let us show that if a is as in 2.2.9, then $a = 0$. Pick a ψ as in 2.2.9. In D, $B(X) = I$, and so $\int_D B \nabla u \nabla \psi = \int_D \nabla u \cdot \nabla \psi = \lim_{\varepsilon \to 0} \int_{D_\rho^\varepsilon} \nabla u \cdot \nabla \psi$, where $D_\rho^\varepsilon = \{(x,y) : |(x,y)| < \rho,\ y > \phi(x) + \varepsilon\}$, and ρ is large. The right-hand side equals $\lim_{\varepsilon \to 0} \int_{\partial D_\rho^\varepsilon} \frac{\partial u}{\partial N} = \lim_{\varepsilon \to 0} \int_{\partial D_\rho^\varepsilon} [\psi - 1] \frac{\partial u}{\partial N}$, since, by the harmonicity of u, $\int_{\partial D_\rho^\varepsilon} \frac{\partial u}{\partial N} = 0$. Let $\partial D_{\rho,1}^\varepsilon = \{(x,y) \in \partial D_\rho^\varepsilon : y > \phi(x) + \varepsilon\}$, and $\partial D_{\rho,2}^\varepsilon = \partial D_\rho^\varepsilon \backslash \partial D_{\rho,1}^\varepsilon$. Then, $\lim_{\varepsilon \to 0} \int_{\partial D_\rho^\varepsilon} [\psi - 1] \frac{\partial u}{\partial N} = \lim_{\varepsilon \to 0} \int_{\partial D_{\rho,1}^\varepsilon} [\psi - 1] \frac{\partial u}{\partial N}$ $\lim_{\varepsilon \to 0} \int_{\partial D_{\rho,2}^\varepsilon} [\psi - 1] \frac{\partial u}{\partial N} = \int_{\partial D} [\psi - 1] a = \int_{\partial D} \psi a - \int_{\partial D} a = \int_{\partial D} \psi a = 0$, since $\psi = 0$ on supp a. Moreover, $\int_{D^-} B \nabla u \nabla \psi = \int_D \nabla u \cdot \nabla \psi_*$, where $\psi_* = \psi \circ \Phi$, by our construction of B. The last term is also 0 by the same argument, and so $a = 0$. We now show that u (and hence $\tilde{u}$) is bounded. We will assume that $n \ge 4$ for simplicity. Since $\|a\|_{L^2(\Lambda)} < C$, we know that $u(X) = c_n \int_{\partial D} \frac{f(Q)}{|X-Q|^{n-2}} d\sigma(Q)$, with $\|f\|_{L^2(\Lambda)} < C$. Now, for $X \in D_1 = \{(x,y) : y > \phi(x) + 1\}$, $\frac{1}{|X-Q|^{n-2}} \le \frac{C}{1 + |Q|^{n-2}} \in L^2(\Lambda)$, and so $u \in L^\infty(D_1)$.

Let now B be any ball in R^n so that $2B \subset R^n \backslash B^*$, B is of unit size, and such that a fixed fraction of B is contained in D_1. Since $N(\nabla u) \in L^2(\Lambda)$, with norm less than C, $\int_{2B \cap D} |\nabla u|^2 \le C$, and moreover on $B \cap D_1$, $|u(X)| \le C$. Therefore, by the Poincaré inequality

$\int_{2B} |\tilde{u}|^2 \leq C$. But, since $\tilde{u}$ solves $L\tilde{u} = 0$, $\max_B |\tilde{u}| \leq C(\int_{2B} |\tilde{u}|^2)^{1/2} \leq C$ ([26]). Therefore, $\tilde{u} \in L^\infty(R^n \backslash B^*)$, $\|\tilde{u}\|_{L^\infty(R^n\backslash B^*)} \leq C$. Hence, since $\alpha = 0$, $\nabla u = \nabla v$, and $|v(x,y)| \leq C/(|x|+|y|)^{n-2+\nu}$, $\nu > 0$. For $R \geq R_0 =$ diam B^*, set $b(R) = \int_{A_R} N(\nabla u)^2$, where, $A_R = \{(x, \phi(x)) : R < |X| < 2R\}$. For each fixed R, let $N_1(\nabla u)(x) = \sup\{|\nabla u(x,y)| : (z,y) \in \Gamma_i(x)$, dist$((z,y), \partial D) \leq \delta R\}$, $N_2(\nabla u)(x) = \sup\{|\nabla u(z,y)| : (z,y) \in \Gamma_i(x)$, dist$((z,y), \partial D) \geq \delta R\}$. In the set where the sup in N_2 is taken, u is harmonic, and the distance of any point X to the boundary is comparable to $|X|$. Thus, using our bound on v, we see that $N_2(\nabla u)(x) \leq C/|X|^{n-1+\nu} \approx C/R^{n-1+\nu}$, and so $\int_{A_R} N_2(\nabla u)^2 \leq CR^{1-n-2\nu}$. Let now $\Omega_\tau = \{(x,y) : \phi(x) < y < \phi(x) + CR, \tau R < |x| < \tau^{-1}R\}$, $\tau \in \left(\frac{1}{4}, \frac{1}{2}\right)$. By the L^2-Neumann theory in Ω_τ, $\int_{A_R} N_1(\nabla u)^2 d\sigma \leq C \int_{\partial\Omega_\tau} |\nabla u|^2 d\sigma$. Integrating in τ from 1/4 to 1/2 gives

$$\int_{A_R} N_1(\nabla u)^2 d\sigma \leq \frac{C}{R} \int_{\Omega_{1/4}\backslash\Omega_{1/2}} |\nabla u|^2 dX \leq \frac{C}{R^3} \int_{C_1R<|X|<C_2R} \tilde{u}^2 ,$$

since $L\tilde{u} = 0$ (see [26] for example). The right-hand side is bounded by $\frac{C}{R^3} \frac{1}{R^{2(n-2)-2\nu}} R^n = CR^{1-n-2\nu}$, and hence $b(R) \leq CR^{1-n-2\nu}$. Then, $\int_{A_R} N(\nabla u) \leq C(\int_{A_R} N(\nabla u)^2)^{1/2} R^{\frac{n-1}{2}} \leq CR^{-\nu}$. Choosing now $R = 2^j$, and adding in j, we obtain the desired estimate.

We now turn to the case $1 < p < 2$ of 2.2.6. We need a further definition.

DEFINITION 2.2.10. A function a is an H_1^1 atom if $\vec{A} = \nabla_t a$ satisfies (a) supp $\vec{A} \subset B$, a surface ball, (b) $\|\vec{A}\|_{L^\infty} \leq 1/\sigma(B)$, (c) $\int \vec{A}\, d\sigma = 0$.

We will use the following interpolation result:

THEOREM 2.2.11. *Let* T *be a linear operator such that* $\|Tf\|_{L^2(\Lambda)} \leq C\|f\|_{L_1^2(\Lambda)}$ *and* $\|Ta\|_{L^1(\Lambda)} \leq C$ *for all* H_1^1 *atoms* a. *Then, for* $1 < p < 2$, $\|Tf\|_{L^p(\Lambda)} \leq C\|f\|_{L_1^p(\Lambda)}$.

Hence, all we need to show is that if $\Delta u = 0$, $\nabla_t u = \nabla_t a$, and a is a unit size H_1^1 atom, $N(\nabla u) \in L^1(\Lambda)$. But note that if we let

$$\tilde{u}(x,y) = \begin{cases} u(x,y) & (x,y) \in D \\ -u^*(x,y) & (x,y) \in D^- \end{cases},$$

then $\tilde{u}$ is a weak solution of $L\tilde{u} = 0$ in $R^n \backslash B^*$, since $u|_{\partial D \backslash B^*} = 0$. Then, $\tilde{u} = \tilde{u}_\infty + \alpha g + v$, but $\alpha = 0$ since $\tilde{u} - \tilde{u}_\infty$ must change sign at ∞. The argument is then identical to the one given before.

Before we pass to the case $2 < p < 2+\varepsilon$, we would like to point out that using the techniques described above, one can develop the Stein-Weiss [32] Hardy space theory on an arbitrary Lipschitz domain in R^n. This generalizes the results for $n = 2$ obtained in [20] and [21], and the results for C^1 domains in [12].

Some of the results one can obtain are the following: Let $H_{at}^1(\partial D) = \{\Sigma \lambda_i a_i : \Sigma|\lambda_i| < \infty,\ a_i$ is an atom$\}$, $H_{1,at}^1(\partial D) = \{\Sigma\lambda_i a_i : \Sigma|\lambda_i| < +\infty,\ a_i$ is an H_1^1 atom$\}$.

THEOREM 2.2.12. a) *Given* $f \in H_{at}^1(\partial D)$, *there exists a unique harmonic function* u, *which tends to* 0 *at* ∞, *such that* $N(\nabla u) \in L^1(\partial D)$, *and such that* $N_Q \nabla u(X) \to f(Q)$ *non-tangentially a.e. Moreover,* $u(X) = S(g)(X)$, $g \in H_{at}^1$. *Also,* $u|_{\partial D} \in H_{1,at}^1(\partial D)$. b) *Given* $f \in H_{1,at}^1$, *there exists a unique (modulo constants) harmonic function* u, *such that* $N(\nabla u) \in L^1(\partial D)$, *and such that* $\nabla_t u|_{\partial D} = \nabla_t f$ *a.e. Moreover,* $u = S(g)$, $g \in H_{at}^1$, *and* $\frac{\partial u}{\partial N} \in H_{at}^1(\partial D)$. c) *If* u *is harmonic, and* $N(\nabla u) \in L^1(\partial D)$, *then* $\frac{\partial u}{\partial N} \in H^1_{at}(\partial D)$, $u|_{\partial D} \in H^1_{1,at}(\partial D)$. d) $f \in H^1_{at}(\partial D)$ *if and only if* $N(\nabla Sf) \in L^1(\partial D)$, *if and only if* $\left(\frac{1}{2} I - K^*\right) f \in H^1_{at}(\partial D)$.

We turn now to the L^p theory, $2 < p < 2+\varepsilon$. In this case, the results are obtained as automatic real variable consequences of the fact that the L^2 results hold for all Lipschitz domains. We will now show that $\|N(\nabla u)\|_{L^p(\Lambda)} \leq C \|\frac{\partial u}{\partial N}\|_{L^p(\Lambda)}$ for $2 < p < 2+\varepsilon$.

The geometry will be clearer if we do it in R^n_+, and then we transfer it to D by the bi-Lipschitzian mapping $\Phi : R^n_+ \to D$, $\Phi(x,y) = (x, y+\phi(x))$. We will systematically ignore the distinction between sets in R^n_+ and their images under Φ.

Let $\gamma = \{(x,y) \in R^n_+ : |x| < y\}$, $\gamma^* = \{(x,y) \in R^n_+ : a|x| < y\}$, where a is a small constant to be chosen. Let $m(x) = \sup_{(z,y)\in x+\gamma} |\nabla u(z,y)|$, and $m^*(x) = \sup_{(z,y)\in x+\gamma^*} |\nabla u(z,y)|$. Our aim is to show that there is a small $\varepsilon_0 > 0$ such that $\int m^{2+\varepsilon} dx \leq c \int |f|^{2+\varepsilon} dx$, for all $0 < \varepsilon \leq \varepsilon_0$, where $f = \frac{\partial u}{\partial N}$. Let $h = M(f^2)^{1/2}$, where M denotes the Hardy-Littlewood maximal operator. Let $E_\lambda = \{x \in R^{n-1} : m^*(x) > \lambda\}$. We claim that $\int_{\{m^*>\lambda;h\leq\lambda\}} m^2 \leq C\lambda^2|E_\lambda| + Ca \int_{\{m^*>\lambda\}} m^2$. Let us assume the claim, and prove the desired estimate. First, note that

$$\int_{E_\lambda} m^2 \leq \int_{\{m^*>\lambda;h\leq\lambda\}} m^2 + \int_{\{h>\lambda\}} m^2 \leq C\lambda^2|E_\lambda| + Ca \int_{\{m^*>\lambda\}} m^2 + \int_{\{h>\lambda\}} m^2,$$

by the claim. Choose now and fix a so that $C\cdot a < 1/2$. Then, $\int_{E_\lambda} m^2 \leq C\lambda^2|E_\lambda| + C \int_{\{h\geq\lambda\}} m^2$. For $\varepsilon > 0$, $\int m^{2+\varepsilon} = \varepsilon \int_0^\infty \lambda^{\varepsilon-1} \int_{\{m>\lambda\}} m^2 d\lambda \leq \varepsilon \int_0^\infty \lambda^{\varepsilon-1} \int_{E_\lambda} m^2 d\lambda \leq C\varepsilon \int_0^\infty \lambda^{1+\varepsilon} |\{m^*>\lambda\}| d\lambda + C\varepsilon \int_0^\infty \lambda^{\varepsilon-1} (\int_{h>\lambda} m^2) d\lambda$.

By a well-known inequality (see [14] for example), $|E_\lambda| \leq C_\alpha |\{m>\lambda\}|$. Thus, $\int m^{2+\varepsilon} \leq C\varepsilon \int_0^\infty \lambda^{1+\varepsilon} |\{m>\lambda\}| d\lambda + C\varepsilon \int_0^\infty \lambda^{\varepsilon-1}(\int_{h>\lambda} m^2) d\lambda \leq C\varepsilon \int m^{2+\varepsilon} + C \int m^2 h^\varepsilon$. If we now choose ε_0 so that $C\varepsilon_0 < 1/2$, for $\varepsilon < \varepsilon_0$, $\int m^{2+\varepsilon} \leq C \int m^2 h^\varepsilon$. If we now use Holder's inequality with exponents $\frac{2+\varepsilon}{2}$ and $\frac{2+\varepsilon}{\varepsilon}$, we see that $\int m^{2+\varepsilon} \leq C(\int m^{2+\varepsilon})^{\frac{2}{2+\varepsilon}}(\int M(f^2)^{\frac{2+\varepsilon}{2}})^{\frac{\varepsilon}{2+\varepsilon}}$, and the desired inequality follows from the Hardy-Littlewood maximal theorem.

It remains to establish the claim. Let $\{Q_k\}$ be a Whitney decomposition of the set $E_\lambda = \{m^*>\lambda\}$, such that $3Q_k \subset E_\lambda$, and $\{3Q_k\}$ has bounded overlap. Fix k, we can assume that there exists $x \in Q_k$ such that $h(x) \leq \lambda$, and hence, $\int_{2Q_k} f^2 \leq C\lambda^2|Q_k|$. For $1 \leq \tau \leq 2$, let $Q_{k,\tau} = \tau Q_k$, and $\widetilde{Q}_{k,\tau} = \{(x,y) : x \in \tau Q_k,\ 0 < y < \tau \text{ length } (Q_k)\}$. $\widetilde{Q}_{k,\tau}$ (and $\Phi(\widetilde{Q}_{k,\tau})$) is a Lipschitz domain, uniformly in k,τ. Also, by construction of Q_k, there exists x_k with $\operatorname{dist}(x_k,Q_k) \approx \text{length } (Q_k)$, and such that $m^*(x_k) \leq \lambda$. Let

$$A_{k,\tau} = \partial\widetilde{Q}_{k,\tau} \cap x_k + \gamma^*$$

$$B_{k,\tau} = \partial\widetilde{Q}_{k,\tau} \cap R^n_+ \backslash A_{k,\tau},$$

so that $\partial\widetilde{Q}_{k,\tau} = Q_{k,\tau} \cup A_{k,\tau} \cup B_{k,\tau}$. Note that the height of $B_{k,\tau}$ is dominated by $C\alpha$ length (Q_k), and that $|\nabla u| \leq \lambda$ on $A_{k,\tau}$. Let m_1 be the maximal function of ∇u, corresponding to the domain $Q_{k,\tau}$ (i.e. where the cones are truncated at height $\approx \ell(Q_k)$). Then, for $x \in Q_k$, $m(x) \leq m_1(x) + \lambda$. Also,

$$\int_{Q_k} m_1^2 \leq \int_{\partial\widetilde{Q}_{k,\tau}} m_1^2 \leq (\text{using the } L^2\text{-theory on } \widetilde{Q}_{k,\tau}) \leq$$

$$C \int_{B_{k,\tau}} |\nabla u|^2 d\sigma + c \int_{A_{k,\tau}} |\nabla u|^2 d\sigma + c \int_{2Q_k} f^2 \leq C \int_{B_{k,\tau}} |\nabla u|^2 d\sigma + c\lambda^2 |Q_k| .$$

Integrating in τ between 1 and 2, we see that

$$\int_{Q_k} m_1^2 \le \frac{C}{\text{length}(Q_k)} \int_0^{\alpha\,\text{length}(Q_k)} \int_{2Q_k} |\nabla u|^2 + c\lambda^2|Q_k| \le C\alpha \int_{2Q_k} m^2 + C\lambda^2|Q_k| .$$

Thus, $\int_{Q_k} m^2 \le C\alpha \int_{2Q_k} m^2 + C\lambda^2|Q_k|$. Adding in k, we see that $\int_{\{m^*>\lambda, h\le\lambda\}} \le C\lambda^2|E_\lambda| + C\alpha \int_{\{m^*>\lambda\}} m^2$, which is the claim. Note also that the same argument gives the estimate $\|N(\nabla u)\|_p \le C\|\nabla_t u\|_p$, $2 < p < 2+\varepsilon$, and the L^p theory is thus completed.

§3. *Systems of equations on Lipschitz domains*

(a) *The systems of elastostatics.*

In this part we will sketch the extension of the L^2 results for the Laplace equation to the systems of linear elastostatics on Lipschitz domains. These results are joint work of B. Dahlberg, C. Kenig and G. Verchota, and will be discussed in detail in a forthcoming paper ([8]). Here we will describe some of the main ideas in that work. For simplicity here we restrict our attention to domains D above the graph of a Lipschitz function $\phi : \mathbf{R}^2 \to \mathbf{R}$.

Let $\lambda > 0$, $\mu \ge 0$ be constants (Lame moduli). We will seek to solve the following boundary value problems, where $\vec{u} = (u_1, u_2, u_3)$

(3.1.1)
$$\begin{cases} \mu\Delta\vec{u} + (\lambda+\mu)\nabla \operatorname{div} \vec{u} = 0 \text{ in } D \\ \vec{u}|_{\partial D} = \vec{f} \in L^2(\partial D, d\sigma) \end{cases}$$

(3.1.2)
$$\begin{cases} \mu\Delta\vec{u} + (\lambda+\mu)\nabla \operatorname{div} \vec{u} = 0 \text{ in } D \\ \lambda(\operatorname{div}\vec{u})N + \mu\{\nabla\vec{u} + (\nabla\vec{u})^t\}N|_{\partial D} = \vec{f} \in L^2(\partial D, d\sigma) . \end{cases}$$

(3.1.1) corresponds to knowing the displacement vector $\vec{u}$ on the boundary of D, while (3.1.2) corresponds to knowing the surface stresses on the boundary of D. We seek to solve (3.1.1) and (3.1.2) by the method of layer potentials. In order to do so, we introduce the Kelvin matrix of fundamental solutions (see [24] for example), $\Gamma(X) = (\Gamma_{ij}(X))$, where

$$\Gamma_{ij}(X) = \frac{A}{4\pi}\frac{\delta_{ij}}{|X|} + \frac{C}{4\pi}\frac{X_iX_j}{|X|^3}, \text{ and } A = \frac{1}{2}\left[\frac{1}{\mu} + \frac{1}{2\mu+\lambda}\right], \; C = \frac{1}{2}\left[\frac{1}{\mu} - \frac{1}{2\mu+\lambda}\right].$$

We will also introduce the stress operator T, where $T\vec{u} = \lambda(\operatorname{div}\vec{u})N + \mu\{\nabla\vec{u} + \nabla\vec{u}^{\,t}\}N$.

The double layer potential of a density $\vec{g}(Q)$ is then given by $\vec{u}(X) = \mathcal{K}\vec{g}(X) = \int_{\partial D}\{T(Q)\Gamma(X-Q)\}^t\vec{g}(Q)\,d\sigma(Q)$, where the operator T is applied to each column of the matrix Γ.

The single layer potential of a density $\vec{g}(Q)$ is

$$\vec{u}(X) = S\vec{g}(X) = \int_{\partial D}\Gamma(X-Q)\cdot\vec{g}(Q)\,d\sigma(Q).$$

Our main results here parallel those of Section 2, part a). They are

THEOREM 3.1.3. (a) *There exists a unique solution of problem 3.1.1 in* D, *with* $N(\vec{u}) \in L^2(\partial D, d\sigma)$. *Moreover, the solution* $\vec{u}$ *has the form* $\vec{u}(X) = \mathcal{K}\vec{g}(X)$, $\vec{g} \in L^2(\partial D, d\sigma)$.

(b) *There exists a unique solution of (3.1.2) in* D, *which is* 0 *at infinity, with* $N(\nabla\vec{u}) \in L^2(\partial D, d\sigma)$. *Moreover the solution* $\vec{u}$ *has the form* $\vec{u}(X) = S\vec{g}(X)$, $\vec{g} \in L^2(\partial D, d\sigma)$.

(c) *If the data* $\vec{f}$ *in 3.1.1 belongs to* $L^2_1(\partial D, d\sigma)$, *then we can solve (3.1.1), with* $N(\nabla\vec{u}) \in L^2(\partial D, d\sigma)$. *Moreover, we can take*

$$\vec{u}(X) = S\vec{g}(x), \quad \vec{g} \in L^2(\partial D, d\sigma).$$

The proof of Theorem 3.1.3 starts out following the pattern we used to prove 2.1.1, 2.1.2 and 2.1.14. We first show, as in Theorem 2.1.3. that the following lemma holds:

LEMMA 3.1.4. *Let* $\mathcal{K}\vec{g}$, $S\vec{g}$ *be defined as above, so that they both solve* $\mu\Delta\vec{u} + (\lambda+\mu)\nabla \operatorname{div} \vec{u} = 0$ *in* $R^3\backslash\partial D$. *Then*:

(a) $$\|N(\mathcal{K}\vec{g})\|_{L^p(\partial D, d\sigma)} \le C\|\vec{g}\|_{L^p(\partial D, d\sigma)},$$

$$\|N(\nabla S\vec{g})\|_{L^p(\partial D, d\sigma)} \le C\|\vec{g}\|_{L^p(\partial D, d\sigma)}, \quad \textit{for } 1 < p < \infty .$$

(b) $$(\mathcal{K}\vec{g})^{\pm}(P) = \pm\frac{1}{2}\,\vec{g}(P) + K\vec{g}(P)$$

$$\left(\frac{\partial}{\partial X_i}(S\vec{g})_j\right)^{\pm} = \pm\left\{\frac{A+C}{2}\,n_i(P)g_j(P) - n_i(P)\,n_j(P)<N(P), \vec{g}(P)>\right\} +$$

$$+\left(\text{p.v.}\int_{\partial D}\frac{\partial}{\partial P_i}\Gamma(P-Q)\,\vec{g}(Q)\,d\sigma(Q)\right)_j ,$$

where $K\vec{g}(P) = \text{p.v.}\int_{\partial D}\{T(Q)\Gamma(P-Q)\}^t\,\vec{g}(Q)\,d\sigma(Q)$, *and* A, C *are the constants in the definition of the fundamental solution.*

Thus, just as in Section 2, part (a) is reduced to proving the invertibility on $L^2(\partial D, d\sigma)$ of $\pm\frac{1}{2}I+K$, $\pm\frac{1}{2}I+K^*$, and the invertibility from $L^2(\partial D, d\sigma)$ onto $L_1^2(\partial D, d\sigma)$ of S. Just as before, using the jump relations, it suffices to show that if $\vec{u}(X) = S\vec{g}(X)$, then $\|T\vec{u}\|_{L^2(\partial D, d\sigma)} \approx \|\nabla_t\vec{u}\|_{L^2(\partial D, d\sigma)}$. Before explaining the difficulties in doing so, it is very useful to explain the stress operator T (and thus the boundary value problem 3.1.2), from the point of view of the theory of constant coefficient second order elliptic systems. We go back to working on R^n, and use the summation convention.

Let a_{ij}^{rs}, $1 \le r, s \le m$, $1 \le i, j \le n$ be constants satisfying the ellipticity condition

$$a_{ij}^{rs}\,\xi_i\,\xi_j\,\eta^r\eta^s \ge c|\xi|^2|\eta|^2$$

and the symmetry condition $a_{ij}^{rs} = a_{ji}^{sr}$. Consider vector valued functions $\vec{u} = (u^1, \cdots, u^m)$ on R^n satisfying the divergence from system $\frac{\partial}{\partial X_i} a_{ij}^{rs} \frac{\partial}{\partial X_j} u^s = 0$ in D. From variational considerations, the most natural boundary conditions are to Dirichlet condition $(\vec{u}|_{\partial D} = f)$ or the Neumann type conditions, $\frac{\partial \vec{u}}{\partial \nu} = n_i\, a_{ij}^{rs} \frac{\partial u^s}{\partial X_j} = f_r$. The interpretation of problem (2) in this context is that we can find constants a_{ij}^{rs}, $1 \le i, j \le 3$, $1 \le r, s \le 3$, which satisfy the ellipticity condition and the symmetry condition, and such that $\mu \Delta \vec{u} + (\lambda + \mu) \nabla \operatorname{div} \vec{u} = 0$ in D if and only if $\frac{\partial}{\partial X_i} a_{ij}^{rs} \frac{\partial u^s}{\partial X_j} = 0$ in D, and with $T\vec{u} = \frac{\partial}{\partial \nu} \vec{u}$. In order to obtain the equivalence between the tangential derivatives and the stress operator we need an identity of the Rellich type. Such identities are available for general constant coefficient systems (see [29], [27]).

LEMMA 3.1.5 (The Rellich, Payne-Weinberger, Nečas identities). *Suppose that* $\frac{\partial}{\partial X_i} a_{ij}^{rs} \frac{\partial}{\partial X_j} u^s = 0$ *in* D, $a_{ij}^{rs} = a_{ji}^{sr}$, $\vec{h}$ *is a constant vector in* R^n, *and* $\vec{u}$ *and its derivatives are suitably small at* ∞. *Then,*

$$\int_{\partial D} h_\ell n_\ell\, a_{ij}^{rs} \frac{\partial u^r}{\partial X_i} \frac{\partial u^s}{\partial X_j}\, d\sigma = 2 \int_{\partial D} h_i \frac{\partial u^r}{\partial X_i}\, n_\ell a_{\ell j}^{rs} \frac{\partial u^s}{\partial X_j}\, d\sigma .$$

Proof. Apply the divergence theorem to the formula

$$\frac{\partial}{\partial X_\ell}\left[(h_\ell a_{ij}^{rs} - h_i a_{\ell j}^{rs} - h_j a_{i\ell}^{rs}) \frac{\partial u^r}{\partial X_i} \cdot \frac{\partial u^s}{\partial X_j}\right] = 0 .$$

REMARK 1. Note that if we are dealing with the case $m = 1$, $a_{ij} = I$, and we choose $\vec{h} = e_n$, we recover the identity we used before for Laplace's equation.

REMARK 2. Note that if we had the stronger ellipticity assumption that $a_{ij}^{rs}\xi_i^r\xi_j^s \geq C \sum_{\ell,t} |\xi_\ell^t|^2$, we would have, if $\partial D = \{(x, \phi(x)) : \phi : R^{n-1} \to R, \|\nabla\phi\|_\infty \leq M\}$, that $\|\nabla_t \vec{u}\|_{L^2(\partial D, d\sigma)} \approx \left\|\frac{\partial \vec{u}}{\partial \nu}\right\|_{L^2(\partial D, d\sigma)}$. In fact, if we take $\vec{h} = e_n$, then we would have

$$\sum_r \int_{\partial D} |\nabla u^r|^2 d\sigma \leq C \int_{\partial D} h_\ell n_\ell a_{ij}^{rs} \frac{\partial u^r}{\partial X_i} \frac{\partial u^s}{\partial X_j} d\sigma =$$

$$= 2C \int_{\partial D} h_i \frac{\partial u^r}{\partial X_i} \cdot n_\ell a_{\ell j}^{rs} \frac{\partial u^s}{\partial X_j} d\sigma \leq 2C \left(\sum_r \int_{\partial D} |\nabla u^r|^2 d\sigma\right)^{1/2} \left(\int_{\partial D} \left|\frac{\partial \vec{u}}{\partial \nu}\right|^2 d\sigma\right)^{1/2}$$

Thus, $\sum \int_{\partial D} |\nabla u^r|^2 d\sigma \leq C \int_{\partial D} \left|\frac{\partial \vec{u}}{\partial \nu}\right|^2 d\sigma$.

For the opposite inequality, observe that, for each r,s,j fixed, the vector $h_i n_\ell a_{\ell j}^{rs} - h_\ell n_\ell a_{ij}^{rs}$ is perpendicular to N. Because of Lemma 3.1.5,

$$\int_{\partial D} h_\ell n_\ell a_{ij}^{rs} \frac{\partial u^r}{\partial X_i} \cdot \frac{\partial u^s}{\partial X_j} d\sigma = 2 \int_{\partial D} (h_\ell n_\ell a_{ij}^{rs} - h_i n_\ell a_{\ell j}^{rs}) \frac{\partial u^\ell}{\partial X_i} \cdot \frac{\partial u^s}{\partial X_j} d\sigma .$$

Hence, $\int_{\partial D} |\nabla \vec{u}|^2 d\sigma \leq C(\int_{\partial D} |\nabla_t \vec{u}|^2 d\sigma)^{1/2} (\int_{\partial D} |\nabla \vec{u}|^2 d\sigma)^{1/2}$, and so

$$\int_{\partial D} \left|\frac{\partial \vec{u}}{\partial \nu}\right|^2 d\sigma \leq c \int_{\partial D} |\nabla \vec{u}|^2 d\sigma \leq C \left(\int_{\partial D} |\nabla_t \vec{u}|^2\right)^{1/2} .$$

REMARK 3. In the case in which we are interested, i.e. the case of the systems of elastostatics,

$$a^{rs}_{ij}\frac{\partial u^s}{\partial X_i}\cdot\frac{\partial u^r}{\partial X_j} = \lambda(\text{div } u)^2 + \frac{\mu}{2}\sum_{i,j}\left(\frac{\partial u^j}{\partial X_i} + \frac{\partial u^i}{\partial X_j}\right)^2,$$

which clearly does not satisfy $a^{rs}_{ij}\,\xi^r_i\xi^s_j \geq C\sum_{\ell,t}|\xi^t_\ell|^2$, since the quadratic form involves only the symmetric part of the matrix (ξ^r_i). In this case, of course $\frac{\partial\vec{u}}{\partial\nu} = T\vec{u} = \lambda(\text{div } \vec{u}\,N + \mu\{\nabla\vec{u} + \nabla\vec{u}^t\}N$.

REMARK 4. The inequality $\|\nabla\vec{u}\|_{L^2(\partial D, d\sigma)} \leq C\|\nabla_t\vec{u}\|_{L^2(\partial D, d\sigma)}$ holds in the general case, directly from Lemma 3.1.5, by a more complicated algebraic argument. In fact, as in Remark 2, $\int_{\partial D} h_\ell n_\ell a^{rs}_{ij}\frac{\partial u^r}{\partial X_i}\frac{\partial u^s}{\partial X_j}d\sigma = 2\int_{\partial D}(h_\ell n_\ell a^{rs}_{ij} - h_i n_\ell a^{rs}_{\ell j})\frac{\partial u^\ell}{\partial X_i}\cdot\frac{\partial u^s}{\partial X_j}d\sigma$, and for fixed r, s, j, $(h_\ell n_\ell a^{rs}_{ij} - h_i n_\ell a^{rs}_{\ell j})$ is a tangential vector. Thus, $\int_{\partial D} h_\ell n_\ell a^{rs}_{ij}\frac{\partial u^r}{\partial X_i}\frac{\partial u^s}{\partial X_j}d\sigma \leq C\left(\int_{\partial D}|\nabla_t\vec{u}|^2 d\sigma\right)^{1/2}\left(\int_{\partial D}|\nabla\vec{u}|^2 d\sigma\right)^{1/2}$. Consider now the matrix $d_{rs} = (a^{rs}_{ij}\,n_i n_j)^{-1}$. This is a strictly positive matrix, since $a^{rs}_{ij}\,\xi_i\xi_j\eta^r\eta^s \geq C|\xi|^2|\eta|^2$. Moreover, $d_{rs}\left(\frac{\partial\vec{u}}{\partial\nu}\right)_r\left(\frac{\partial\vec{u}}{\partial\nu}\right)_s - a^{rs}_{ij}\frac{\partial u^r}{\partial X_i}\frac{\partial u^s}{\partial X_j} = d_{rs}n_i a^{rt}_{ij}\frac{\partial u^t}{\partial X_j}\cdot n_\ell a^{sm}_{\ell k}\frac{\partial u}{\partial X}$ $a^{rs}_{ij}\frac{\partial u^r}{\partial X_i}\frac{\partial u^s}{\partial X_j} = d_{rs}n_k a^{rt}_{k\ell}\frac{\partial u^t}{\partial X_\ell}\cdot n_m a^{st}_{mv}\frac{\partial u^\tau}{\partial X_v} - a^{t\tau}_{v\ell}\frac{\partial u^t}{\partial X_v}\frac{\partial u^\tau}{\partial X_\ell} = d_{rs}n_k a^{rt}_{kv}\frac{\partial u^t}{\partial X_v}\cdot n_m a^{s\tau}_{m\ell}\frac{\partial u^\tau}{\partial X_\ell} - a^{t\tau}_{v\ell}\frac{\partial u^t}{\partial X_v}\frac{\partial u^\tau}{\partial X_\ell} = \{d_{rs}n_k a^{rt}_{kv}n_m a^{st}_{m\ell} - a^{t\tau}_{v\ell}\}\frac{\partial u^t}{\partial X_v}\frac{\partial u^\tau}{\partial X_\ell}$. Now, note that for t, τ, ℓ fixed, $\{d_{rs}n_k a^{rt}_{kv}n_m a^{st}_{m\ell} - a^{t\tau}_{v\ell}\}$ is perpendicular to N, by our definition of d_{rs}, and the symmetry of a^{rs}_{ij}: $d_{rs}n_k a^{rt}_{kv}n_m a^{st}_{m\ell}n_v - a^{t\tau}_{v\ell}n_v = a^{rt}_{kv}n_k n_v d_{rs}a^{st}_{m\ell}n_m - a^{t\tau}_{m\ell}n_m = a^{tr}_{vk}n_v n_k d_{rs}a^{s\tau}_{m\ell}n_m - a^{t\tau}_{m\ell}n_m = \delta_{ts}a^{st}_{m\ell}n_m - a^{t\tau}_{m\ell}n_m = a^{t\tau}_{m\ell}n_m - a^{t\tau}_{m\ell}n_m = 0$. Therefore, $\int_{\partial D} h_\ell n_\ell d_{rs}\left(\frac{\partial\vec{u}}{\partial\nu}\right)_r\left(\frac{\partial\vec{u}}{\partial\nu}\right)_s d\sigma \leq$

$C(\int_{\partial D} |\nabla_t \vec{u}|^2 d\sigma)^{1/2} (\int_{\partial D} |\nabla \vec{u}|^2 d\sigma)^{1/2}$. Now, $\frac{\partial \vec{u}}{\partial \nu}_r - a^{rs}_{kj} n_k n_j \frac{\partial u^s}{\partial N} =$ $n_i a^{rs}_{ij} \frac{\partial u^s}{\partial X_j} - a^{rs}_{kj} n_k n_j n_i \frac{\partial u^s}{\partial X_i} = n_i a^{rs}_{ij} \frac{\partial u^s}{\partial X_j} - a^{rs}_{ki} n_k n_j n_i \frac{\partial u^s}{\partial X_j} = \{ n_i a^{rs}_{ij} -$ $a^{rs}_{ki} n_k n_i n_j \} \frac{\partial u^s}{\partial X_j} = \{ n_i a^{rs}_{ij} - a^{rs}_{ik} n_k n_i n_j \} \frac{\partial u^s}{\partial X_j}$. But, for i, r, s fixed, $a^{rs}_{ij} -$ $a^{rs}_{ik} n_k n_j$ is perpendicular to N, and so $\int_{\partial D} h_\ell n_\ell d_{rs} \left\{ a^{rt}_{kj} n_k n_j \frac{\partial u^t}{\partial N} \right\}$ $a^{s\tau}_{i\ell} n_i n_\ell \frac{\partial u^\tau}{} \; d\sigma \le C \{ (\int_{\partial D} |\nabla_t \vec{u}|^2 d\sigma)^{1/2} (\int_{\partial D} |\nabla \vec{u}|^2 d\sigma)^{1/2} + \int_{\partial D} |\nabla_t \vec{u}|^2 d\sigma \}$.

We now choose $\vec{h} = e_n$, so that $h_\ell n_\ell \ge C$, and recall that (d_{rs}) and $(a^{rt}_{kj} n_k n_j)$ are strictly positive definite matrices. We then see that

$$\int_{\partial D} \left| \frac{\partial \vec{u}}{\partial N} \right|^2 d\sigma \le C \left\{ \left(\int_{\partial D} |\nabla_t \vec{u}|^2 d\sigma \right)^{1/2} \left(\int_{\partial D} |\nabla \vec{u}|^2 d\sigma \right)^{1/2} + \int_{\partial D} |\nabla_t \vec{u}|^2 d\sigma \right\}.$$

Now, as $|\nabla \vec{u}|^2 = |\nabla_t \vec{u}|^2 + \left| \frac{\partial \vec{u}}{\partial N} \right|^2$, the remark follows.

REMARK 5. In order to show that $\int_{\partial D} |\nabla_t u|^2 d\sigma \le C \int_{\partial D} |T\vec{u}|^2 d\sigma$, it suffices to show that $\int_{\partial D} |\nabla \vec{u}|^2 d\sigma \le C \int_{\partial D} |\lambda (\operatorname{div} \vec{u}) I + \mu \{ \nabla \vec{u} + \nabla \vec{u}^t \}|^2 d\sigma$. In fact, if this inequality holds, we would clearly have that $\int_{\partial D} |\nabla \vec{u}|^2 d\sigma \le C \int_{\partial D} |\nabla \vec{u} + \nabla \vec{u}^t|^2 d\sigma$ (Korn type inequality at the boundary). The Rellich-Payne-Weinberger-Nečas identity is, in this case (with $h = e_n$),

$$\int_{\partial D} n_n \left\{ \frac{\mu}{2} |\nabla \vec{u} + \nabla \vec{u}^t|^2 + \lambda (\operatorname{div} \vec{u})^2 \right\} d\sigma = 2 \int_{\partial D} \frac{\partial \vec{u}}{\partial y} \cdot \{ \lambda (\operatorname{div} \vec{u}) N + \mu \{ \nabla \vec{u} + \nabla \vec{u}^t \} N \} d\sigma.$$

But then, $\int_{\partial D} |\nabla \vec{u}|^2 d\sigma \le C (\int_{\partial D} |\nabla \vec{u}|^2 d\sigma)^{1/2} (\int_{\partial D} |\lambda (\operatorname{div} \vec{u}) N + \mu \{ \nabla \vec{u} + \nabla \vec{u}^t \} N|^2 d\sigma)^{1/2}$. The rest of part (a) is devoted to sketching the proof of the above inequality.

THEOREM 3.1.6. *Let* $\vec{u}$ *solve* $\mu \Delta \vec{u} + (\lambda + \mu) \nabla \operatorname{div} \vec{u} = 0$ *in* D, $\vec{u} = S(\vec{g})$, *where* $\vec{g}$ *is nice. Then, there exists a constant* C, *which depends only on the Lipschitz constant of* ϕ *so that*

$$\int_{\partial D} |\nabla\vec{u}|^2 d\sigma \leq C \int_{\partial D} |\lambda(\operatorname{div}\vec{u})I + \mu\{\nabla\vec{u} + \nabla\vec{u}^t\}|^2 d\sigma .$$

The proof of the above theorem proceeds in two steps. They are:

LEMMA 3.1.7. *Let* $\vec{u}$ *be as in Theorem 3.1.6. Then,*

$$\int_{\partial D} N(\nabla\vec{u})^2 d\sigma \leq C \int_{\partial D} N(\lambda(\operatorname{div}\vec{u})I + \mu\{\nabla\vec{u} + \nabla\vec{u}^t\})^2 d\sigma .$$

LEMMA 3.1.8. *Let* $\vec{u}$ *be as in Theorem 3.1.6. Then,*

$$\int_{\partial D} N(\lambda(\operatorname{div}\vec{u})I + \mu\{\nabla\vec{u} + \nabla\vec{u}^t\})^2 d\sigma \leq C \int_{\partial D} |\lambda(\operatorname{div}\vec{u})I + \mu\{\nabla\vec{u} + \nabla\vec{u}^t\}|^2 d\sigma .$$

Lemma 3.1.7 is proved by first doing so in the case when the Lipschitz constant is small, and then passing to the general case by using the ideas of G. David ([9]). Lemma 3.1.8 is proved by observing that if $\vec{v}$ is any row of the matrix $\lambda(\operatorname{div}\vec{u})I + \mu\{\nabla\vec{u} + \nabla\vec{u}^t\}$, then $\vec{v}$ is a solution of the Stokes system

$$(S)\begin{cases} \Delta\vec{v} = \nabla p \text{ in } D \\ \operatorname{div}\vec{v} = 0 \text{ in } D \\ \vec{v}|_{\partial D} = \vec{f} \in L^2(\partial D, d\sigma) \end{cases} .$$

This is checked directly by using the system of equations $\mu\Delta\vec{u} + (\lambda+\mu)\nabla \operatorname{div}\vec{u} = 0$. One then invokes the following Theorem of E. Fabes, C. Kenig and G. Verchota, whose proof will be presented in the next section.

THEOREM 3.1.9. *Given* $\vec{f} \in L^2(\partial D, d\sigma)$, *there exists a unique solution* $(\vec{v}, p)$ *to system* (S) *with* p *tending to* 0 *at* ∞, *and* $N(\vec{v}) \in L^2(\partial D, d\sigma)$. *Moreover* $\|N(\vec{v})\|_{L^2(\partial D, d\sigma)} \leq C\|\vec{f}\|_{L^2(\partial D, d\sigma)}$.

We now turn to a sketch of the proof of Lemma 3.1.7. We will need the following unpublished real variable lemma of G. David ([10]).

LEMMA 3.1.10. *Let* $F : R \times R^n \to R$ *be a function of two variables* $t \in R$, $x = (x_1, \cdots, x_n) \in R^n$. *Assume that for each* x, *the function* $t \mapsto F(t,x)$ *is Lipschitz, with Lipschitz constant less than or equal to* M, *and for each* i, $1 \leq i \leq n$, *the function* $x_i \mapsto F(t,x)$ *is Lipschitz, with Lipschitz constant less than or equal to* M_i, *for any choice of the other variables. Given an interval* $I \times J = I \times J_1 \times \cdots \times J_n$, *where the* J_i's *and* I *are* 1 *dimensional compact intervals, there exists a function* $G(t,x) : R \times R^n \to R$ *with the following properties*:

(a) $G(t,x) \geq F(t,x)$ *on* $I \times J$.

(b) *If* $E = \{(t,x) \in I \times J : F(t,x) = G(t,x)\}$, *then* $|E| \geq \frac{3}{8} |I| \, |J|$.

(c) *For each* i, *the function* $G(t, x_1, x_2, \cdots, x_{i-1}, -, x_{i+1}, \cdots, x_n)$ *is Lipschitz, with Lipschitz constant less than or equal to* M_i, *and one of the following statements is true*:

$$\textit{Either for each } x, \quad -M \leq \frac{\partial G}{\partial t}(t,x) \leq \frac{4M}{5}, \quad \textit{or}$$

$$\textit{for each } x \qquad , \quad -\frac{4M}{5} \leq \frac{\partial G}{\partial t}(t,x) \leq M .$$

The proof of this lemma is the same as in the 1 dimensional case, treating x as a parameter (see [9]).

Before we proceed with the proof of Lemma 3.1.7, we would like to point out that in the analogue of Lemma 3.1.7 for bounded domains, a normalization is necessary since if $\vec{u}(X)$ solves the systems of elastostatics, so does $\vec{u}(X) + \vec{a} + BX$, where $\vec{a}$ is a constant vector, while B is any antisymmetric 3×3 matrix. The right-hand side of the inequality in the lemma of course remains unchanged, while the left-hand side increases if B 'increases.' The most convenient normalization is that for some fixed point X^* in the domain $\nabla\vec{u}(X^*) - \nabla\vec{u}(X^*)^t = 0$. This gives uniqueness modulo constants to problem 3.1.2 in bounded domains.

We now need to introduce some definitions. Let $D_0 \subset R^n_+$ be a fixed, C^∞ domain with $\{(x,0) : |||x||| = \max |x_i| \leq 1\} \subset \partial D_0$, $\{(x,y) : 0 < y < 1, |||x||| \leq 1\} \subset D_0 \subset \{(x,y) : 0 < y < 2, |||x||| < 2\}$. If $\phi : R^{n-1} \to R$ is Lipschitz, with $|||\nabla\phi||| \leq M$, we construct the mapping $T_\phi : R^n_+ \to R^n$ by $T_\phi(x,y) = (x, cy + \eta_y * \phi(x))$ where $\eta \in C_0^\infty(R^{n-1})$ is radial, $\int \eta = 1$, and $C = C(M)$ is chosen so that $T_\phi(R^n_+) \subset \{(x,y) : y > \phi(x)\}$, and so that T_ϕ is a bi-Lipschitzian mapping. Also, it is clear that T_ϕ is smooth for (x,y) with $y > 0$, and $T_\phi(x,0) = (x, \phi(x))$. We will denote by A_ϕ the point $T_\phi(0,1)$. Lemma 3.1.7 is an easy consequence of

LEMMA 3.1.11. *Given* $M > 0$ *and* ϕ *with* $|||\nabla\phi||| \leq M$, *there exists a constant* $C = C(M)$ *such that for all functions* $\vec{u}$ *in* D_ϕ, *which are Lipschitz in* $\bar{D}_\phi$, *which satisfy* $\mu\Delta\vec{u} + (\lambda+\mu)\nabla \operatorname{div} \vec{u} = 0$ *in* D_ϕ *and* $\nabla\vec{u}(A_\phi) = \nabla\vec{u}(A_\phi)^t$, we have $\|N_\phi(\nabla\vec{u})\|_{L^2(\partial D, d\sigma)} \leq C\|N_\phi(\lambda(\operatorname{div}\vec{u})I + \mu\{\nabla\vec{u} + \nabla\vec{u}^t\}\|_{L^2(\partial D, d\sigma)}$. *Here* N_ϕ *is the non-tangential maximal operator corresponding to the domain* D_ϕ.

This lemma will be proved by a series of propositions. Before we proceed we need to introduce one more definition. We say that Proposition (M,ε) holds if whenever ϕ is such that $|||\nabla\phi||| \leq M$, and there exists a constant vector $\vec{a}$ with $|||\vec{a}||| \leq M$ so that $|||\nabla\phi - \vec{a}||| \leq \varepsilon$, then, for all Lipschitz functions $\vec{u}$ on D_ϕ with $\mu\Delta\vec{u} + (\lambda+\mu)\nabla \operatorname{div}\vec{u} = 0$ in D_ϕ, with $\nabla\vec{u}(A_\phi) = \nabla\vec{u}^t(A_\phi)$ we have

$$\|N_\phi(\nabla\vec{u})\|_{L^2(\partial D, d\sigma)} \leq C\|N_\phi(\lambda(\operatorname{div}\vec{u})I + \mu\{\nabla\vec{u} + \nabla\vec{u}^t\}\|_{L^2(\partial D, d\sigma)},$$

where $C = C(M,\varepsilon)$.

Note that if Proposition (M,ε) holds, then the corresponding estimates automatically hold for all translates, rotates or dilates of the domains D_ϕ, when ϕ satisfies the conditions in Proposition (M,ε). In the rest of this section, a coordinate chart will be a translate, rotate or dilate of a domain D_ϕ. The bottom B_ϕ of ∂D_ϕ will be $T_\phi(\partial D_0 \cup (x,0) : x \in R^{n-1})$.

PROPOSITION 3.1.12. *Given* $M > 0$, *there exists* $\varepsilon = \varepsilon(M)$ *so that Proposition* (M,ε) *holds.*

We will not give the proof of Proposition 3.1.12 here. We will just make a few remarks about its proof. First, in this case the stronger estimate $\|N_\phi(\nabla\vec{u})\|_{L^2(\partial D, d\sigma)} \leq C\|\lambda(\text{div}\,\vec{u})N + \mu\{\nabla\vec{u} + \nabla\vec{u}^t\}N\|_{L^2(\partial D, d\sigma)}$ holds. This is because in this case, the domain D_ϕ is a small perturbation of the smooth domain $D_{\vec{a}X}$. For the smooth domain $D_{\vec{a}X}$, we can solve problem 3.1.2 by the method of layer potentials (see [24], for example). If ε is small, a perturbation analysis based on the theorem of Coifman-McIntosh-Meyer ([2]) shows that this is still the case. This easily gives the estimate claimed above.

PROPOSITION 3.1.13. *For all* $M > 0$, $\varepsilon > 0$, $\alpha \in (0,0.1)$, *if Proposition* (M,ε) *holds, then Proposition* $((1-\alpha)M, 1.1\varepsilon)$ *holds.*

We postpone the proof of Proposition 3.1.13, and show first how Proposition 3.1.12 and Proposition 3.1.13 yield Lemma 3.1.11.

Proof of Lemma 3.1.11. We will show that Proposition (M,ε) holds for any M,ε. Fix M,ε, and choose R so large that if $\varepsilon(10M)$ is as in Proposition 3.1.12, then $(1.1)^R\,\varepsilon(10M) \geq \varepsilon$. Pick now $\alpha_j > 0$ so that $\prod_{j=1}^{R}(1-\alpha_j) = 1/10$.

Then, since Proposition $(10M, \varepsilon(10M))$ holds, by Proposition 3.1.12, applying Proposition 3.1.13 R times we see that Proposition (M,ε) holds. We will not sketch the proof of Proposition 3.1.13. We first note that it suffices to show that

$$\|N_\phi(\nabla\vec{u})\|_{L^2(\partial D, d\sigma)} \leq C\|\tilde{N}_\phi(\lambda(\text{div}\,\vec{u})I + \mu\{\nabla\vec{u} + \nabla\vec{u}^t\}\|_{L^2(\partial D, d\sigma)}$$

where $\tilde{N}_\phi$ is the nontangential maximal operator with a wider opening of the non-tangential region. This follows because of classical arguments relating non-tangential maximal functions with different openings (see [14] for example). Pick now ϕ with $|||\nabla\phi-\vec{a}\,||| \le 1{\cdot}1\varepsilon$, $|||\vec{a}\,||| \le (1-\alpha)M$. We will choose $\widetilde{N}_\phi$ as follows: Since $\partial D_\phi \setminus B_\phi$ is smooth, it is easy to see that we can find a finite number of coordinate charts (i.e. rotates, translates and dilates of D_ψ), which are entirely contained in D_ϕ, such that their bottoms B_ψ are contained in ∂D_ϕ, such that $T_\psi((x,0): |\,\|x\|\,| < 1/2)$ cover ∂D_ϕ, and such that the ψ's involved satisfy $|||\nabla\psi||| \le \left(1-\frac{\alpha}{2}\right)M$, and there exist $\vec{a}_\psi$ such that $|||\vec{a}_\psi||| \le 1.11\varepsilon$. The non-tangential region defining $\widetilde{N}_\phi$, on $T_\psi\left((x,0): |||x||| < \frac{1}{2}\right)$ is defined as follows: let $F \subset \left\{(x,0): |||x||| < \frac{1}{2}\right\}$ be a closed set. Consider the cone on R^n_+, $\gamma = \{(x,y) \in R^n_+ : b|x| < y\}$, where b is a small constant. Consider now the domain D_F on R^n_+, given by $D_F = \bigcup_{x\in F} ((x,0)+\gamma)$. Then D_F is the domain above the graph of a Lipschitz function θ, for which $|||\nabla\theta||| \le Cb$, for some absolute constant C (independent of F). It is also easy to see that we can take now b so small (depending only on M and ε) that $T_\psi(D_F)$ is the domain above the graph of a Lipschitz function $\tilde{\psi}$, with $\tilde{\psi} \ge \psi$, and which satisfies $|||\nabla\tilde{\psi}||| \le \left(1-\frac{\alpha}{10}\right)M$, $|||\nabla\psi-\vec{a}_\psi||| \le 1.111\varepsilon$. The non tangential region defining $\widetilde{N}_\phi$, for $Q \in T_\psi\left((x,0): |||x||| < \frac{1}{2}\right)$ is then the image under T_ψ of $(x,0)+\gamma$, with b chosen as above, suitably truncated, and where $Q = T_\psi((x,0))$. Let now, to lighten notation, $m = N_\phi(\nabla\vec{u})$, $\overline{m} = \widetilde{N}_\phi(\lambda(\operatorname{div}\vec{u})I + \mu\{\nabla\vec{u}+\nabla\vec{u}^{\,t}\})$.

For $t > 0$, consider the open-set $E_t = \{m > t\}$. We now produce a Whitney type decomposition of E_t into a family of disjoint sets $\{U_j\}$ with the property that each U_j is contained in $T_\psi\left((x,0): |||x||| < \frac{1}{2}\right)$ for a coordinate chart D_ψ, each U_j contains $T_\psi(I_j)$, where I_j is a cube in $|||x||| < \frac{1}{2}$, and is contained in $T_\psi(\overline{I}_j)$, where $\overline{I}_j$ is a fixed multiple of I_j. Finally, we can also assume that there exists a constant η_0 such that if $\operatorname{diam}(U_j) \le \eta_0$, there exists a point Q_j in ∂D_ϕ, with $\operatorname{dist}(Q_j,U_j) \approx \operatorname{diam} U_j$, such that $m(Q_j) \le t$. Let now $\beta > 1$ be given. We claim that there exists $\delta > 0$ so small that if $E_j = U_j \cap \{m > \beta t, \overline{m} \le \delta t\}$

then $\sigma(E_j) \le (1-\eta_M)\sigma(U_j)$ where $\eta_M > 0$. Assume the claim for the time being. Then

$$\int_{\partial D_\phi} m^2 d\sigma = 2\int_0^\infty t\sigma(E_t)dt = 2\beta^2\int_0^\infty t\sigma(E_{\beta t})dt = \sum_j 2\beta^2\int_0^\infty t\sigma(U_j \cap E_{\beta t})dt \le$$

$$\le \sum_j 2\beta^2\int_0^\infty t\sigma(E_j)\,dt + 2\beta^2\int_0^\infty t\sigma(\overline{m} > \delta t)\,dt \le \sum_j 2\beta^2(1-\eta_M)\int_0^\infty t\sigma(U_j)\,dt +$$

$$+ 2\frac{\beta^2}{\delta^2}\int_0^\infty t\sigma\{\overline{m} > t\}dt = \beta^2\cdot(1-\eta_M)\int_{\partial D_\phi} m^2 d\sigma + \frac{\beta^2}{\delta^2}\int_{\partial D_\phi} \overline{m}^2 d\sigma.$$ Thus, if

we choose $\beta > 1$, but so that $\beta^2\cdot(1-\eta_M) < 1$, the desired result follows. It remains to establish the claim. We argue by contradiction. Suppose not, then $\sigma(E_j) > (1-\eta_M)\sigma(U_j)$. Let $\widetilde{E}_j = T_\psi^{-1}(E_j)$. If η_M is chosen sufficiently small, we can guarantee that $|\widetilde{E}_j \cap I_j| \ge .99|\underset{\sim}{I_j}|$. Let now $F_j = \widetilde{E}_j \cap I_j$, and construct now the Lipschitz function ψ corresponding to it, as in the definition of $\widetilde{N}_\phi$. Thus, $\widetilde{\psi} \ge \psi$, $|||\nabla\widetilde{\psi}||| \le \left(1-\frac{a}{10}\right)M$, $|||\nabla\widetilde{\psi} - \vec{a}_\psi||| \le 1.111\varepsilon$. We now apply Lemma 3.1.10 to $\widetilde{\psi}$, one variable at a time, to find a Lipschitz function f, with $f \ge \widetilde{\psi}$ on I_j, such that if $\overline{F}_j = \{x \in I_j : f = \widetilde{\psi}\}$, then $|\overline{F}_j \cap F_j| \ge C\,\sigma(U_j)$, with $|||\nabla f||| \le \left(1-\frac{a}{10}\right)M$, and such that there exists $\vec{a}_f$, with $|||\vec{a}_f||| \le \left(1-\frac{a}{10}\right)M$ so that $|||\nabla f - a_f||| \le \frac{4}{5}\,1.111\varepsilon < \varepsilon$. We can also arrange the truncation of our nontangential regions in such a way that on the appropriate rotate, translate and dilate of D_f (which of course is contained in the corresponding coordinate chart associated to D_ψ, which is contained in D_ϕ), $|\lambda(\text{div } u)I + \mu\{\nabla u + \nabla u^t\}| \le \delta t$. To lighten the exposition, we will still denote by D_f the translate, rotate and dilate of D_f. Note that Proposition (M,ε) applies to it. We divide the sets U_j into two types. Type I are those with diam $U_j \geq \eta_0$, and type II those for which diam $U_j \leq \eta_0$. We first deal with the U_j of type I. In this case, D_f has diameter of the order of 1. Because of the solvability of problem 3.1.2 for balls, and our normalization, we see that on a ball $B \subset D_\phi$, diam $B \approx 1$, $A_\phi \in B$, we have $\int_B |\nabla\vec{u}|^2 \leq C \int_B |\lambda \operatorname{div} \vec{u} I + \mu\{\nabla\vec{u} + \nabla\vec{u}^t\}|^2$. Joining A_f to A_ϕ by a finite number of balls, and using interior regularity results for the system $\mu\Delta\vec{u} + (\lambda+\mu)\nabla \operatorname{div} \vec{u} = 0$, we see that $|\nabla\vec{u}(A_f)| \leq C\delta t$, for some absolute constant C. Then

$$C\,\sigma(U_j)\beta^2 t^2 \leq \int_{T_\psi(\bar{F}_j \cap F_j)} m^2 d\sigma \leq C \int_{\partial D_f} N_f^2(\nabla\vec{u})\, d\sigma \leq$$

$$\leq C\,\sigma(U_j)\delta^2 t^2 + C \int_{\partial D_f} N_f^2\left(\nabla\vec{u} - \left[\frac{\nabla\vec{u}(A_f) - \nabla\vec{u}^{\,t}(A_f)}{2}\right]\right) d\sigma \leq$$

$$\leq C\,\sigma(U_j)\delta^2 t^2 + C \int_{\partial D_f} N_f^2(\lambda(\operatorname{div}\vec{u})I + \mu\{\nabla\vec{u} + \nabla\vec{u}^{\,t}\})^2\, d\sigma\,,$$

by (M,ε). The last quantity is also bounded by $C\,\sigma(U_j)\delta^2 t^2$, which is a contradiction for small δ.

Now, assume that U_j is of type II. Note that in this case there exists $Q_j \in \partial D_\phi$, with $\operatorname{dist}(Q_j, U_j) \approx \operatorname{diam} U_j$, and such that $|\nabla\vec{u}(X)| \leq t$ for all X in the nontangential region associated to Q_j. Because of this, it is easy to see, using the arguments we used to bound $|\nabla\vec{u}(A_f)|$ in case I, that for all X in a neighborhood of A_f and also on the top part of D_f, we have that $|\nabla u(X)| \leq t + C\delta t$. Since for $Q \in T_\psi(\bar{F}_j \cap F_j)$, $m(Q) \geq \beta t$, and $\beta > 1$, if δ is small enough, we see that we must have $N_f(Q) \geq m(Q)$. Hence, $N_f\left(\nabla\vec{u} - \left[\frac{\nabla\vec{u}(A_f) - \nabla\vec{u}^{\,t}(A_f)}{2}\right]\right)(Q) \geq (\beta - 1 - C\delta)t \geq$

$\frac{(\beta-1)}{2}t$ if δ is small and $Q \in T_\psi(\bar{F}_j \cap F_j)$. Thus, applying (M,ε) to D_f, we see that $\alpha(\beta-1)^2t^2\sigma(U_j) \le \int_{T_\psi(\bar{F}_j\cap F_j)} N_f\left(\nabla\vec{u} - \left[\frac{\nabla\vec{u}(A_f) - \nabla\vec{u}^t(A_f)^f}{2}\right]\right)^2 d\sigma \le \int_{\partial D_f} N_f\left(\nabla\vec{u} - \left[\frac{\nabla\vec{u}(A_f) - \nabla\vec{u}^t(A_f)}{2}\right]\right)^2 d\sigma \le C\sigma(U_j)\delta^2t^2$, a contradiction if δ is small. This finishes the proof of Proposition 3.1.13, and hence of Lemma 3.1.11.

(b) *The Stokes system of linear hydrostatics*

In this part I will sketch the proof of the L^2 results for the Stokes system of hydrostatics. These results are joint work of E. Fabes, C. Kenig and G. Verchota ([13]). We will keep using the notation introduced in part (a).

We seek a vector valued function $\vec{u} = (u^1,u^2,u^3)$ and a scalar valued function p satisfying

$$(3.2.1)\qquad \begin{cases} \Delta\vec{u} = \nabla p \ \text{ in } D \\ \operatorname{div} \vec{u} = 0 \ \text{ in } D \\ \vec{u}|_{\partial D} = \vec{f} \in L^2(\partial D, d\sigma) \ \text{ in the non-tangential sense.} \end{cases}$$

THEOREM 3.2.2 (also Theorem 3.1.9). *Given* $f \in L^2(\partial D, d\sigma)$, *there exists a unique solution* $(\vec{u},p)$ *to* (3.2.1), *with* p *tending to* 0 *at* ∞, *and* $N(\vec{u}) \in L^2(\partial D, d\sigma)$. *Moreover,* $\vec{u}(X) = \mathcal{K}\vec{g}(X)$, *with* $\vec{g} \in L^2(\partial D, d\sigma)$.

In order to sketch the proof of 3.2.2 we introduce the matrix $\Gamma(X)$ of fundamental solutions (see the book of Ladyzhenskaya [25]), $\Gamma(X) = (\Gamma_{ij}(X))$, where $\Gamma_{ij}(X) = \frac{1}{8\pi}\frac{\delta_{ij}}{|X|} + \frac{1}{8\pi}\frac{X_iX_j}{|X|^3}$, and its corresponding pressure vector $q(X) = q^i(X))$, where $q^i(X) = \frac{X_i}{4\pi|X|^3}$. Our solution of (3.2.2) will be given in the form of a double layer potential, $\vec{u}(X) = \mathcal{K}\vec{g}(X) = \int_{\partial D} \{H'(Q)\Gamma(X-Q)\}\vec{g}(Q)\,d\sigma(Q)$, where $(H'(Q)\Gamma(X-Q))_{i\ell} = \delta_{ij}q^\ell(X-Q)\,n_j(Q) +$

$\frac{\partial \Gamma_{i\ell}}{\partial Q_j}(X-Q)\,n_j(Q)$. We will also use the single layer potential $\vec{u}(X) = S\vec{g}(X) = \int_{\partial D} \Gamma(X-Q)\vec{g}(Q)\,d\sigma(Q)$.

In the same way as one establishes 3.1.4, one has:

LEMMA 3.2.3. *Let* $\mathcal{K}\vec{g}$, $S\vec{g}$ *be defined as above, with* $\vec{g} \in L^2(\partial D, d\sigma)$. *Then, they both solve* $\Delta\vec{u} = \nabla p$ *in* D, *and* D^-, $\operatorname{div}\vec{u} = 0$ *in* D *and* D^-. *Also*

(a) $\|N(\mathcal{K}\vec{g})\|_{L^2(\partial D, d\sigma)} \le C\|\vec{g}\|_{L^2(\partial D, d\sigma)}$,

(b) $(\mathcal{K}\vec{g})^{\pm}(P) = \pm\frac{1}{2}\vec{g}(P) - \text{p.v.}\int_{\partial D}\{H'(Q)\Gamma(P-Q)\}\vec{g}(Q)\,d\sigma(Q)$

(c) $\|N(\nabla S\vec{g})\|_{L^2(\partial D, d\sigma)} \le C\|\vec{g}\|_{L^2(\partial D, d\sigma)}$

(d) $$\frac{\partial}{\partial X_i}(S\vec{g})_j^{\pm}(P) = \pm\left\{\frac{n_i(P)\,g_j(P)}{2} - \frac{n_i(P)\,n_j(P)}{2}\langle N(P), \vec{g}(P)\rangle\right\}$$
$$+\ \text{p.v.}\int_{\partial D}\frac{\partial}{\partial P_i}\Gamma(P-Q)\vec{g}(Q)\,d\sigma(Q)$$

(e) $(HS\vec{g})^{\pm}(P) = \pm\frac{1}{2}\vec{g}(P) + \text{p.v.}\int_{\partial D}\{H(P)\Gamma(P-Q)\}\vec{g}(Q)\,d\sigma(Q)$,

where $(H(X)\Gamma(X-Q))_{i\ell} = n_j(x)\frac{\partial \Gamma_{i\ell}}{\partial X_j}(X-Q) - \delta_{ij}\,q^{\ell}(X-Q)\,n_j(X)$.

For the proof of this lemma in the case of smooth domains, see [25].

The proof of Theorem 3.2.2 (at least the existence part of it), reduces to the invertibility in $L^2(\partial D, d\sigma)$ of the operator $\frac{1}{2}I + K$, where $K\vec{g}(P) = -\text{p.v.}\int_{\partial D}\{H'(Q)\Gamma(P-Q)\}\vec{g}(Q)\,d\sigma(Q)$. As in previous cases, it is enough to show

(3.2.4) $$\left\|\left(\tfrac{1}{2}I - K^*\right)\vec{g}\right\|_{L^2(\partial D, d\sigma)} \approx \left\|\left(\tfrac{1}{2}I + K^*\right)\vec{g}\right\|_{L^2(\partial D, d\sigma)}.$$

This is shown by using the following two integral identities.

LEMMA 3.2.5. *Let* $\vec{h}$ *be a constant vector in* R^n, *and suppose that* $\Delta\vec{u} = \nabla p$, $\operatorname{div}\vec{u} = 0$ *in* D, *and that* $\vec{u}, p$ *and their derivatives are suitably small at* ∞. *Then,*

$$\int_{\partial D} h_\ell n_\ell \frac{\partial u^s}{\partial X_j} \cdot \frac{\partial u^s}{\partial X_j}\, d\sigma = 2\int_{\partial D} \frac{\partial u^s}{\partial N} \cdot h_\ell \frac{\partial u^s}{\partial X_\ell}\, d\sigma - 2\int_{\partial D} p\, n_s h_\ell \frac{\partial u^s}{\partial X_\ell}\, d\sigma .$$

LEMMA 3.2.6. *Let* $\vec{h}, p$ *and* $\vec{u}$ *be as in 3.2.5. Then,*

$$\int_{\partial D} h_\ell n_\ell p^2\, d\sigma = 2\int_{\partial D} h_r \frac{\partial u^r}{\partial N} p\, d\sigma - 2\int_{\partial D} h_r \frac{\partial u^r}{\partial X_i} \frac{\partial u^i}{\partial N}\, d\sigma +$$

$$+ 2\int_{\partial D} h_r n_s \frac{\partial u^s}{\partial X_j} \cdot \frac{\partial u^r}{\partial X_j}\, d\sigma .$$

The proofs of 3.2.5 and 3.2.6 are simple applications of the properties of $\vec{u}, p$, and the divergence theorem.

Choosing $\vec{h} = e_3$, we see that, from 3.2.6 we obtain

COROLLARY 3.2.7. *Let* $\vec{u}, p$ *be as in 3.2.5. Then,*

$$\int_{\partial D} p^2\, d\sigma \le C \int_{\partial D} |\nabla\vec{u}|^2\, d\sigma ,$$

where C *depends only on* M.

A consequence of Corollary 3.2.7 and Lemma 3.2.5, is that, if $\frac{\partial\vec{u}}{\partial\nu} = \frac{\partial\vec{u}}{\partial N} - p\cdot N$, then we have

COROLLARY 3.2.8. *Let* $\vec{u}, p$ *be as in 3.2.5. Then,*

$$\int_{\partial D} \left|\frac{\partial \vec{u}}{\partial \nu}\right|^2 d\sigma \approx \int_{\partial D} |\nabla_t \vec{u}|^2 d\sigma + \sum_j \int_{\partial D} \left| n_s \frac{\partial u^s}{\partial X_j}\right|^2 d\sigma ,$$

where the constants of equivalence depend only on M.

Proof. 3.2.5 clearly implies, by Schwartz's inequality, that $\int_{\partial D} |\nabla \vec{u}|^2 d\sigma \leq C \int_{\partial D} \left|\frac{\partial \vec{u}}{\partial \nu}\right|^2 d\sigma$. Moreover, arguing as in the second part of the Remark 2 after 3.1.5, we see that 3.2.5 shows that

$$\int_{\partial D} |\nabla \vec{u}|^2 d\sigma \leq C \int_{\partial D} |\nabla_t \vec{u}|^2 d\sigma + \left| \int_{\partial D} p\, n_s h_\ell \frac{\partial u^s}{\partial X_\ell} d\sigma \right| .$$

By Corollary 3.2.7, the right-hand side is bounded by

$$C \left(\int_{\partial D} |\nabla \vec{u}|^2 d\sigma \right)^{1/2} \left(\sum_j \int_{\partial D} \left| n_s \frac{\partial u^s}{\partial X_j}\right|^2 d\sigma \right)^{1/2} .$$

3.2.8 follows now, using 3.2.7 once more.

To prove 3.2.4, let $\vec{u} = S(\vec{g})$. By d) in 3.2.3, $\nabla_t \vec{u}$ and $n_s \frac{\partial u^s}{\partial X_j}$ are continuous across ∂D. Using this fact, 3.2.3 e) and Corollary 3.2.8, 3.2.4 follows.

In closing we would like to point out another boundary value problem for the Stokes system, which is of physical significance, the so-called slip boundary condition

$$(3.2.9)\qquad \begin{cases} \Delta\vec{u} = \nabla p \ \text{ in } D \\ \operatorname{div} \vec{u} = 0 \ \text{ in } D \\ \{(\nabla\vec{u} + \nabla\vec{u}^{\,t})N - p\cdot N)\}\big|_{\partial D} = \vec{f} \in L^2(\partial D, d\sigma) . \end{cases}$$

This problem is very similar to (3.1.2). Using the techniques introduced in part (a), together with the observation that if $\Delta\vec{u} = \nabla p$, $\operatorname{div}\vec{u} = 0$ in D, the same is true for each row v of the matrix $[\nabla\vec{u} + \nabla\vec{u}^{\,t} - p\cdot I]$, we have obtained

THEOREM 3.2.10. *Given* $f \in L^2(\partial D, d\sigma)$ *there exists a unique solution* $(\vec{u},p)$ *to (3.2.9), which tends to* 0 *at* ∞, *and with* $N(\nabla\vec{u}) \in L^2(\partial D, d\sigma)$. *Moreover,* $\vec{u}(X) = S(\vec{g})(X)$, *with* $\vec{g} \in L^2(\partial D, d\sigma)$.

DEPARTMENT OF MATHEMATICS
UNIVERSITY OF CHICAGO
CHICAGO, ILLINOIS 60637

REFERENCES

[1] A. P. Calderon, Cauchy integrals on Lipschitz curves and related operators, Proc. Nat. Acad. Sc. U.S.A. 74 (1977), 1324-1327.

[2] R. R. Coifman, A. McIntosh and Y. Meyer, L'integrale de Cauchy definit un operateur borne sur L^2 pour les courbes lipschitziennes, Annals of Math. 116 (1982), 361-387.

[3] R. R. Coifman and G. Weiss, Extensions of Hardy spaces and their use in analysis, Bull. AMS 83 (1977), 569-645.

[4] B.E.J. Dahlberg, On estimates of harmonic measure, Arch. Rational Mech. and Anal. 65 (1977), 272-288.

[5] ________, On the Poisson integral for Lipschitz and C^1 domains, Studia Math. 66 (1979), 13-24.

[6] B.E.J. Dahlberg and C. E. Kenig, Hardy spaces and the L^p Neumann problem for Laplace's equation in a Lipschitz domain, to appear, Annals of Math.

[7] ________, Area integral estimates for higher order boundary value problems on Lipschitz domains, to appear.

[8] D.E.J. Dahlberg, C.E. Kenig and G.C. Verchota, Boundary value problems for the systems of elastostatics on a Lipschitz domain, in preparation.

[9] G. David, Operateurs integraux singuliers sur certaines courbes du plan complex, Ann. Sci. del'Ecole Norm. Sup. 17 (1984), 157-189.

[10] ________, personal communication, 1983.

[11] E. Fabes, M. Jodeit, Jr., and N. Riviere, Potential techniques for boundary value problems on C^1 domains, Acta Math. 141, (1978), 165-186.

[12] E. Fabes and C. E. Kenig, On the Hardy space H^1 of a C^1 domain, Ark. Mat. 19 (1981), 1-22.

[13] E. Fabes, C. E. Kenig and G. C. Verchota, The Stokes system on a Lipschitz domain, in preparation.

[14] C. Fefferman and E. Stein, H^p spaces of several variables, Acta Math. 129 (1972), 137-193.

[15] A. Gutierrez, Boundary value problems for linear elastostatics on C^1 domains, University of Minnesota preprint, 1980.

[16] D. S. Jerison and C. E. Kenig, An identity with applications to harmonic measure, Bull. AMS Vol. 2 (1980), 447-451.

[17] ________, The Dirichlet problem in non-smooth domains, Annals of Math. 113 (1981), 367-382.

[18] ________, The Neumann problem on Lipschitz domains, Bull. AMS Vol. 4 (1981), 203-207

[19] ________, Boundary value problems on Lipschitz domains, MAA Studies in Mathematics, Vol. 23, Studies in Partial Differential Equations, W. Littmann, editor (1982), 1-68.

[20] C. E. Kenig, Weighted H^p spaces on Lipschitz domains, Amer. J. of Math. 102 (1980), 129-163.

[21] ________, Weighted Hardy spaces on Lipschitz domains, Proceedings of Symposia in Pure Mathematics, Vol. 35, Part 1, (1979), 263-274.

[22] ________, Boundary value problems of linear elastostatics and hydrostatics on Lipschitz domains, Seminaire Goulaouic-Meyer-Schwartz, 1983-84, Expose no. XXI, Ecole Polytechnique, Palaiseau, France.

[23] ________, Recent progress on boundary value problems on Lipschitz domains, to appear, Proceedings of Symposia in Pure Mathematics, Proceedings of the Notre Dame Conference on Pseudodifferential Operators, Volume 43 (1985), 175-205.

[24] V. D. Kupradze, Three dimensional problems of the mathematical theory of elastocity and thermoelasticity, North Holland, New York, 1979.

[25] O. A. Ladyzhenskaya, The mathematical theory of viscous incompressible flow, Gordon and Breach, New York, 1963.

[26] J. Moser, On Harnack's theorem for elliptic differential equations, Comm. Pure Appl. Math. Vol. XIV (1961), 577-591.

[27] J. Nečas, Les methodes directes en theorie des equations elliptiques, Academia, Prague, 1967.

[28] L. Payne and H. Weinberger, New bounds in harmonic and biharmonic problems, J. Math. Phys. 33 (1954), 291-307.

[29] ________, New bounds for solutions of second order elliptic partial differential equations, Pacific J. of Math. 8 (1958), 551-573.

[30] F. Rellich, Darstellung der Eigenwerte von $\Delta u + \lambda u$ durch ein Randintegral, Math Z. 46 (1940), 635-646.

[31] J. Serrin and H. Weinberger, Isolated singularities of solutions of linear elliptic equations, Amer. J. of Math. 88 (1966), 258-272.

[32] E. Stein and G. Weiss, On the theory of harmonic functions of several variables, I, Acta Math. 103 (1960), 25-62.

[33] G. C. Verchota, Layer potentials and boundary value problems for Laplace's equation in Lipschitz domains, Thesis, University of Minnesota, (1982), also, J. of Functional Analysis, 59 (1984), 572-611.

INTEGRAL FORMULAS IN COMPLEX ANALYSIS

Steven G. Krantz*

§1. *Three basic methods for obtaining integral formulas*

We begin by discussing three ways to think about integral formulas on domains in $\mathbb{C}^1$, with a view to finding techniques which might generalize to $\mathbb{C}^n$. These are

(I) Exploit symmetry of the domain;

(II) Use differential forms and Stokes's theorem;

(III) Use functional analysis.

Discussion of I. Let $\Delta = \{z \in \mathbb{C} : |z| < 1\}$. For $n \in \mathbb{Z}$, define $\phi_n(re^{i\theta}) = r^{|n|}e^{in\theta}$. Then direct calculation shows that

$$\phi_n(0) = \frac{1}{2\pi} \int_0^{2\pi} \phi_n(e^{i\theta})\, d\theta, \quad \text{all } n. \tag{1.1}$$

Now if f is harmonic on a neighborhood of $\bar{\Delta}$, then f has an L^2 convergent Fourier expansion

$$f(re^{i\theta}) = \sum_{n=-\infty}^{\infty} a_n \phi_n(re^{i\theta}). \tag{1.2}$$

By linearity, (1.1) and (1.2) yield

*Work supported in part by the National Science Foundation. The splendid lecture notes prepared by Li Hui Ping, Li Xin Min and Ye Ke Ying greatly simplified the task of writing this paper.

$$f(0) = \frac{1}{2\pi} \int_0^{2\pi} f(e^{i\theta})\, d\theta . \tag{1.3}$$

Of course formula (1.3) holds in particular for holomorphic f; then we rewrite (1.3) as

$$f(0) = \frac{1}{2\pi i} \int_0^{2\pi} \frac{f(e^{i\theta})}{e^{i\theta} - 0} (ie^{i\theta} d\theta) = \frac{1}{2\pi i} \oint_\gamma \frac{f(\zeta)}{\zeta - 0} d\zeta \tag{1.4}$$

where $\gamma(\theta) = e^{i\theta}$, $0 \leq \theta < 2\pi$.

Now (1.4) is the Cauchy integral formula on the disc for the point $z = 0$. In order to obtain the general Cauchy formula, we exploit more symmetry. Recall that if $z \in \Delta$ is fixed then the function

$$\phi(\zeta) = \phi_z(\zeta) = \frac{\zeta + z}{1 + \bar{z}\zeta},$$

called a *Möbius transformation*, has the following properties:

(a) $\phi : \Delta \to \Delta$ is *biholomorphic* (i.e. holomorphic, one-to-one, and onto, with a holomorphic inverse).

(b) $\phi(0) = z$

(c) ϕ, ϕ^{-1} are smooth on a neighborhood of $\bar{\Delta}$.

If f is holomorphic on a neighborhood of $\bar{\Delta}$, define $g(\xi) = f \circ \phi(\xi)$. Then (1.4) applied to g yields

$$f(z) = g(0) = \frac{1}{2\pi i} \oint_\gamma \frac{g(\xi)}{\xi} d\xi = \frac{1}{2\pi i} \int_\gamma \frac{f(\phi(\xi))}{\xi} d\xi .$$

Change variables by $\xi = \phi^{-1}(\zeta) = (\zeta - z)/(1 - \bar{z}\zeta)$. Then

$$f(z) = \frac{1}{2\pi i} \oint_\gamma f(\zeta) \left[\frac{(\phi^{-1})'(\zeta)}{\phi^{-1}(\zeta)} \right] d\zeta$$

and logarithmic differentiation of ϕ^{-1} gives

$$f(z) = \frac{1}{2\pi i} \oint_\gamma f(\zeta)\left[\frac{1}{\zeta - z} + \frac{\bar{z}}{1-\bar{z}\zeta}\right] d\zeta \tag{1.5}$$

$$= \frac{1}{2\pi i} \oint_\gamma \frac{f(\zeta)}{\zeta - z}\, d\zeta + \frac{1}{2\pi i} \oint_\gamma \frac{(\bar{z}\zeta f(\zeta)/(1-\bar{z}\zeta))}{\zeta - 0}\, d\zeta \,' \tag{1.6}$$

Now the numerator of the second integrand in (1.6) is a holomorphic function of ζ which vanishes at 0. By (1.4), the second integral vanishes. Formula (1.6) now becomes

$$f(z) = \frac{1}{2\pi i} \oint_\gamma \frac{f(\zeta)}{\zeta - z}\, d\zeta$$

which is the Cauchy Integral Formula for the disc.

REMARK 1. If f is only assumed to be harmonic, then we cannot argue that the second integral in (1.6) vanishes. Instead, a little algebra applied to (1.5) gives

$$f(z) = \frac{1}{2\pi} \int_0^{2\pi} f(e^{i\theta}) \frac{1-|z|^2}{|e^{i\theta}-z|^2}\, d\theta \,.$$

This is the Poisson Integral Formula.

REMARK 2. Among bounded domains in $\mathbb{C}$, only the disc (and domains biholomorphic to it) has a transitive group of biholomorphic self maps. (This follows from the Uniformization Theorem; see also [32].) Thus approach I has serious limitations in $\mathbb{C}^1$. In $\mathbb{C}^n$ the limitations are even more severe. Indeed in Section 7, after we develop a lot of machinery, we shall return to the concept of symmetry in $\mathbb{C}^n$ and gain some new insights.

Discussion of II. We need some notation. Recall that in real differential analysis on $\mathbb{R}^2$ we use the basis $\frac{\partial}{\partial x}$, $\frac{\partial}{\partial y}$ for the tangent space (i.e. all linear first order differential operators are linear combinations of these) and dx, dy for the cotangent space. We have the pairings

$$< \frac{\partial}{\partial x}, dx > = < \frac{\partial}{\partial y}, dy > = 1 ,$$

$$< \frac{\partial}{\partial x}, dy > = < \frac{\partial}{\partial y}, dx > = 0 .$$

In complex analysis, it is convenient to define differential operators

$$\frac{\partial}{\partial z} = \frac{1}{2}\left(\frac{\partial}{\partial x} - i\frac{\partial}{\partial y}\right), \quad \frac{\partial}{\partial \bar{z}} = \frac{1}{2}\left(\frac{\partial}{\partial x} + i\frac{\partial}{\partial y}\right).$$

The motivation for this notation is twofold. First,

$$\frac{\partial}{\partial z} z = \frac{\partial}{\partial \bar{z}} \bar{z} = 1, \quad \frac{\partial}{\partial z} \bar{z} = \frac{\partial}{\partial \bar{z}} z = 0 .$$

Secondly, if $f(z) = u(z) + iv(z)$ is a C^1 function, with u and v real valued, then

$$\frac{\partial f}{\partial \bar{z}} \equiv 0 \Longleftrightarrow \left(\frac{\partial u}{\partial x} \equiv \frac{\partial v}{\partial y} \text{ and } \frac{\partial u}{\partial y} \equiv -\frac{\partial v}{\partial x}\right), \tag{1.7}$$

which is the Cauchy-Riemann equations. Thus $\frac{\partial f}{\partial \bar{z}} = 0$ means that f is holomorphic.

We also define

$$dz = dx + idy , \quad d\bar{z} = dx - idy .$$

It is immediate that

$$< \frac{\partial}{\partial z}, dz > = < \frac{\partial}{\partial \bar{z}}, d\bar{z} > = 1 ,$$

$$< \frac{\partial}{\partial z}, d\bar{z} > = < \frac{\partial}{\partial \bar{z}}, dz > = 0 .$$

An arbitrary 1-form is written

$$u(z) = a(z)\,dz + b(z)\,d\bar{z} \tag{1.8}$$

and we define the exterior differentials

$$\partial u = \frac{\partial b}{\partial z}\,dz \wedge d\bar{z}\,, \quad \bar{\partial} u = \frac{\partial a}{\partial \bar{z}}\,d\bar{z} \wedge dz\,. \tag{1.9}$$

Clearly $du = \partial u + \bar{\partial} u$. Recall

STOKES' THEOREM. *If* $\Omega \subset\subset \mathbf{R}^n$ *is a bounded domain with smooth boundary and* u *is a smooth form on* $\bar{\Omega}$ *then*

$$\int_{\partial\Omega} u = \int_{\Omega} du\,.$$

In our new notation, if $\Omega \subseteq \mathbf{C}^1 \approx \mathbf{R}^2$ and u is a 1-form as in (1.8), (1.9), then Stokes' Theorem becomes

$$\begin{aligned}\int_{\partial\Omega} u &= \int_{\Omega} \partial u + \bar{\partial} u \\ &= \int_{\Omega} \left(\frac{\partial b}{\partial z} - \frac{\partial a}{\partial \bar{z}}\right) dz \wedge d\bar{z}\,.\end{aligned} \tag{1.10}$$

Now we can prove

THEOREM. *If* $\Omega \subseteq \mathbf{C}$ *is smoothly bounded and* f *is holomorphic on a neighborhood of* $\bar{\Omega}$ *then*

$$f(z) = \frac{1}{2\pi i} \int_{\partial\Omega} \frac{f(\zeta)}{\zeta - z}\,d\zeta\,, \quad \textit{all } z \in \Omega\,.$$

Proof. Fix $z \in \Omega$. Let $\varepsilon <$ distance $(z, \partial\Omega)$. Define $D(z,\varepsilon) = \{\zeta \in \mathbb{C} : |\zeta - z| < \varepsilon\}$ and $\tilde{\Omega} = \Omega \setminus \overline{D(z,\varepsilon)}$. We apply Stokes' Theorem to the 1-form $u(\zeta) = \frac{f(\zeta)}{\zeta - z} d\zeta$ on the domain $\tilde{\Omega}$ (note that u has smooth coefficients on a neighborhood of the closure of $\tilde{\Omega}$, but *not* on all of Ω). Thus, by (1.10),

$$\int_{\partial\tilde{\Omega}} u(\zeta) d\zeta = \int_{\tilde{\Omega}} du = -\int_{\tilde{\Omega}} \frac{\partial}{\partial\bar{\zeta}}\left(\frac{f(\zeta)}{\zeta - z}\right) d\zeta \wedge d\bar{\zeta} . \qquad (1.11)$$

This last is 0 by (1.7). Thus, since $\partial\tilde{\Omega} = \partial\Omega \cup \partial D(z,\varepsilon)$ (with suitable orientations), we have

$$\int_{\partial\Omega} u(\zeta) = \int_{\partial D(z,\varepsilon)} u(\zeta) .$$

Parametrizing $\partial D(z,\varepsilon)$ by $\zeta = z + \varepsilon e^{i\theta}$, $0 \leq \theta < 2\pi$, we obtain

$$\int_{\partial\Omega} \frac{f(\zeta)}{\zeta - z} d\zeta = \int_0^{2\pi} f(z + \varepsilon e^{i\theta}) i d\theta \to 2\pi i f(z)$$

as $\varepsilon \to 0^+$. This completes the proof. □

REMARK 1. In the proof of the theorem, if we assume that f is smooth but not necessarily holomorphic, then the right side of (1.11) does not vanish; instead it equals $\int_{\tilde{\Omega}} \frac{(\partial f/\partial\bar{\zeta})}{\zeta - z} d\bar{\zeta} \wedge d\zeta$. Thus the proof of the theorem yields

$$f(z) = \frac{1}{2\pi i} \int_{\partial\Omega} \frac{f(\zeta)}{\zeta - z} d\zeta - \frac{1}{2\pi i} \int_{\Omega} \frac{(\partial f/\partial\bar{\zeta})}{\zeta - z} d\bar{\zeta} \wedge d\zeta . \qquad (1.12)$$

This formula, valid for all $f \in C^1(\bar{\Omega})$, will be valuable later on.

REMARK 2. As it stands, the method of Stokes' theorem will not generalize to $\mathbb{C}^n$. Namely, we used the fact that $\frac{f(\cdot)}{\cdot - z}$ is holomorphic with an isolated singular point at z. In $\mathbb{C}^n$, $n \geq 2$, holomorphic functions *never* have isolated singularities. A more sophisticated approach will therefore be needed.

Discussion of III. If $\Omega \subseteq \mathbb{C}$ is a smoothly bounded domain and if $\varepsilon > 0$, define $\Omega_\varepsilon = \{z \in \Omega : \mathrm{dist}(z, \partial\Omega) > \varepsilon\}$. If ε is sufficiently small, say $0 < \varepsilon < \varepsilon_0$, then Ω_ε will also be smoothly bounded. Define

$$H^2(\Omega) = \left\{ f \text{ holomorphic on } \Omega : \sup_{0<\varepsilon<\varepsilon_0} \int_{\partial\Omega_\varepsilon} |f(\zeta)|^2 ds(\zeta) < \infty \right\}.$$

(Here ds is the element of arc length.) That this definition is unambiguous is a technical matter (see [31, Ch. 8]). We shall need

PROPOSITION ([31]). *Each* $f \in H^2(\Omega)$ *has associated to it a unique* $\tilde{f} \in L^2(\partial\Omega)$. *The Poisson integral of* $\tilde{f}$ *is* f. *Also*

$$\sup_{0<\varepsilon<\varepsilon_0} \int_{\partial\Omega_\varepsilon} |f(\zeta)|^2 d\sigma(\zeta)^{1/2} \approx \|\tilde{f}\|_{L^2(\partial\Omega)} .$$

If $\pi_\varepsilon : \partial\Omega_\varepsilon \to \partial\Omega$ *is normal projection then*

$$(f|_{\partial\Omega_\varepsilon}) \circ (\pi_\varepsilon)^{-1} \longrightarrow \tilde{f} \text{ in } L^2(\partial\Omega) .$$

Thus we may make H^2 *into an inner product space by defining*

$$\langle f,g \rangle = \int_{\partial\Omega} \tilde{f}\,\overline{\tilde{g}}\,d\sigma \text{ for } f,g \in H^2(\Omega) . \tag{1.13}$$

BASIC LEMMA. *If* $K \subset \Omega$ *is compact then there is a constant* $C = C(K)$ *such that*

$$\sup_{z \in K} |f(z)| \leq C\|f\|_{H^2}, \text{ all } f \in H^2(\Omega). \tag{1.14}$$

Proof. Fix $z \in K$. Let $\varepsilon_1 < \frac{1}{10}$ (distance $(K, \partial\Omega)$). Then

$$|f(z)| = \left| \frac{1}{2\pi i} \oint_{\partial D(z,\varepsilon_1)} \frac{f(\zeta)}{\zeta - z} d\zeta \right|$$

(Stokes)

$$= \left| \frac{1}{2\pi i} \oint_{\partial\Omega_{\varepsilon_1}} \frac{f(\zeta)}{\zeta - z} d\zeta \right|$$

$$\leq \frac{1}{2\pi\varepsilon_1} \int_{\partial\Omega_{\varepsilon_1}} |f| \, ds$$

(Schwartz)

$$\leq C(\varepsilon_1) \cdot \int_{\partial\Omega_{\varepsilon_1}} |f(\zeta)|^2 ds(\zeta)^{1/2}$$

$$\leq C(K) \|f\|_{H^2} . \quad \square$$

Now (1.13) and (1.14) imply that $H^2(\Omega)$ is a Hilbert space. Fix $z \in \Omega$ and define the functional

$$\phi_z : H^2(\Omega) \to \mathbb{C}$$

$$f \mapsto f(z) .$$

Then the lemma, with $K = \{z\}$, shows that ϕ_z is continuous. The Riesz Representation Theorem yields a unique $k_z \in H^2(\Omega)$ such that

$$\phi_z(f) = \langle f, k_z \rangle, \quad \text{all } f \in H^2 .$$

In other words,

$$f(z) = \int_{\partial\Omega} \tilde{f}(\zeta)\overline{\tilde{k}_z(\zeta)}\,ds(\zeta), \quad \text{all } f \in H^2(\Omega), \quad z \in \Omega . \tag{1.15}$$

Formula (1.12) is called the *Szëgo formula* and $\overline{k_z(\zeta)} \equiv S(z,\zeta)$ the *Szëgo kernel* (see [31, Ch. 1] for further details).

REMARK 1. Formula (1.15) has the advantage of working on any smoothly bounded domain (even in $\mathbb{C}^n$), and the disadvantage of being non-explicit. As an exercise, check that when $\Omega = \Delta \subseteq \mathbb{C}$ and $z = 0$ then $S(z,\zeta) \equiv \frac{1}{2\pi}$. Then use Möbius transformations to calculate $S(z,\zeta)$ for any $z \in \Delta$. You will rediscover the Cauchy formula!

§2. *The Cauchy-Fantappié Formula*

Now we begin to consider integral formulas in $\mathbb{C}^n$. For purposes of differential analysis we introduce the notation

$$\frac{\partial}{\partial z_j} = \frac{1}{2}\left(\frac{\partial}{\partial x_j} - i\,\frac{\partial}{\partial y_j}\right), \quad \frac{\partial}{\partial \bar{z}_j} = \frac{1}{2}\left(\frac{\partial}{\partial x_j} + i\,\frac{\partial}{\partial y_j}\right),$$

$$dz_j = dx_j + i\,dy_j\,, \quad d\bar{z}_j = dx_j - i\,dy_j\,, \quad j = 1,\cdots,n .$$

It is easily checked that

$$\frac{\partial}{\partial z_j} z_j = \frac{\partial}{\partial \bar{z}_j}\bar{z}_j = 1\,, \quad \langle \frac{\partial}{\partial z_j}, dz_j \rangle = \langle \frac{\partial}{\partial \bar{z}_j}, d\bar{z}_j \rangle = 1\,, \quad j = 1,\cdots,n\,,$$

and all other pairings are 0.

If $\alpha = (\alpha_1, \cdots, \alpha_k)$, $\beta = (\beta_1, \cdots, \beta_\ell)$ are tuples of non-negative integers (*multi-indices*) then we write

$$dz^\alpha = dz_{\alpha_1} \wedge \cdots \wedge dz_{\alpha_k}, \quad d\bar{z}^\beta = d\bar{z}_{\beta_1} \wedge \cdots \wedge d\bar{z}_{\beta_\ell}.$$

A differential form is written

$$u = \sum_{\alpha,\beta} a_{\alpha\beta}\, dz^\alpha \wedge d\bar{z}^\beta \tag{2.1}$$

with smooth coefficients $a_{\alpha\beta}$. (If $0 \le p,q \in \mathbf{Z}$ and the sum in (2.1) ranges over $|\alpha| = p$, $|\beta| = q$ only, then u is called a form of *type* (p,q).) We then define

$$\partial u = \sum_{\alpha,\beta,j} \frac{\partial a_{\alpha\beta}}{\partial z_j} dz_j \wedge dz^\alpha \wedge d\bar{z}^\beta, \quad \bar{\partial} u = \sum_{\alpha,\beta,j} \frac{\partial a_{\alpha\beta}}{\partial \bar{z}_j} d\bar{z}_j \wedge dz^\alpha \wedge d\bar{z}^\beta.$$

By a calculation (or functoriality), $du = \partial u + \bar{\partial} u$. A C^1 function f is called *holomorphic* if $\bar{\partial} u \equiv 0$. (Note: this means that $\frac{\partial f}{\partial \bar{z}_j} \equiv 0$, $j = 1, \cdots, n$, so f is holomorphic in the one variable sense in each variable separately.)

Finally, we introduce two special forms: if $w = (w_1, \cdots, w_n)$ is an n-tuple of smooth functions then we define the *Leray form* to be

$$\eta(w) = \sum_{j=1} (-1)^{j+1} w_j \wedge dw_1 \wedge \cdots \wedge dw_{j-1} \wedge dw_{j+1} \wedge \cdots \wedge dw_n.$$

Likewise

$$\omega(w) = dw_1 \wedge \cdots \wedge dw_n.$$

We define a constant

$$W(n) = \int_{B(0,1)} \omega(\overline{\zeta}) \wedge \omega(\zeta) .$$

Here $B(z,r) = \{\zeta \in \mathbb{C}^n : |\zeta - z| < r\}$.

Now we may formulate a generalization of approach II in Section 1.

THEOREM (The Cauchy-Fantappié formula). *Let* $\Omega \subseteq \mathbb{C}^n$ *be a smoothly bounded domain. Assume that* $w = (w_1, \cdots, w_n) \in C^\infty(\overline{\Omega} \times \overline{\Omega} \setminus \Delta)$, $w_j = w_j(z, \zeta)$, and

$$\sum_{j=1}^{n} w_j(z,\zeta) \cdot (\zeta_j - z_j) = 1 \quad \text{on } \overline{\Omega} \times \overline{\Omega} \setminus \Delta . \tag{2.2}$$

If $f \in C^1(\overline{\Omega})$ *is holomorphic on* Ω, *then for any* $z \in \Omega$ *we have*

$$f(z) = \frac{1}{nW(n)} \int_{\partial\Omega} f(\zeta)\, \eta(w) \wedge \omega(\zeta) . \tag{2.3}$$

Before proving this result, we make some detailed remarks.

REMARK 1. In case $n = 1$, then $w = w_1 = \dfrac{1}{\zeta - z}$ (of necessity). So

$$\eta(w) = \frac{1}{\zeta - z}, \quad \omega(\zeta) = d\zeta, \quad \frac{1}{nW(n)} = \frac{1}{2\pi i} .$$

The Cauchy-Fantappié formula becomes

$$f(z) = \frac{1}{2\pi i} \int_{\partial\Omega} \frac{f(\zeta)}{\zeta - z}\, d\zeta ,$$

which is just Cauchy's formula.

REMARK 2. As soon as $n \geq 2$, the condition (2.2) no longer uniquely determines w. However an interesting example is given by

$$w(z,\zeta) = (w_1(z,\zeta),\cdots,w_n(z,\zeta)) = \left(\frac{\bar{\zeta}_1-\bar{z}_1}{|\zeta-z|^2},\ldots,\frac{\bar{\zeta}_n-\bar{z}_n}{|\zeta-z|^2}\right).$$

Let us calculate what the theorem says for this w in case $n = 2$. Now

$$\eta(w) \wedge \omega(\zeta) = (w_1 dw_2 - w_2 dw_1) \wedge d\zeta_1 \wedge d\zeta_2$$

$$= \left(w_1 \frac{\partial w_2}{\partial \bar{\zeta}_1} d\bar{\zeta}_1 + w_1 \frac{\partial w_2}{\partial \bar{\zeta}_2} d\bar{\zeta}_2\right.$$

$$\left. - w_2 \frac{\partial w_1}{\partial \bar{\zeta}_1} d\bar{\zeta}_1 - w_2 \frac{\partial w_1}{\partial \bar{\zeta}_2} d\bar{\zeta}_2\right) \wedge d\zeta_1 \wedge d\zeta_2$$

which by direct calculation

$$= \left(-\frac{\bar{\zeta}_2-\bar{z}_2}{|\zeta-z|^4} d\bar{\zeta}_1 + \frac{\bar{\zeta}_1-\bar{z}_1}{|\zeta-z|^4} d\bar{\zeta}_2\right) \wedge d\zeta_1 \wedge d\zeta_2 \tag{2.4}$$

Thus, by the theorem, we get a form of the *Bochner-Martinelli formula*:

$$f(z) = \frac{1}{zW(2)} \int_{\partial\Omega} f(\zeta) \frac{\eta(\bar{\zeta}_1-\bar{z}_1, \bar{\zeta}_2-\bar{z}_2)}{|\zeta-z|^4} \wedge d\zeta_1 \wedge d\zeta_2$$

for $f \in C^1(\bar{\Omega})$, holomorphic on Ω.

Now we turn to the proof of the Cauchy-Fantappié formula. For simplicity, we restrict attention to $n = 2$. Let

$$\mathcal{F} = \left\{ a = (a_1, a_2) \in C^\infty(\bar{\Omega} \times \bar{\Omega} \setminus \Delta) : \sum_{j=1}^{2} a_j(z,\zeta)(\zeta_j - z_j) \equiv 1 \right\}.$$

If $\alpha^1, \alpha^2 \in \mathfrak{F}$ then we define

$$B(\alpha^1, \alpha^2) = \det(\alpha^1, \bar{\partial}\alpha^2) \equiv \sum_{\sigma \in S_2} \varepsilon(\sigma) a^1_{\sigma(1)} \wedge \bar{\partial}_\zeta a^2_{\sigma(2)} .$$

We claim that B has three key properties:

(1) $B(\alpha^1, \alpha^2)$ does not depend on α^1 ;

(2) $\bar{\partial}_\zeta B(\alpha^1, \alpha^2) \equiv 0$;

(3) If $\alpha^1, \alpha^2, \beta^1, \beta^2 \in \mathfrak{F}$ then $B(\alpha^1, \alpha^2) - B(\beta^1, \beta^2)$ is $\bar{\partial}_\zeta$ exact.

Assuming (1), (2) and (3) for the moment, let us complete the proof (note that (1) is used only to prove (3)). Letting

$$\alpha^1 = \alpha^2 = \begin{pmatrix} w_1 \\ w_2 \end{pmatrix} \in \mathfrak{F}$$

$$\beta^1 = \beta^2 = \begin{pmatrix} (\bar{\zeta}_1 - \bar{z}_1)/|\zeta - z|^2 \\ \\ (\bar{\zeta}_2 - \bar{z}_2)/|\zeta - z|^2 \end{pmatrix} \in \mathfrak{F}$$

we have (observe that $B(\alpha^1, \alpha^2) = \eta(w)$)

$$\int_{\partial\Omega} f(\zeta)\, \eta(w) \wedge \omega(\zeta) = \int_{\partial\Omega} f(\zeta)\, B(\alpha^1, \alpha^2) \wedge \omega(\zeta) . \tag{2.5}$$

Note that, by (2),

$$\begin{aligned} &d_\zeta(f(\zeta)\, B(\alpha^1, \alpha^2) \wedge \omega(\zeta)) \\ &= \bar{\partial}_\zeta f(\zeta) \wedge B(\alpha^1, \alpha^2) \wedge \omega(\zeta) + f(\zeta)\, \bar{\partial}_\zeta B(\alpha^1, \alpha^2) \wedge \omega(\zeta) \\ &= 0 + 0 = 0 . \end{aligned}$$

Hence, letting $0 < \varepsilon < \operatorname{dist}(z, \partial\Omega)$, we have by Stokes' Theorem that (2.5)

$$= \int_{\partial B(z,\varepsilon)} f(\zeta) B(\alpha^1, \alpha^2) \wedge \omega(\zeta)$$

$$= \int_{\partial B(z,\varepsilon)} f(\zeta) \{B(\alpha^1, \alpha^2) - B(\beta^1, \beta^2)\} \wedge \omega(\zeta) \tag{2.6}$$

$$+ \int_{\partial B(z,\varepsilon)} B(\beta^1, \beta^2) \wedge \omega(\zeta) .$$

But the first integral

$$= \int_{\partial B(z,\varepsilon)} f(\zeta)\, dA \wedge \omega(\zeta)$$

(for some A, by (3))

$$\begin{aligned} &= \int_{\partial B(z,\varepsilon)} d(f(\zeta) \wedge A \wedge \omega(\zeta)) \\ &= 0 , \end{aligned} \tag{2.7}$$

by Stokes' Theorem, since $\partial\partial B(z,\varepsilon) = 0$. Thus (2.5) - (2.7) give

$$\int_{\partial\Omega} f(\zeta)\, \eta(w) \wedge \omega(\zeta) = \int_{\partial B(z,\varepsilon)} f(\zeta) B(\beta^1, \beta^2) \wedge \omega(\zeta)$$

which, by (2.4),

$$= \int_{\partial B(z,\varepsilon)} f(\zeta) \frac{\eta(\bar{\zeta}_1 - \bar{z}_1, \bar{\zeta}_2 - \bar{z}_2)}{|\zeta - z|^4} \wedge \omega(\zeta)$$

$$= \frac{1}{\varepsilon^4} \int_{\partial B(z,\varepsilon)} f(\zeta)\, \eta(\bar{\zeta} - \bar{z}) \wedge \omega(\zeta) \,.$$

$$= \frac{1}{\varepsilon^4} \int_{\partial B(z,\varepsilon)} f(z)\, \eta(\bar{\zeta} - \bar{z}) \wedge \omega(\zeta) + 0(\varepsilon)$$

(Stokes)

$$= \frac{1}{\varepsilon^4} f(z) \int_{B(z,\varepsilon)} 2\omega(\bar{\zeta} - \bar{z}) \wedge \omega(\zeta) + 0(\varepsilon)$$

$$= \frac{1}{\varepsilon^4} f(z)\, \varepsilon^4 \int_{B(0,1)} 2\omega(\bar{\zeta}) \wedge \omega(\zeta) + 0(\varepsilon)$$

$$= f(z) \cdot 2W(2) + 0(\varepsilon) \,.$$

Letting $\varepsilon \to 0$ yields the desired result. □

We conclude this section by proving (1) - (3). For (1), we have

$$B(\alpha^1, \alpha^2) = \det \begin{pmatrix} a_1^1 & \bar{\partial}_\zeta a_1^2 \\ a_2^1 & \bar{\partial}_\zeta a_2^2 \end{pmatrix}$$

$$= \frac{1}{(\zeta_1 - z_1)(\zeta_2 - z_2)} \det \begin{pmatrix} (\zeta_1 - z_1) a_1^1 & \bar{\partial}_\zeta ((\zeta_1 - z_1) a_1^2) \\ (\zeta_2 - z_2) a_2^1 & \bar{\partial}_\zeta ((\zeta_2 - z_2) a_2^2) \end{pmatrix}$$

which, by adding row 2 to row 1,

$$= \frac{1}{(\zeta_1 - z_1)(\zeta_2 - z_2)} \det \begin{pmatrix} 1 & \bar\partial_\zeta 1 \\ (\zeta_2 - z_2) a_2^1 & \bar\partial_\zeta((\zeta_2 - z_2) a_2^2) \end{pmatrix}$$

$$= \frac{1}{(\zeta_1 - z_1)} \bar\partial_\zeta a_2^2 .$$

This calculation is correct for $\zeta_1 \neq z_1$, $\zeta_2 \neq z_2$. The full result follows by continuity.

For (2), imitate the proof of (1).

For (3), use (1) to write

$$B(\alpha^1, \alpha^2) - B(\beta^1, \beta^2) = B(\alpha^1, \alpha^2 - \beta^2)$$

$$= \det \begin{pmatrix} a_1^1 & \bar\partial(a_1^2 - b_1^2) \\ a_2^1 & \bar\partial(a_2^2 - b_2^2) \end{pmatrix} .$$

Now an easy calculation, as in (1), shows that this last equals $\bar\partial A$ where

$$A = \det \begin{pmatrix} a_1^1 & a_1^2 - b_1^2 \\ a_2^1 & a_2^2 - b_2^2 \end{pmatrix} .$$

This completes our discussion of the Cauchy-Fantappié formula.

§3. *Introduction to the $\bar\partial$ problem*

One of the principal problems in complex function theory is the construction of holomorphic functions with specified properties. In one dimension, there are a number of highly developed techniques: Runge and Mergelyan theorems, power series, infinite products, integral formulas, and so on. In several variables, these techniques are either unavailable, much less useful, or much less accessible.

The two most prevalent techniques for constructing functions in several complex variables are sheaf theory and the inhomogeneous Cauchy-Riemann equations. The latter interact strongly with the subject of integral formulas, and in any case are a more flexible technique than the former. To these we now turn our attention.

The setup for our study is that, for a given form $f = \sum_{j=1}^{n} f_j(z)\, d\bar{z}_j$, we seek a function u such that $\bar{\partial} u = f$. Notice that since $0 = d^2 = \partial^2 + (\partial\bar{\partial} + \bar{\partial}\partial) + \bar{\partial}^2$, linear independence considerations yield that $\bar{\partial}^2 = 0$. Hence $\bar{\partial} u = f$ *necessitates* $\bar{\partial} f = 0$.

A simple calculation shows that, for $n \geq 2$, this compatibility condition is equivalent to $\dfrac{\partial f_j}{\partial \bar{z}_k} = \dfrac{\partial f_k}{\partial \bar{z}_j}$, all j, k. Notice, however, that when $n = 1$ the condition $\bar{\partial} f = 0$ is always vacuously satisfied. This difference can be explained in part by the fact that the equation $\bar{\partial} u = f$ is really n equations $\left(\dfrac{\partial u}{\partial \bar{z}_j} = f_j\right)$ in one unknown (namely u). For $n > 1$ the system is then "over-determined" and a compatibility condition is necessary. For $n = 1$ the system is *not* over-determined.

The three basic considerations about a PDE are existence, uniqueness and regularity. It is easy to check that $\bar{\partial}$ is elliptic on functions in the interior of a given domain; hence, if u exists, it will be smooth whenever f is (we will see this in a more elementary fashion later). So interior regularity is not a problem. Also, since the kernel of $\bar{\partial}$ consists of all holomorphic functions, uniqueness is out of the question. So, for us, existence is the main issue.

The following example shows that the compatibility condition $\bar{\partial} f = 0$ does not by itself guarantee existence of u.

EXAMPLE. Let $\Omega \subseteq \mathbf{C}^2$ be given by

$$\Omega = (B(0,4) \setminus B(0,2)) \cup B\left((2,0), \tfrac{3}{2}\right).$$

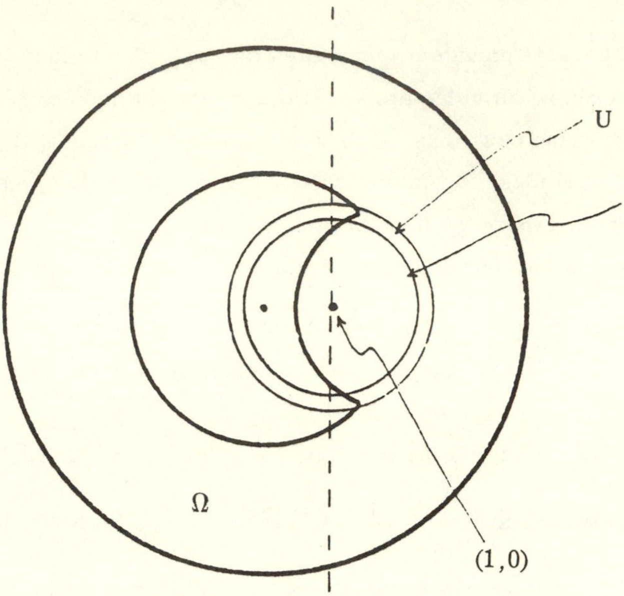

Let $U = B\left((1,0), \frac{3}{2}\right)$ and $V = B\left((1,0), \frac{5}{4}\right)$ as shown. Let $\eta \in C_c^\infty(U)$ satisfy $\eta \equiv 1$ on V. Finally, let

$$f = \frac{\bar{\partial}\eta}{z_1 - 1} .$$

Then f is smooth and $\bar{\partial}$ closed *on* Ω since $\frac{1}{z_1-1}$ is well-defined and holomorphic on $\operatorname{supp}(\bar{\partial}\eta) \cap \Omega$. If there existed a u satisfying $\bar{\partial}u = f$ then the function $h \equiv u - \frac{\eta}{z_1 - 1}$ would be holomorphic $(\bar{\partial}h = 0)$ on $\Omega \setminus \left(B\left((1,0), \frac{5}{4}\right) \cap \{z_1 = 1\}\right)$. But u would necessarily be smooth near $(1,0)$ (since f is) hence h has a singularity at, for instance, $(1,0)$. Thus we have created a function holomorphic on $B(0,4) \setminus B(0,2)$ which does not continue analytically to $(1,0)$. This contradicts the Hartogs extension phenomenon (an independent proof of this phenomenon will be given momentarily). □

Now that we know that $\bar{\partial}u = f$ is not always solvable, let us turn to an example where it is useful to be able to solve the $\bar{\partial}$ equation.

EXAMPLE. Consider the following question for an open domain $\Omega \subset\subset \mathbb{C}^n$:

$$(3.1)\quad \begin{cases} \text{If } \omega \equiv \Omega \cap \{z_n = 0\} \neq \emptyset \text{ and if } g \text{ is holomorphic} \\ \text{on } \omega \text{ (in an obvious sense), can we find } G \\ \text{holomorphic on } \Omega \text{ such that } G|_\omega = g\,? \end{cases}$$

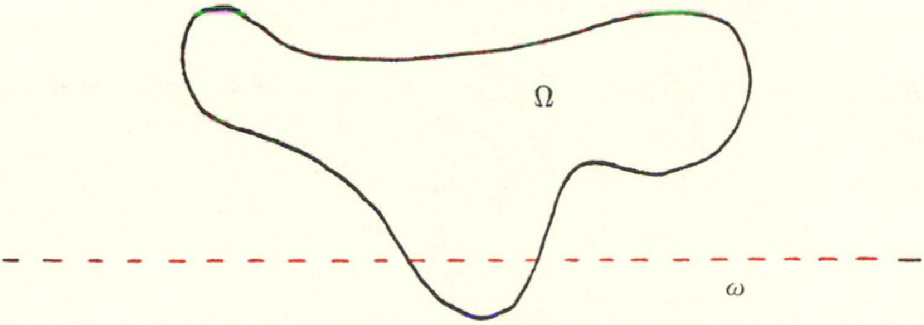

If Ω is the unit ball, then the trivial extension $G(z_1, z_2, \cdots, z_n) \equiv g(z_1, \cdots, z_{n-1}, 0)$ will suffice. However if $\Omega = B(0,2) \setminus \overline{B(0,1)} \subseteq \mathbb{C}^2$ then $g(z_1, 0) = 1/z_1$ is holomorphic on ω but could not have an extension G (else the Hartogs extension phenomenon would be contradicted). □

THEOREM. *Suppose that* $\omega \subset \mathbb{C}^n$ *is a connected open set such that whenever* f *is a smooth* $\bar\partial$*-closed* $(0,1)$ *form on* Ω *then the equation* $\bar\partial u = f$ *has a smooth solution. Then the answer to (3.1) is "yes."*

Proof. Let $\pi : \mathbb{C}^n \to \mathbb{C}^n$ be given by $\pi(z_1, \cdots, z_n) = (z_1, \cdots, z_{n-1}, 0)$. Let $B = \{z \in \Omega : \pi z \notin \omega\}$. Then B, ω are disjoint relatively closed subsets of Ω, so there is a C^∞ function ϕ on Ω such that $\phi \equiv 1$ on a relative neighborhood of ω and $\phi \equiv 0$ on a relative neighborhood of B. Define $\tilde{F}$ on Ω by

$$\tilde{F}(z) = \begin{cases} \phi(z) \cdot f(\pi(z)) & \text{if} \quad z \in \operatorname{supp} \phi \\ 0 & \text{else.} \end{cases}$$

Then $\tilde{F}$ gives a C^∞ (but certainly not holomorphic) extension of f to Ω.

We now seek a v such that $\tilde{F} + v$ is holomorphic and $\tilde{F} + v|_{\omega} = f$. With this in mind, we take v of the form $z_n \cdot u$ and we want

$$\bar{\partial}(\tilde{F} + v) = 0$$

or

$$\bar{\partial}\phi \cdot (f \circ \pi) + \phi \cdot \bar{\partial}(f \circ \pi) + (\bar{\partial} z_n) \cdot u + z_n \cdot \bar{\partial} u = 0 .$$

Now $f \circ \pi$ is holomorphic on $\operatorname{supp} \phi$ and z_n is holomorphic so all that remains is

$$\bar{\partial}\phi \cdot (f \circ \pi) + z_n \cdot \bar{\partial} u = 0$$

or

$$\bar{\partial} u = \frac{(-\bar{\partial}\phi) \cdot (f \circ \pi)}{z_n} . \tag{3.2}$$

The critical fact is that, by construction, $\bar{\partial}\phi = 0$ in a neighborhood of $\Omega \cap \{z_n = 0\}$ so the right-hand side of (3.2) is smooth on Ω. Also it is easily checked to be $\bar{\partial}$ closed. Thus our hypothesis is satisfied and a u satisfying (3.2) exists. Therefore $F \equiv \tilde{F} + v$ has all the desired properties. □

Our two examples show that solving the $\bar{\partial}$ equation is (i) subtle and (ii) useful. Thus we have ample motivation to prove our next result.

LEMMA. *Let* $\phi \in C_c^k(\mathbb{C})$, $k \geq 1$, *and define* $f = \phi(z) d\bar{z}$. *Then*

$$u(z) \equiv -\frac{1}{2\pi i} \iint_{\mathbb{C}} \frac{\phi(\zeta)}{\zeta - z} \, d\bar{\zeta} \wedge d\zeta$$

satisfies $u \in C^k(\mathbb{C})$ *and* $\bar{\partial} u = f$.

Proof. We have

$$\frac{\partial u}{\partial \bar{z}} = \frac{\partial}{\partial \bar{z}} \left(-\frac{1}{2\pi i} \iint_{\mathbf{C}} \frac{\phi(\zeta+z)}{\zeta} \, d\bar{\zeta} \wedge d\zeta \right)$$

$$= -\frac{1}{2\pi i} \iint_{\mathbf{C}} \frac{(\partial\phi/\partial\bar{z})(\zeta+z)}{\zeta} \, d\bar{\zeta} \wedge d\zeta$$

$$= -\frac{1}{2\pi i} \iint_{D(0,R)} \frac{(\partial\phi/\partial\bar{\zeta})(\zeta)}{\zeta - z} \, d\bar{\zeta} \wedge d\zeta$$

where $D(0,R)$ is a disc which contains $\operatorname{supp} \phi$. We apply Remark 1 of II in Section 1 to ϕ on $D(0,R)$ to obtain that the last line

$$= \phi(z) - \frac{1}{2\pi i} \int_{\partial D(0,R)} \frac{\phi(\zeta)}{\zeta - z} \, d\zeta .$$

The integral vanishes since $\phi = 0$ on $\partial D(0,R)$, hence $\bar{\partial} u = f$. Observe finally that $u \in C^k$ by differentiation under the integral sign. □

REMARK. In general, the u given by the lemma will not be compactly supported. Indeed if $\iint \phi \neq 0$ and if u were compactly supported (say $u \subseteq D(0,R)$) then a contradiction arises as follows:

$$0 = \int_{\partial D(0,R)} u \, d\zeta = \int_{D(0,R)} \phi \, d\bar{\zeta} \wedge d\zeta \neq 0 .$$

The supports of solutions to the $\bar{\partial}$ problem explain many phenomena in one and several complex variables. We explore this theme later. Meanwhile, contrast the Lemma and Remark with the following result.

THEOREM. *Let* $n > 1$ *and let* $\Phi(z) = \phi_1 d\bar{z}_1 + \cdots + \phi_n d\bar{z}_1$ *be* $\bar{\partial}$*-closed on* C^n *and suppose each* $\phi_j \in C_c^k(C^n)$. *Then for any* $1 \le j \le n$ *the function*

$$u^j(z) = -\frac{1}{2\pi i} \iint_{C} \frac{\phi_j(z_1, \cdots, z_{j-1}, \zeta, z_{j+1}, \cdots, z_n)}{\zeta - z_j} d\bar{\zeta} \wedge d\zeta$$

satisfies $u^j \in C_c^k(C^n)$ *and* $\bar{\partial} u^j = \Phi$. *Moreover* $u^j = u^\ell$ *for all* j, ℓ.

Proof. Fix $1 \le m \le n$. We need to check that $\frac{\partial u^j}{\partial \bar{z}_m} = \phi_j$, $1 \le m \le n$. If $m = j$ then the result follows from the lemma. If $m \ne j$ then use the compatibility condition $\frac{\partial \phi_j}{\partial \bar{z}_m} = \frac{\partial \phi_m}{\partial \bar{z}_j}$ to write

$$\frac{\partial}{\partial \bar{z}_m} u^j(z) = \frac{1}{2\pi i} \iint_{C} \frac{\frac{\partial \phi_j}{\partial \bar{z}_m}(z_1, \cdots, z_{j-1}, \zeta, z_{j+1}, \cdots, z_n)}{\zeta - z_j} d\bar{\zeta} \wedge d\zeta$$

$$= -\frac{1}{2\pi i} \iint_{C} \frac{\frac{\partial \phi_m}{\partial \bar{z}_j}(z_1, \cdots, z_{j-1}, \zeta, z_{j+1}, \cdots, z_n)}{\zeta - z_j} d\bar{\zeta} \wedge d\zeta .$$

By the lemma, this last equals ϕ_m. Thus $\bar{\partial} u^j = \Phi$. Notice that, if $\ell \ne j$, then $u^j = 0$ for z_ℓ large (since then $\phi_j = 0$). Also u^j is holomorphic for z_ℓ large (since $\bar{\partial} u = \Phi$ is then 0). So, by analytic continuation, $u \equiv 0$ off a compact set. Next, $u^j - u^\ell \equiv 0$ since it is compactly supported and holomorphic. Finally, $u^j \in C_c^k(C^n)$ by differentiation under the integral sign. □

REMARK. The proof actually shows that u is zero on the unbounded component of ${}^c(\bigcup_{j=1}^{n} \operatorname{supp} \phi_j)$. It also shows that there is *at most* one compactly supported solution to $\bar{\partial}u = f$. In the present case, one exists and is given by an integral formula.

§4. *The Hartogs Extension phenomenon and more on the* $\bar{\partial}$ *problem*

We have cited the Hartogs extension phenomenon in the examples of Section 3. The reader will want to check that the proof of it that we now give is independent of those examples.

THEOREM (The Hartogs Extension Phenomenon). *Let* $\Omega \subseteq \mathbb{C}^n$, $n > 1$, *be a bounded, connected open set. Let* $K \subseteq \Omega$ *be compact. Assume that* $\Omega \setminus K$ *is connected. If* f *is holomorphic on* $\Omega \setminus K$ *then there is a holomorphic* F *on* Ω *such that* $F|_{\Omega \setminus K} = f$.

Proof. Choose $\phi \in C^\infty(\Omega)$ such that $\phi \equiv 1$ in a neighborhood of $\partial\Omega$ and $\phi \equiv 0$ in a neighborhood of K. Define

$$\tilde{F}(z) = \begin{cases} \phi(z) \cdot f(z) & \text{if} \quad z \in \Omega \setminus K \\ 0 & \text{if} \quad z \in K . \end{cases}$$

Then $\tilde{F}$ is a C^∞ extension of f to Ω, but it is not in general holomorphic. We now seek v such that $\tilde{F} + v$ is holomorphic on Ω (and $\tilde{F} + v|_{\Omega \setminus K} = f$). Thus we seek v satisfying

$$\bar{\partial}(\tilde{F} + v) = 0$$

or

$$\bar{\partial}(\phi f) + \bar{\partial}v = 0$$

or

$$\bar{\partial}v = (-\bar{\partial}\phi) \cdot f \tag{4.1}$$

since f is holomorphic on $\operatorname{supp} \phi$. Now $\phi \equiv 1$ near $\partial\Omega$ so $(-\bar{\partial}\phi) \cdot f$ is smooth and compactly supported in Ω. The theorem of Section 3 now guarantees that there exists a v satisfying (4.1). Moreover, the remark following the theorem guarantees that $v \equiv 0$ near $\partial\Omega$. Thus $F + v$ is holomorphic and, near $\partial\Omega$, $\widetilde{F}+v=\widetilde{F}=\phi\cdot f=f$. By analytic continuation, $\widetilde{F}+v=f$ on U and the result follows. □

Notice how the hypothesis $n>1$ was used in the proof to control supp v. The Hartogs extension phenomenon has several interesting consequences:

(i) A holomorphic function f in $\mathbf{C}^n$, $n\geq 2$, cannot have an isolated singularity. If it did, say at P, then f would be holomorphic on $B(P,2\varepsilon)\setminus\overline{B}(P,\varepsilon)$ for ε small hence, by the Hartogs phenomenon, on $B(P,2\varepsilon)$, and hence at P. That is a contradiction

(ii) A holomorphic function f in $\mathbf{C}^n$, $n\geq 2$, cannot have an isolated zero. If it did, say at P, then apply (i) to $1/f$ to obtain a contradiction.

(iii) If $U\subseteq\mathbf{C}^n$ is open, $E\subseteq U$, f is holomorphic on $U\setminus E$, and E is a complex manifold of complex codimension at least 2, then f continues analytically to all of U. To see this, notice that for $n=2$ the set E is discrete and the result follows from (i). For $n>2$, the result follows from the case $n=2$ by considering $f|_{(\Omega\setminus E)\cap\ell}$ ranging over all two dimensional complex affine spaces $\ell\subseteq\mathbf{C}^n$.

Now we return to discussion of the $\overline{\partial}$ operator. There are essentially four aspects to this matter:

(1) Existence of solutions;

(2) Support of $\overline{\partial}$ data and $\overline{\partial}$ solutions;

(3) Choosing a *good solution*, where "good" means smooth or bounded;

(4) Estimates and regularity.

Regarding (4), we have noted that ellipticity considerations imply that when $\overline{\partial}v=g$ then v is smooth wherever g is. As an exercise, use the theorem of Section 3 to give another proof of this assertion. (Hint: if g is smooth on $B(P,r)$, then let $\phi\in C_c^\infty(B,(P,r))$ satisfy $\phi\equiv 1$ on $B\left(P,\frac{r}{2}\right)$. Let $\psi\in C_c^\infty\left(B\left(P,\frac{r}{2}\right)\right)$ satisfy $\psi\equiv 1$ on $B\left(P,\frac{r}{4}\right)$. Define $u=\phi\cdot v$ and $f=\overline{\partial}(\phi\cdot v)=\overline{\partial}\phi\cdot v+\phi\cdot g$. Then $\overline{\partial}u=f$. Apply the theorem of Section 3, decomposing f as $f = (\psi \cdot \bar{\partial}\phi \cdot v) + (1-\psi)(\bar{\partial}\phi \cdot v) + \phi \cdot g$.) This is all that we shall say about interior regularity.

Topic (2) has been discussed vis à vis the Hartogs phenomenon. Topics (1) and (3) are more subtle. First note that if $\bar{\partial}u = f$ then also $\bar{\partial}(u+h) = f$ for any holomorphic h. Given f, one cannot expect *all* $u + h$ to be nice (i.e. bounded, or L^2, or C^∞ up to the boundary). How does one find a nice solution?

An idea from Hodge theory is to study the solution u to $\bar{\partial}u = f$ which is orthogonal to the kernel of $\bar{\partial}$, i.e. which is orthogonal to holomorphic functions. This solution has been studied by Kohn [25], [27], [28], Catlin [3], [4], Greiner-Stein [16], and others. It is often called the *Kohn solution* or *canonical solution* to the $\bar{\partial}$ equation.

We now briefly review some of what is known about the Kohn solution, and other solutions, to the $\bar{\partial}$ problems on domains in $\mathbb{C}^n$. (In this section we shall take "pseudoconvex" and "strongly pseudoconvex" as undefined terms. These terms will be discussed in detail in Section 5; for now, a pseudoconvex domain is a domain of existence for the $\bar{\partial}$ operator.)

(a) If Ω is a bounded pseudoconvex domain in $\mathbb{C}^n$ and $f = \Sigma f_j d\bar{z}_j$ is a $\bar{\partial}$-closed form with all $f_j \in L^2(\Omega)$, then there is a $u \in L^2(\Omega)$ with $\bar{\partial}u = f$. Also $\|u\|_{L^2} \leq C(\Omega) \sum_j \|f_j\|_{L^2}$. See [20]. (Exercise: the canonical solution also satisfies this estimate.)

(b) If $\Omega \subseteq \mathbb{C}^n$ is smoothly bounded and pseudoconvex and $f = \Sigma f_j d\bar{z}_j$ is a $\bar{\partial}$-closed (0,1) form with all $f_j \in C^\infty(\bar{\Omega})$ then there is a $u \in C^\infty(\bar{\Omega})$ satisfying $\bar{\partial}u = f$. See [26]. *It is not known* whether the canonical solution has this property.

(c) If $\Omega \subseteq \mathbb{C}^n$ is strongly pseudoconvex with C^2 boundary and if $f = \Sigma f_j d\bar{z}_j$ is a $\bar{\partial}$-closed (0,1) form with bounded coefficients, then there is a u satisfying $\bar{\partial}u = f$ and $\|u\|_{L^\infty} \leq C\Sigma \|f_j\|_{L^\infty}$. The canonical solution has this property. See [11], [17], [22], [16], [35]. Sibony [39] has shown that there are smooth pseudoconvex domains on which uniform estimates for $\bar{\partial}$ do not hold. It is not known on which parameters the uniform estimates depend (however see [13]). Range [38], Henkin [17], and others have proved uniform estimates on certain weakly pseudoconvex domains.

(d) Complete, and sharp, estimates have been computed on strongly pseudoconvex domains in Lipschitz, Sobolev, Besov and other norms. See [16], [30]. These estimates hold for the canonical solution. One feature of the theory is that the operator assigning the canonical solution Kf to a $\bar{\partial}$-closed (0,1) form f is compact in these norms. This compactness is best expressed as a "subelliptic estimate" (see [27]). Catlin [4] has announced a characterization of those domains on which $\bar{\partial}$ satisfies a subelliptic estimate.

Many times an estimate tells us how to choose the right solution to $\bar{\partial}u = f$. We conclude this section with an example of how estimates can be useful.

DEFINITION. Let $\Omega \subseteq \mathbf{C}^n$ be a domain and $P \in \partial\Omega$. A holomorphic function $f : \Omega \to \mathbf{C}$ is called *singular* at P if for every $\varepsilon > 0$, $f|_{\Omega \cap B(P,\varepsilon)}$ is unbounded.

It is useful to be able to construct singular functions. Often we can *nearly* do this in the sense that we can find a neighborhood U of P and a holomorphic function on $U \cap \Omega$ which is singular at P (this is called a *local* singular function). Then the problem reduces to extending local singular functions to global ones.

LEMMA. *Let $\Omega \subseteq \mathbf{C}^n$ be a domain on which the $\bar{\partial}$ operator satisfies uniform estimates. If $P \in \partial\Omega$ and there exists a local singular function at P which is bounded off any $B(P,\varepsilon)$ then there exists a global one.*

Outline of proof. Let $g : U \cap \Omega \to \mathbb{C}$ be a local singular function at P. Let V be an open neighborhood of P such that $\overline{V} \subseteq U$. Let $\phi \in C^\infty(U)$ satisfy $\phi \equiv 1$ near P and $\phi \equiv 0$ off V. Set $f = \phi \cdot g + u$ and solve a $\bar{\partial}$ problem to find a bounded u. Then f is a global singular function at P.

Fix a strongly pseudoconvex domain Ω. We shall prove later that (i) uniform estimates for the $\bar{\partial}$ operator hold on Ω and (ii) local singular functions satisfying the hypotheses of the lemma exist for each $P \in \partial\Omega$. By taking a suitable root of the local singular function and applying the lemma, we may construct for each $P \in \partial\Omega$ a singular function F_P at P which is in $L^2(\Omega)$. Now we will prove that there is an L^2 holomorphic function F on Ω that cannot be holomorphically continued past *any* boundary point. This shows that Ω is a *domain* of *holomorphy* and essentially solves the *Levi problem* (see [31]).

For the construction of F, let $\{P_i\}_{i=1}^\infty$ be a countable dense set in $\partial\Omega$. Let H_{ij} be the L^2 holomorphic functions on $\Omega \cup B\left(P_i, \frac{1}{j}\right)$, $j = 1,2,\cdots$. Let $A^2(\Omega)$ be the L^2 holomorphic functions on Ω. Consider the restriction map $\gamma_{ij} : H_{ij} \to A^2(\Omega)$. Define $X_{ij} = \text{image } \gamma_{ij} \subseteq A^2(\Omega)$. Because F_{P_i} exists for each i, $X_{ij} \neq A^2(\Omega)$ for all i,j. We claim that $\bigcup_{i,j} X_{ij} \neq A^2(\Omega)$. Assume the claim for now. Take $F \in A^2(\Omega) \setminus \bigcup_{i,j} X_{ij}$. This is the F we seek. The claim now follows from:

PROPOSITION. *Let* X *and* Y *be Banach spaces, and* $T : X \to Y$ *a continuous linear map. Then the following are equivalent:*

(1) T(X) *is not of first category in* Y.

(2) T *is an open mapping.*

(3) T *is onto.*

Proof. This a variant of the Open Mapping Theorem for Banach spaces. (I am grateful to R. Huff for this proposition.) □

§5. *Convexity and pseudoconvexity*

Let $\Omega \subseteq \mathbf{R}^N$ be an open set. Then Ω is called *geometrically convex* if whenever $P, Q \in \Omega$ and $0 \le t \le 1$ then $(1-t)P + tQ \in \Omega$. In calculus, however, a C^2 function $y = f(x)$ is called convex if $f'' \ge 0$. How are these ideas related?

If Ω has smooth boundary, then we may think of Ω as given by

$$\Omega = \{x \in \mathbf{R}^N : \rho(x) < 0\}$$

for a smooth function ρ with $\nabla\rho \neq 0$ on $\partial\Omega$ (Exercise: use the implicit function theorem). The function ρ is called a *defining function* for Ω. If $P \in \partial\Omega$, let

$$T_P(\partial\Omega) = \left\{(\alpha_1, \cdots, \alpha_N) \in \mathbf{R}^N : \Sigma \frac{\partial \rho}{\partial x_j}(\rho) \cdot \alpha_j = 0\right\}.$$

Then $T_P(\partial\Omega)$ is the (real) *tangent space* to $\partial\Omega$ at P. If $\Omega = \{\rho < 0\}$ we say that Ω is *convex* at $P \in \partial\Omega$ if

$$\sum_{j,k=1}^{n} \frac{\partial^2 \rho}{\partial x_j \partial x_j}(P)\alpha_j \alpha_k \ge 0 \quad A\alpha \in T_P(\partial\Omega)\,. \tag{5.1}$$

We say that Ω is *strongly convex* at P if strict inequality obtains in (5.1) when $0 \neq \alpha \in T_P(\partial\Omega)$. A domain is convex (strongly convex) if each boundary point is.

EXERCISES (see [31]):

(i) For a smoothly bounded domain, geometric convexity is equivalent to convexity.

(ii) If Ω is smoothly bounded and convex, then Ω can be written as an increasing union of strongly convex domains.

(iii) The above definitions are independent of the choice of ρ.

In order to understand the role of convexity in complex analysis, we need to discuss inner products. If $z, w \in \mathbf{C}^n$, we define the *Hermitian inner product*

$$<z,w>_H \equiv \sum_{j=1}^{n} z_j \bar{w}_j$$

and if we identify

$$z = (z_1,\cdots,z_n) = (x_1+iy_1,\cdots,x_n+iy_n) \approx (x_1,y_1,\cdots,x_n,y_n)$$
$$\equiv (t_1,\cdots,t_{2n})$$

and likewise

$$w = (w_1,\cdots,w_n) \approx (s_1,\cdots,s_{2n})$$

then we define the *real inner product*

$$<z,w>_{Re} \equiv \sum_{j=1}^{2n} t_j s_j \ .$$

Notice the following facts:

(1) $<z,w>_{Re} = \mathrm{Re}\,(<z,w>_H)$;

(2) If $\Omega = \{z \in \mathbb{C}^n : \rho(z) < 0\}$ has smooth boundary, $P \in \partial\Omega$, then $T_P(\partial\Omega) = \{\alpha \in \mathbb{C}^n : <\alpha, \bar{\partial}\rho(P)>_{Re} = 0\}$.

(3) With Ω, P as in (2), let $\mathcal{T}_P(\partial\Omega) = \{\alpha \in \mathbb{C}^n : <\alpha, \bar{\partial}\rho(P)>_H = 0\}$. We call $\mathcal{T}_P(\partial\Omega)$ the *complex tangent space* to $\partial\Omega$ at P. If $\alpha \in \mathcal{T}_P(\partial\Omega)$ then $i\alpha \in \mathcal{T}_P(\partial\Omega)$. Also $\mathcal{T}_P(\partial\Omega) \subseteq T_P(\partial\Omega)$ and it is the largest subspace of $T_P(\partial\Omega)$ which is closed under multiplication by i.

Now if $\Omega = \{z \in \mathbb{C}^n : \rho(z) < 0\}$ is smooth and convex and $P \in \partial\Omega$ then Ω lies on one side of $T_P(\partial\Omega)$. Thus if we define $f_P(z) = <z-P, \bar{\partial}\rho(P)>_H$ then the zero set $\mathcal{Z}(f_P)$ of f_P lies in $P + T_P(\partial\Omega)$. In particular $\mathcal{Z}(f_P) \cap \bar{\Omega} \subseteq \partial\Omega$ (and if Ω is strongly convex then $\mathcal{Z}(f_P) \cap \partial\Omega = \{P\}$). Thus $1/f_P$ is singular at P. Also $\mathrm{Re}\, f_P < 0$ on Ω so we may choose $0 < N \in \mathbb{Z}$ such that $1/(f_P)^{1/N}$ is holomorphic on Ω and in $L^2(\Omega)$.

Thus each $P \in \partial\Omega$ has an L^2 singular function. By the argument at the end of Section 4, Ω supports an L^2 holomorphic function which cannot be analytically continued to any large open set. So any convex domain is a domain of holomorphy.

If we want to understand domains of holomorphy, convexity will not tell the whole story. For convexity is not a biholomorphic invariant: consider $\phi : \Delta \to \mathbb{C}$ given by $\phi(z) = (z+3)^3$. What is needed is a new notion called pseudoconvexity:

DEFINITION. If $\Omega = \{z \in \mathbb{C}^n : \rho(z) < 0\}$ is smoothly bounded we say that Ω is (Levi) *pseudoconvex* at $P \in \partial\Omega$ if

$$\sum_{j,k=1}^{n} \frac{\partial^2 \rho}{\partial z_j \partial \bar{z}_k}(P) w_j \bar{w}_k \geq 0 \quad \forall w \in \mathcal{T}_P(\partial\Omega) . \tag{5.2}$$

We call Ω *strongly pseudoconvex* at P if strict inequality holds in (5.2) for all $0 \neq w \in \mathcal{T}_P(\partial\Omega)$. The domain is pseudoconvex (strongly pseudoconvex) if each boundary point is.

EXERCISE. If $P \in \partial\Omega$ is strongly pseudoconvex prove that if $\lambda > 0$ is sufficiently large and $\tilde{\rho}(z) \equiv (e^{\lambda\rho(z)}-1)/\lambda$ then

$$\sum_{j,k=1}^{n} \frac{\partial^2 \tilde{\rho}}{\partial z_j \partial \bar{z}_k}(P) w_j \bar{w}_k \geq C|w|^2, \forall P \in \partial\Omega, \forall w \in \mathcal{T}_P(\partial\Omega) . \tag{5.3}$$

See [31] for details.

The rather technical notion of pseudoconvexity is vindicated by the following deep theorem (see [31]):

THEOREM. *If $\Omega \subseteq \mathbb{C}^n$ is smoothly bounded then the following are equivalent:*

(i) *Ω is pseudoconvex*

(ii) *Ω is a domain of holomorphy*

(iii) *the equation $\bar{\partial}u = f$, f a $\bar{\partial}$ closed (p,q) form, is always solvable.*

This theorem means that pseudoconvex domains are the natural arena for complex function theory. Also (exercise, or see [31]) any pseudoconvex domain is the increasing union of smooth strongly pseudoconvex domains. So strongly pseudoconvex domains, in a certain sense, are generic.

In order to unify and illustrate the ideas introduced so far in this section we prove

LEMMA. *If* $\Omega \subseteq \mathbf{C}^n$ *is smoothly bounded and* $P \in \partial\Omega$ *is a point of convexity then* P *is a point of pseudoconvexity.*

Proof. Let ρ be a defining function for Ω. Let $\alpha \in \mathcal{T}_P(\partial\Omega)$. Writing the definition of convexity in complex notation we have

$$\sum_{j,k=1}^{n} \frac{\partial^2\rho}{\partial z_j \partial z_k}(P)\alpha_j\alpha_k + \sum_{j,k=1}^{n} \frac{\partial^2\rho}{\partial \bar{z}_j \partial \bar{z}_k}(P)\bar{\alpha}_j\bar{\alpha}_k$$

$$+\frac{1}{2}\sum_{j,k=1}^{n} \frac{\alpha^2\rho}{\partial z_j \partial \bar{z}_k}(P)\alpha_j\bar{\alpha}_k \geq 0 .$$

But a similar inequality also obtains for $i\alpha \in \mathcal{T}_P(\partial\Omega)$. Adding the two inequalities yields the result. □

§6. *Solutions for the* $\bar{\partial}$ *problem*

We briefly describe the Hilbert space setup for Hörmander's L^2 theory of the $\bar{\partial}$ problem. We fix a smoothly bounded $\Omega \subseteq \mathbf{C}^n$ and introduce the notation

$$\begin{array}{ccccc} L^2(\Omega) & \xrightarrow{\bar{\partial}_{(0,0)} \equiv T} & L^2_{(0,1)}(\Omega) & \xrightarrow{\bar{\partial}_{(0,1)} \equiv S} & L^2_{(0,1)}(\Omega) \\ ||| & & ||| & & ||| \\ H_1 & & H_2 & & H_3 \end{array}$$

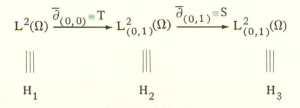

The operators T,S are of course unbounded, but they are densely defined (since C_c^∞ is dense in L^2). It is easy to check that T, S are closed; also $S \circ T = 0$ so if $F = \ker S$ then Range $T \subseteq F$.

An existence theorem for the $\bar{\partial}$ *equation amounts to proving that* Range $T = F$. Moreover, it is an exercise in functional analysis to check that this is equivalent to proving an inequality of the form

$$\|y\|_{H_2} \leq C\|T^*y\|_{H_1}, \quad \forall y \in \mathcal{D}_{T^*} \cap F. \tag{6.1}$$

See [20] for details. Rather than prove (6.1), it is more convenient to study the symmetric inequality

$$\|y\|_{H_2} \leq C(\|T^*y\|_{H_1} + \|Sy\|_{H_3}) \quad \forall y \in \mathcal{D}_{T^*} \cap \mathcal{D}_S. \tag{6.2}$$

Notice that when $y \in F$, (6.2) reduces to (6.1).

Unfortunately this program, as stated, fails. The difficulty is that the computation of T^* gives rise to boundary terms which, in general, cannot be controlled. (However, in the strongly pseudoconvex case and on weakly pseudoconvex domains satisfying a non-degeneracy condition, there are delicate techniques for handling the boundary terms. See [25].)

Hörmander's idea [20] was to work not in Euclidean L^2 but rather in L^2 of the measure space $e^{-\phi}dx$. If ϕ is chosen to have certain convexity properties and to blow up rapidly at $\partial\Omega$, then the boundary is effectively suppressed (Ω becomes a complete Riemannian manifold) and the Hilbert space program outlined above works. After the existence problem is thus tamed, the weights can be eliminated (provided Ω is bounded) and one obtains an existence theorem in $L^2(\Omega, dx)$. A leisurely exposition of all these ideas can be found in [31]. We now formulate a version of Hörmander's result, which we will use freely in what follow:

THEOREM (Hormander). *If* $\Omega \subseteq \mathbb{C}^n$ *is smoothly bounded and pseudoconvex, and if* f *is a* $\bar{\partial}$ *closed* (0,1) *from on* Ω *with* $L^2(\Omega,dx)$ *coefficients, then there exists* $u \in L^2(\Omega,dx)$ *such that* $\bar{\partial}u = f$.

REMARK. Hörmander's theorem can be used to prove most of the theorems at the end of Section 5 characterizing domains of holomorphy. See [31] for details.

Our next main goal is to obtain an integral formula for a solution to the $\bar{\partial}$ problem on a convex domain. We first need:

THEOREM (Bochner-Martinelli). *Let* Ω *be a domain in* $\mathbb{C}^n$ *with* C^1 *boundary. Let* $f \in C^1(\bar{\Omega})$. *Then for all* $z \in \Omega$ *we have*

$$f(z) = \frac{1}{nW(n)} \left\{ \int_{\partial\Omega} f(\zeta)\eta\left(\frac{\overline{\zeta}-\overline{z}}{|\zeta-z|^2}\right) \wedge \omega(\zeta) \right.$$

$$\left. - \int_{\Omega} \bar{\partial} f(\zeta) \wedge \eta\left(\frac{\overline{\zeta}-\overline{z}}{|\zeta-z|^2}\right) \wedge \omega(\zeta) \right\} .$$

Proof. Apply Stokes' theorem to the form

$$\mu(\zeta) = f(\zeta)\eta\left(\frac{\overline{\zeta}-\overline{z}}{|\zeta-z|^2}\right) \wedge \omega(\zeta)$$

on the domain $\Omega \setminus B(z,\varepsilon)$. Imitate the proof of the Cauchy Integral Formula (or see [31]).

Now fix Ω a bounded, C^2, convex domain. Choose $\varepsilon > 0$ so small that $\hat{\Omega} \equiv \{z \in \mathbb{C}^n : \text{dist}(z,\Omega) < \varepsilon\}$ is convex (hence pseudoconvex). Let f be a smooth, $\bar{\partial}$-closed (0,1) form on $\hat{\Omega}$. By Hörmander's theorem, there is a smooth u on $\hat{\Omega}$ such that $\bar{\partial}u = f$. We apply the Bochner-Martinelli formula to u (which is certainly in $C^1(\bar{\Omega})$). Thus

$$u(z) = \frac{1}{nW(n)} \int_{\partial\Omega} u(\zeta)\eta\left(\frac{\bar{\zeta}-\bar{z}}{|\zeta-z|^2}\right) \wedge \omega(\zeta)$$

$$- \frac{1}{nW(n)} \int_{\Omega} \bar{\partial}u(\zeta) \wedge \eta\left(\frac{\bar{\zeta}-\bar{z}}{|\zeta-z|^2}\right) \wedge \omega(\zeta) \tag{6.3}$$

$$= \frac{1}{nW(n)} \int_{\partial\Omega} u(\zeta)\eta\left(\frac{\bar{\zeta}-\bar{z}}{|\zeta-z|^2}\right) \wedge \omega(\zeta) - \frac{1}{nW(n)} \int_{\Omega} f(\zeta) \wedge \eta\left(\frac{\bar{\zeta}-\bar{z}}{|\zeta-z|^2}\right) \wedge \omega(\zeta)\,.$$

The first term on the right is not useful, since it involves u, so we will remedy matters by subtracting an appropriate holomorphic function from the right side of (6.3) (see the discussion in Section 4 on choosing a good solution). The Cauchy-Fantappié formalism now comes into play:

If $\Omega = \{\rho < 0\}$, let

$$w(z,\zeta) = \left(\frac{-\frac{\partial\rho}{\partial\zeta_1}(\zeta)}{\Phi(z,\zeta)}, \ldots, \frac{-\frac{\partial\rho}{\partial\zeta_n}(\zeta)}{\Phi(z,\zeta)}\right)$$

with $\Phi(z,\zeta) = \sum_{j=1}^{n} \frac{\partial}{\partial\zeta_j}(\zeta)(z_j-\zeta_j)$ and observe that

$$H(z) = \frac{1}{nW(n)} \int_{\partial\Omega} u(\zeta)\eta(w(z,\zeta)) \wedge \omega(\zeta)$$

is well defined and *holomorphic* in z (because w is).

Now let $v(z) \equiv u(z) - h(z)$. Then certainly $\bar{\partial}v = \bar{\partial}u = f$ and, by (6.3),

$$v(z) = \frac{1}{nW(n)} \left\{ \int_{\partial\Omega} u(\zeta)\eta\left(\frac{\bar\zeta-\bar z}{|\zeta-z|^2}\right) \wedge \omega(\zeta) - \int_{\partial\Omega} u(\zeta)\eta(w) \wedge \omega(\zeta) \right\}$$

$$- \frac{1}{nW(n)} \int_{\Omega} f(\zeta) \wedge \eta\left(\frac{\bar\zeta-\bar z}{|\zeta-z|^2}\right) \wedge \omega(\zeta) \equiv I - II .$$

Let $G = \Omega \times [0,1]$ and define $g(z,\zeta,\lambda) = (1-\lambda)\dfrac{\bar\zeta-\bar z}{|\zeta-z|^2} + \lambda w(z,\zeta)$ to be a form on G. Then

$$I = \frac{1}{n \cdot W(n)} \int_{\partial G} u(\zeta)\eta(g) \wedge \omega(\zeta) .$$

By Stokes' theorem, *applied on* G, this

$$= \frac{1}{nW(n)} \int_{G} d(u(\zeta)\eta(b) \wedge \omega(\zeta)) .$$

But

$$\begin{aligned} d(u(\zeta)\eta(g) \wedge \omega(\zeta)) &= \bar\partial u \wedge \eta(g) \wedge \omega(\zeta) \\ &\quad + u(\zeta)d(\eta(g)) \wedge \omega(\zeta) \\ &= f \wedge \eta(g) \wedge \omega(\zeta) \end{aligned}$$

(this last equality takes advantage of the special algebraic properties of Cauchy-Fantappié forms). So we finally obtain

HENKIN'S INTEGRAL FORMULA. *If* $\Omega = \{\rho < 0\}$ *is* C^2 *and convex and* f *is a smooth,* $\bar\partial$*-closed* $(0,1)$ *form on a neighborhood of* $\bar\Omega$ *then the function*

$$v(z) = \frac{1}{n \cdot W(n)} \left\{ \int_{\partial\Omega \times [0,1]} f(\zeta) \wedge \eta(g) \wedge \omega(\zeta) - \int_{\Omega} f(\zeta) \wedge \eta \left(\frac{\bar{\zeta}-\bar{z}}{|\rho - z|^2} \right) \wedge \omega(\zeta) \right\}$$

satisfies $\bar{\partial} v = f$ on Ω. Here

$$g(z, \zeta, \lambda) = (1-\lambda) \left(\frac{\bar{\zeta}-\bar{z}}{|\rho - z|^2} \right) + \lambda w(z, \zeta) ,$$

$$w(z, \zeta) = \left(\frac{-\frac{\partial \rho}{\partial \zeta_1}(\zeta)}{\Phi(z, \zeta)}, \cdots, \frac{-\frac{\partial \rho}{\partial \zeta_n}(\zeta)}{\Phi(z, \zeta)} \right)$$

and

$$\Phi(z, \zeta) = \sum_{j=1}^{n} \frac{\partial \rho}{\partial \zeta_j}(\zeta)(z_j - \zeta_j) .$$

The standard reference for Henkin's work is [18]; see also [31]. Similar formulas were derived by Grauert-Lieb [11], Kerzman [22], and Øvrelid [35].

Now we assume that Ω is *strongly convex*, and show how to use Henkin's formula to obtain uniform estimates for solutions to the $\bar{\partial}$ problem. For simplicity we work in $\mathbb{C}^2$ only. So

$$f \wedge \eta(g) \wedge \omega(\zeta) = f \wedge \left[g_1 \frac{\partial g_2}{\partial \lambda} d\lambda - g_2 \frac{\partial g_1}{\partial \lambda} d\lambda \right] \wedge d\zeta_1 \wedge d\zeta_2 .$$

After some algebra, and integrating out λ, the Henkin formula is

$$v(z) = \frac{1}{2W(2)} \int\limits_{\partial\Omega} \frac{-\frac{\partial\rho}{\partial\zeta_1}(\zeta)(\bar{\zeta}_2-\bar{z}_2) + \frac{\partial\rho}{\partial\zeta_2}(\zeta)(\bar{\zeta}_1-\bar{z}_1)}{\Phi(z,\zeta)\,|\zeta-z|^2}$$

$$\wedge (f_1(\zeta)\,d\bar{\zeta}_1 + f_2(\zeta)\,d\bar{\zeta}_2) \wedge d\zeta_1 \wedge d\zeta_2$$

$$- \frac{1}{2W(2)} \int\limits_{\Omega} \frac{f_1(\zeta)(\bar{\zeta}_1-\bar{z}_1) + f_2(\zeta)(\bar{\zeta}_2-\bar{z}_2)}{|\zeta-z|^4}\, d\bar{\zeta}_1 \wedge d\bar{\zeta}_2 \wedge d\zeta_1 \wedge d\zeta_2$$

$$= \frac{1}{2W(2)} \int\limits_{\partial\Omega} f_1(\zeta) A_1(z,\zeta)\, d\bar{\zeta}_1 \wedge d\zeta_1 \wedge d\zeta_2$$

$$+ \frac{1}{2W(2)} \int\limits_{\partial\Omega} f_2(\zeta) A_2(z,\zeta)\, d\bar{\zeta}_2 \wedge d\zeta_1 \wedge d\zeta_2$$

$$+ \frac{1}{2W(2)} \int\limits_{\Omega} f_1(\zeta) B_1(z,\zeta)\, d\bar{\zeta}_1 \wedge d\bar{\zeta}_2 \wedge d\zeta_1 \wedge d\zeta_2$$

$$+ \frac{1}{2W(2)} \int\limits_{\Omega} f_2(\zeta) B_2(z,\zeta)\, d\bar{\zeta}_1 \wedge d\bar{\zeta}_2 \wedge d\zeta_1 \wedge d\zeta_2 .$$

In order to prove an inequality of the form

$$\|v\|_{L^\infty} \leq C\|f\|_{L^\infty}, \tag{6.4}$$

it suffices to check that

$$\int_{\partial\Omega} |A_j(z,\zeta)|\, d\sigma(\zeta) \leq C\,, \quad j = 1,2 \tag{6.5}$$

and

$$\int_{\Omega} |B_j(z,\zeta)|\, dV(\zeta) \leq C\,, \quad j = 1,2 \tag{6.6}$$

with the estimates uniform over $z \in \Omega$. (Here $d\sigma$ is area measure on $\partial\Omega$.) By symmetry we check only $j = 1$.

For B_1, choose $R > 0$ such that $B(z,R) \supseteq \Omega$ for all $z \in \Omega$. Then

$$\int_{\Omega} |B_1(z,\zeta)|\, dV(\zeta) \leq \int_{B(z,R)} \frac{C}{|z-\zeta|^3}\, dV(\zeta)$$

$$\leq C \int_0^R \frac{1}{r^3}\, r^3\, dr = CR < \infty\,.$$

For A_1, we must work harder. We need to know something about the degree to which the denominator $\Phi(z,\zeta)\,|\zeta - z|^2$ of A_1 vanishes. (This is where *strong* convexity plays a role.) Think of $z \in \Omega$ as fixed. Notice that, writing the Taylor expansion for ρ about ζ in complex coordinates, we have

$$\rho(z) = \rho(\zeta) + 2\mathrm{Re}\left\{\frac{\partial\rho}{\partial\zeta_1}(\zeta)(z_1-\zeta_1) + \frac{\partial\rho}{\partial\zeta_2}(\zeta)(z_2-\zeta_2)\right\}$$

$$+ \text{(quadratic terms)} + \text{(error terms)}$$

$$> 0 + 2\mathrm{Re}\,\Phi(z,\zeta) + C|z-\zeta|^2\,, \quad C > 0\,.$$

The last estimate is by strong convexity — see Section 5. Thus

$$|\mathrm{Re}\ \Phi(z,\zeta)| \geq C\{|z-\zeta|^2 + |\rho(z)|\}.$$

Let πz be the normal projection of z to $\partial\Omega$. Let t_1 be a coordinate in the complex normal direction at πz (that is, i times the real normal direction) in $\partial\Omega$. Recall that

$$\sum_{j=1}^{2} \frac{\partial\rho}{\partial\zeta_j}(\zeta)a_j = 0 \Longleftrightarrow a \in \mathcal{T}_P(\Omega)$$

and

$$\mathrm{Re} \sum_{j=1}^{2} \frac{\partial\rho}{\partial\zeta_j}(\zeta)a_j = 0 \Longleftrightarrow a \in T_P(\Omega).$$

It follows that

$$\mathrm{Im}\left(\sum_{j=1}^{2} \frac{\partial\rho}{\partial\zeta_j}(\zeta)(z_j-\zeta_j)\right) = \mathrm{Im}\left(\sum_{j=1}^{2} \frac{\partial\rho}{\partial z_j}(z)(z_j-\zeta_j)\right) + 0(|z-\zeta|^2)$$

measures (essentially) the complex normal component of $z-\zeta$ at πz. As a result

$$|\mathrm{Im}\ \Phi| \geq C|t_1|.$$

Introducing two additional coordinates t_2, t_3 centered at πz, which span the complex tangential directions at πz, we conclude that

$$\begin{aligned}|\Phi(z,\zeta)| &\geq \frac{1}{2}(|\mathrm{Re}\ w(z,\zeta)| + |\mathrm{Im}(z,\zeta)|)\\ &\geq C\cdot\{t_1 + t_2^2 + t_3^2 + |\rho(z)|\}\end{aligned}$$

provided ζ is near πz, say $|\zeta-\pi z| < r_0$. Then

$$\int_{\partial\Omega} A_1(z,\zeta)d\sigma(\zeta) = \int_{B(\pi z,r_0)\cap\partial\Omega} + \int_{\partial\Omega\setminus B(\pi z,r_0)}.$$

The second integral is trivially bounded since when $|z-\zeta| \geq r_0$ then A_1 is bounded. The first is majorized by

$$C \int\limits_{t_1^2+t_2^2+t_3^2 \leq Cr_0^2} \frac{\sqrt{t_1^2+t_2^2+t_3^2+\rho(z)^2}\, dt_1 dt_2 dt_3}{(t_1^2+t_2^2+t_3^2+\rho(z)^2)(|t_1|+t_2^2+t_3^2+|\rho(z)|)}$$

$$= C \int\limits_{\substack{|\rho|+|t_1| \leq t_2^2+t_3^2 \\ t_1^2+t_2^2+t_3^2 \leq Cr_0^2}} + C \int\limits_{\substack{|\rho|+|t_1| > t_2^2+t_3^2 \\ t_1^2+t_2^2+t_3^2 \leq Cr_0^2}}$$

$$\equiv \gamma_1 + \gamma_2 .$$

Now

$$|\gamma_1| \leq C \int\limits_{\substack{|\rho|+|t_1| \leq t_2^2+t_3^2 \\ t_1^2+t_2^2+t_3^2 \leq Cr_0^2}} \frac{1}{\sqrt{t_2^2+t_3^2}\,(t_2^2+t_3^2)}\, dt_1 dt_2 dt_3$$

$$\leq C \int\limits_{t_2^2+t_3^2 \leq Cr_0^2} \frac{1}{\sqrt{t_2^2+t_3^2}}\, dt_2 dt_3$$

$$\leq C \int_0^{C' r_0} \frac{1}{r} \cdot r dr = C \cdot C' r_0 < \infty .$$

Likewise

$$|\gamma_2| \leq C \int\limits_{\substack{|\rho|+|t_1|>t_2^2+t_3^2 \\ t_1^2+t_2^2+t_3^2\leq Cr_0^2}} \frac{dt_1 dt_2 dt_3}{\sqrt{t_2^2+t_3^2}\,(|t_1|+|\rho|)}$$

$$\leq C \int\limits_{t_2^2+t_3^2\leq Cr_0^2} \frac{1}{\sqrt{t_2^2+t_3^2}} (|\log(t_2^2+t_3^2)| + |\log(Cr_0)|)\, dt_2 dt_3$$

$$\leq C \int_0^{C' r_0} \frac{1}{r} (|\log r| + |\log C|)\, r\, dr$$

$$\leq C'' < \infty .$$

Thus

$$\int\limits_{\partial\Omega} |A_1|\, d\sigma \leq C < \infty .$$

The estimates for A_2, B_2 follow by symmetry, as already noted. We have proved

THEOREM. *With* Ω *strongly convex and* f, v *as in Henkin's integral formula for solutions of the* $\bar{\partial}$ *problem,*

$$\|v\|_{L^\infty(\Omega)} \leq C\, \Sigma \|f_j\|_{L^\infty} .$$

REMARKS. 1) To eliminate the hypothesis that f is defined on a neighborhood of $\bar{\Omega}$ we observe that for $\varepsilon > 0$ small enough the domain

$$\Omega_\varepsilon = \{z \in \Omega : \operatorname{dist}(z, \partial\Omega) > \varepsilon\}$$

is smoothly bounded and strongly convex. Moreover, the estimates in the theorem on Ω_ε depend boundedly on ε (by a calculation). Thus estimates can be obtained for f a smooth form on Ω with bounded coefficients by applying a limiting argument to the solution u_ε of $\bar\partial(*) = f$ on Ω_ε.

2) The results of the theorem actually hold on smoothly bounded strongly *pseudoconvex* domains. This is most easily seen by using the following important result:

THE FORNAESS IMBEDDING THEOREM [9]. *Let* $\Omega \subset\subset \mathbb{C}^n$ *be a strongly pseudoconvex domain with* C^2 *boundary. Then there is a neighborhood* $\hat{\Omega}$ *of* $\bar{\Omega}$, *a* $k > 0$, *a* C^2 *strongly convex domain* $U \subseteq \mathbb{C}^{n+k}$, *and a holomorphic imbedding* $F : \hat{\Omega} \to \mathbb{C}^{n+k}$ *such that*

(i) $F(\Omega) \subseteq U$

(ii) $F(\hat{\Omega} \setminus \bar{\Omega}) \subseteq {}^c\bar{U}$

(iii) $F(\partial\Omega) \subseteq \partial U$

(iv) *image* F *is transversal to* ∂U.

The upshot of this theorem is that the Henkin singular function Φ which we know how to construct on U can be pulled back to Ω. The construction of the Henkin solution to the $\bar\partial$ equation and the uniform estimates follow just as before.

3) The singular function Φ can be constructed more directly on a strongly pseudoconvex domain Ω as follows: first write $\Omega = \{\tilde\rho < 0\}$ where

$$\sum \frac{\partial^2 \tilde\rho}{\partial z_j \partial \bar z_k}(\zeta) w_j \bar w_k \ge C|w|^2 \quad \forall \zeta \in \partial\Omega, \; \forall w \in \mathbb{C}^n ,$$

(see (5.3)). For $\zeta \in \partial\Omega$ fixed we define

$$L(z,\zeta) = \sum_{j=1}^n \frac{\partial\tilde\rho}{\partial\zeta_j}(\zeta)(z_j - \zeta_j) + \sum_{j,k=1}^n \frac{\partial^2\tilde\rho}{\partial\zeta_j \partial\zeta_k}(\zeta)(z_j-\zeta_j)(z_k-\zeta_k) .$$

The function L, called the *Levi polynomial*, has the property that there a neighborhood U_ζ of ζ such that $\bar{\Omega} \cap U_\zeta \cap \{z : L(z,\zeta) = 0\} = \{\zeta\}$. One can modify L, by solving a $\bar{\partial}$ problem (see [18] or [31]) to obtain Φ such that $\bar{\Omega} \cap \{z : \Phi(z,\zeta) = 0\} = \{\zeta\}$ *and* $\Phi(z,\zeta) = \sum_{j=1}^{n} P_j(z,\zeta)(z_j - \zeta_j)$ with P_j holomorphic in z. Henkin's program may be carried out using this and $w(z,\zeta) = (-P_1(z,\zeta)/\Phi(z,\zeta), \cdots, -P_n(z,\zeta)/\Phi(z,\zeta))$.

Notice that, by the discussion in the preceding paragraph, $1/L(\cdot,\zeta)$ is a local singular function at ζ. This, together with the uniform estimates for the $\bar{\partial}$ equation which we have obtained, completes the program outlined in Section 4 to show that a strongly pseudoconvex domain is a domain of holomorphy.

§7. *Connections between various integral formulas and applications*

In Section 1 we constructed the Szegö kernel for domains in $\mathbb{C}^1$. However the construction goes through for domains in $\mathbb{C}^n$ once one has the basic lemma, and that follows in $\mathbb{C}^n$ from the Bochner-Martinelli formula. We leave the details of the basic Szegö theory in $\mathbb{C}^n$ as an exercise.

Recall that the Szegö kernel for a domain Ω is the reproducing kernel for $H^2(\Omega)$. Now fix $z \in \Omega$. By construction, $\overline{S(z,\cdot)} \in H^2(\Omega)$. By the reproducing property, it follows that

$$\begin{aligned}\overline{S(z,\zeta)} &= \int_{\partial\Omega} S(\zeta,w)\overline{S(z,w)}\,d\sigma(w) \\ &= \overline{\int_{\partial\Omega} S(z,w)\overline{S(\zeta,w)}\,d\sigma(w)} \\ &= \overline{\overline{S(\zeta,z)}} \\ &= S(\zeta,z)\,.\end{aligned}$$

Thus the operator

$$S : f \mapsto \int S(z,\zeta) f(\zeta) d\sigma(\zeta)$$

has the following three properties:

(a) $S : L^2(\partial\Omega) \to H^2(\Omega)$

(b) S is self-adjoint

(c) S is idempotent.

Therefore S is the Hilbert space projection of $L^2(\partial\Omega)$ onto $H^2(\Omega)$.

It turns out that the Henkin operator on a strongly pseudoconvex domain *very nearly* has properties (a)-(c). First, by a theory of non-isotropic singular integrals developed especially for boundaries of strongly pseudoconvex domains (see [8], [36]), the Henkin operator

$$H : f \mapsto \frac{1}{nW(n)} \int_{\partial\Omega} f(\zeta)\, \eta(w) \wedge \omega(\zeta) \tag{7.1}$$

(with the w produced from the Fornaess theorem as in Section 6) maps $L^2(\partial\Omega)$ onto $H^2(\Omega)$. Also H is idempotent. Now H is not quite self-adjoint, but it is nearly so. To see this, one needs to write (7.1) in the form

$$\frac{1}{nW(n)} \int_{\partial\Omega} f(\zeta) \frac{N(z,\zeta)}{\Phi^n(z,\zeta)} d\sigma(\zeta)$$

where $d\sigma$ is area measure on $\partial\Omega$. This is a straightforward but tedious calculation (see [23]). It turns out that N is real. It also turns out that $\Phi(z,\zeta) - \overline{\Phi(\zeta,z)}$ vanishes to higher order at $z = \zeta$ than does Φ. (Try this when Ω is the ball to see how this works – details are in [23].)

As a result of the preceding observations, the kernel

$$\frac{N(z,\zeta)}{\Phi^n(z,\zeta)} - \overline{\left(\frac{N(\zeta,z)}{\Phi^n(\zeta,z)}\right)}$$

is a kernel which is *less singular* than the original Henkin kernel. Thus $H - H^*$, rather than being a non-isotropic singular integral operator (as is H), is a *smoothing* operator. This observation of Kerzman and Stein is now exploited as follows. Denote $H^* - H = A$.

The reproducing properties of S and H guarantee that

$$(1)\quad S = HS$$

and

$$(2)\quad H = SH .$$

Thus

$$(3)\quad S = S^* = (HS)^* = S^*H^* = SH^* .$$

Subtracting (2) from (3) gives

$$S - H = S(H^* - H) = SA .$$

This is an operator equation on L^2. We may resubstitute the equation into itself as follows:

$$\begin{aligned} S &= H + SA \\ &= H + (H+SA)A \\ &= H + HA + SA^2 \\ &= H + HA + (H+SA)A^2 \\ &= H + HA + HA^2 + SA^3 \\ &= \cdots = H + HA + \cdots HA^k + SA^{k+1} . \end{aligned} \tag{7.2}$$

Now we know that each of the operators $HA, HA^2, \cdots$ are smoothing. If we apply both sides of (7.2) to a sequence $\phi_j \in C_c^\infty(\partial\Omega)$ such that $\phi_j \to \delta_\zeta$

in the weak^{-*} topology on $\partial\Omega$, we obtain an equation relating $S(z,\zeta)$ and $H(z,\zeta)$. In particular, H and S are equal *modulo terms which are less singular.* From this fundamental result, many basic mapping properties of S can be determined (see [36]).

The basic construction of Kerzman and Stein can be used in other contexts. Let us turn now to one of these: the Bergman kernel. Fix a domain $\Omega \subset\subset \mathbb{C}^n$ and define

$$A^2(\Omega) = \left\{ f \text{ holomorphic on } \Omega : \int_\Omega |f(z)|^2 \, dVol(\Omega) < \infty \right\}.$$

(Notice that, for Ω smoothly bounded, $H^2(\Omega)$ is a proper subspace of $A^2(\Omega)$ – Exercise.) Define

$$<f,g> = \int_\Omega f\bar{g} dV$$

$$\|f\| = \int_\Omega |f|^2 dV^{1/2}$$

for $f,g \in A^2(\Omega)$. The basic lemma in this context,

$$\sup_K |f(z)| \le C_K \|f\|$$

for $K \subset\subset \Omega$, is easily derived from the mean value property for holomorphic functions. As in Section 1, the abstract Hilbert space theory yields a reproducing kernel for A^2 which we call the *Bergman kernel.*

Just like the Szegö kernel, the Berman kernel (denoted by the letter K) satisfies $K(z,\zeta) = \overline{K(\zeta,z)}$. Thus the associated operator

$$B : f \mapsto \int_{\Omega} f(\zeta) K(z,\zeta) dV(\zeta)$$

is self-adjoint. It maps onto A^2 (by construction) and is idempotent. So $B : L^2(\Omega) \to A^2(\Omega)$ is Hilbert space projection.

A remarkable construction of S. Bergman (see [31]) is as follows: note that, for $z \in \Omega$ fixed,

$$K(z,z) = \int K(z,w) K(w,z) dV(w)$$

$$= \int |K(z,w)|^2 dV(w) > 0 .$$

Therefore we may set

$$g_{ij}(z) = \frac{\partial^2}{\partial z_i \partial \bar{z}_j} \log K(z,z) .$$

By a calculation (see [31]), the matrix $(g_{ij}(z))$ gives a non-degenerate Kähler metric on Ω (called the *Bergman metric*) which is *invariant under biholomorphic mappings*. In particular it holds that if $\Phi : \Omega_1 \to \Omega_2$ is biholomorphic then

$$\operatorname{dist}^{\Omega_1}_{\mathrm{Berg}}(z,w) = \operatorname{dist}^{\Omega_2}_{\mathrm{Berg}}(\Phi(z), \Phi(w)) .$$

As a result, metric geodesics and curvature are preserved.

The Bergman metric and kernel are potentially powerful tools in function theory, provided we can calculate them. To do so, we exploit the idea of Kerzman and Stein [23] to compare K with the Henkin kernel. However a complication arises: the Henkin integral (7.1) is a boundary integral while the Bergman integral is a solid integral. How can we compare functions with different domains? What we would like to do is apply Stokes' theorem to the Henkin integral and turn it into an integral over Ω. However, for $z \in \Omega$ fixed, Henkin's kernel has a singularity at $\zeta = z$. So Stokes' theorem does not apply.

The remedy to this situation is to use an idea developed in [19], [30], [33]: for each fixed $z \in \Omega$, let

$$\psi_z(\zeta) = \left. \frac{N(z,\zeta)}{\Phi^n(z,\zeta)} \right|_{\zeta \in \partial\Omega} .$$

Now construct a *smooth* extension Ψ_z of ψ_z to $\bar{\Omega}$. The Cauchy-Fantappié formula is still valid with Ψ_z replacing ψ_z (since the integral takes place on the boundary where $\Psi_z = \psi_z$). Thus Stokes' theorem can be applied to the new Henkin formula containing Ψ_z. The resulting solid integral operator on $L^2(\Omega)$ can be compared with the Bergman integral via the program of Kerzman and Stein (details are in [33]). The result is that

$$K(z,\zeta) = \Psi_z(\zeta) + (\text{terms which are less singular}) .$$

As a result, curvature, geodesics, etc. of the Bergman metric may be calculated. Also the dependence of these invariants on deformations of $\partial\Omega$ can be determined (see [12], [13]; it should be noted that the methods of [1] or [6] may be used for the deformation study instead of the Kerzman-Stein technique). The following are the three principal consequences of these calculations for a smoothly bounded strongly pseudoconvex Ω:

(α) As $\Omega \ni z \to \partial\Omega$, the Bergman metric curvature tensor at z converges to the constant Bergman metric curvature tensor of the unit ball. The convergence is uniform over $\partial\Omega$.

(β) The kernel and the curvature vary smoothly with smooth perturbations of $\partial\Omega$.

(γ) Ω, equipped with the Bergman metric, is a complete Riemannian manifold.

Now we conclude this paper by coming full circle and discussing once again the topic of symmetry of domains. The reader should consider that, up to now, all of our effort has been directed at obtaining (α), (β), (γ). Now we use those to derive concrete information about symmetries. If $\Omega \subseteq \mathbf{C}^n$ is a domain, let Aut Ω denote the group of biholomorphic self-mappings. If two domains Ω_1 and Ω_2 are biholomorphic we will write $\Omega_1 \approx \Omega_2$.

THEOREM (Bun Wong [41]). *If* $\Omega \subset\subset \mathbf{C}^n$ *is smoothly bounded and strongly pseudoconvex and if* Aut Ω *acts transitively on* Ω, *then* $\Omega \approx$ ball.

Proof (Klembeck). Let $P_0 \in \Omega$ be any fixed point. Let $\{P_j\} \subseteq \Omega$ satisfy $P_j \to \partial\Omega$. By hypothesis, choose $\phi_j \in$ Aut Ω such that $\phi_j(P_0) = P_j$. Then the holomorphic sectional curvature tensor κ for the Bergman metric satisfies

$$\kappa(P_0) = \kappa(\phi_j(P_0)) = \kappa(P_j) \to \text{(constant curvature tensor of the ball).} \qquad (*)$$

Thus the Bergman metric curvature tensor is constant on Ω. We now use

THEOREM (Lu Qi-Keng [34]). *If* M *is a complete connected Kähler manifold with the constant holomorphic sectional curvature of the ball then* M $\approx$ ball.

This theorem, together with (*), completes the proof. □

THEOREM (Greene-Krantz [13]). *If* $\Omega \subseteq \mathbf{C}^n$ *is smoothly bounded and if* $\partial\Omega$ *is* C^∞ *sufficiently close to the unit ball* B *then either*

(i) $\Omega \approx B$

or

(ii) $\Omega \not\approx B$ *and* Aut Ω *is compact and has a fixed point.*

Proof. Step 1. If $\Omega \not\approx B$ then Aut Ω is compact. For if Aut Ω is not compact then a normal families argument [12] implies that for $P_0 \in \Omega$ $\exists$ $\phi_j \in$ Aut Ω such that $\phi_j(P_0) \to \partial\Omega$. As in the proof of the preceding theorem, it follows that $\Omega \approx$ ball.

Step 2. Recall the following result of Cartan-Hadamard (see [24]):

THEOREM. *If* M *is a complete Riemannian manifold of non-positive curvature and if* K *is a compact group of isometries on* M *then* K *has a fixed point.*

Now we prove the theorem by denying (i) and proving (ii).

Step 3. By a calculation, the ball B has negative (bounded from zero) Bergman metric curvature. But the stability result (β) implies that this statement holds for domains Ω which are C^∞ sufficiently close to B. By Step 1, Aut Ω is compact. So the result follows from (γ) and Step 2. □

Now we turn to a conjecture of Lu Qi-Keng (see [34]):

CONJECTURE. If $\Omega \subseteq \mathbb{C}^n$ is a bounded domain then the Bergman kernel $K(z,\zeta)$ never vanishes on $\Omega \times \Omega$.

On the disc and the ball this conjecture is correct; for the ball in $\mathbb{C}^n$ one can calculate (see [31]) that

$$K(z,\zeta) = \frac{n!}{\pi^n} \frac{1}{(1-z\cdot\bar{\zeta})^{n+1}} \ .$$

However it turns out that in $\mathbb{C}^1$ the conjecture is true if and only if Ω is simply connected (see [40]). From this it follows that the conjecture is not always true in $\mathbb{C}^n$ either. To see this, let $\mathbb{C}^2 \supseteq \Omega \equiv$ disc $\times$ annulus. Then the uniqueness of the Bergman kernel easily implies that the kernel for Ω is the product of those for the disc and annulus. Now $\Omega = \cup\Omega_j$ where $\Omega_j \subseteq \Omega_{j+1}$ and each Ω_j is smooth and strictly pseudoconvex (see [31]). By a theorem of Ramadanov [37], $K_{\Omega_j} \to K_\Omega$ normally.

By Hurwitz's theorem [31], K_{Ω_j} vanishes for j large enough. So there exist smooth strictly pseudoconvex domains with vanishing Bergman kernels. Thus we have the

MODIFIED LU QI-KENG CONJECTURE. If $\Omega \subset\subset \mathbf{C}^n$ is smoothly bounded and diffeomorphic to the ball, then K_Ω never vanishes.

Some results about the modified conjecture may now be formulated. Let us agree to topologize the collection of all smoothly bounded strictly pseudoconvex domains by equipping their defining functions with the C^∞ topology. Then we have (see [12], [13]):

(i) If $\mathcal{C} = \{\Omega : K_\Omega$ never vanishes$\}$ then $\mathcal{C}$ is closed.

(ii) If $\mathcal{U} = \{\Omega : K_\Omega$ is bounded from $0\}$ then $\mathcal{U}$ is open.

Statement (i) follows from Hurwitz's theorem. Statement (ii) follows from (β). Facts (i) and (ii), together with the fact that $\mathcal{C}$, $\mathcal{U}$ are non-empty (since both contain the ball), nearly provide a connectedness argument to verify the modified Lu Qi-Keng conjecture. The conjecture was recently resolved in the negative by Boas and by Catlin.

Now we turn to a semi-continuity result for automorphism groups:

THEOREM (Greene-Krantz [14]). *If* $\Omega_0 \subset\subset \mathbf{C}^n$ *is a smoothly bounded strongly pseudoconvex domain and if* Ω *is a sufficiently small smooth perturbation of* Ω_0 *then*

(i) $\operatorname{Aut}\Omega \underset{\text{subgroup}}{\subseteq} \operatorname{Aut}\Omega_0$

(ii) $\exists \phi : \Omega \to \Omega_0$ *a diffeomorphism such that*

$$\operatorname{Aut}\Omega \ni \alpha \mapsto \phi\circ\alpha\circ\phi^{-1} \in \operatorname{Aut}\Omega_0$$

is a univalent group homomorphism.

Sketch of proof. We may as well suppose that $\Omega_0 \not\approx$ ball, else the result is straightforward. Then normal families arguments show that, for Ω sufficiently near Ω_0, $\Omega \not\approx$ ball. Thus $\operatorname{Aut}\Omega$ is compact and, by averaging the Euclidean metric, one can construct a new metric γ, *smooth* across $\partial\Omega$, which is invariant under Aut Ω. By patching this metric with the Bergman metric, and modifying it near $\partial\Omega$, we can arrange that $\mathrm{Isom}(\gamma) = <\mathrm{Aut}\,\Omega, \overline{\mathrm{Aut}\,\Omega}>$ *and* that γ is a product metric near $\partial\Omega$.

Finally, we construct the metric double M of Ω equipped with γ. Known theorems [5] about semi-continuity of isometry groups of deformations of a compact Riemannian manifold now give the result. □

We introduce our final result by recalling a corollary of the Uniformization Theorem (see [2]):

THEOREM. *If* $\Omega \subset\subset \mathbb{C}$, $P \in \Omega$, *and the isotropy group* I_P *of* Aut Ω *is infinite, then* $\Omega \approx \Delta$.

The generalization of this result to $\mathbb{C}^n$ would require new ideas since, in that context, there is no uniformization theorem. Also, on dimensional grounds, the infinitude of I_P is an insufficient hypothesis when $n > 1$. Instead we have

THEOREM (Greene-Krantz [15]). *If* M *is any* n *dimensional, connected, non-compact complex manifold, and if* I_P *has a compact subgroup* H *which acts transitively on real tangent directions at* P, *then* M *is biholomorphic to either the ball or* $\mathbb{C}^n$.

Idea of proof. First create an H-invariant metric on M by averaging over H. By a continuity argument, we show that metric balls B(P,r) centered at P are biholomorphic to the unit ball in $\mathbb{C}^n$. This last is the heart of the argument: It involves analysis of geodesics and equivariance properties of the Bergman metric on B(P,r) and of the canonical solution to the $\bar{\partial}$ equation. Since (by inspection of the proof), the biholomorphisms of B(P,r) to B vary continuously with r, and since the biholomorphisms match up as r increases, the conclusion follows. □

Let us conclude by briefly reviewing the course we have come. We began by exploiting the many symmetries of the disc to derive an integral reproducing formula on the disc. Since generic domains possess few symmetries, we developed two alternate techniques to find integral formulas – via Stokes' theorem and via Hilbert space theory. The latter method has the advantage of being canonical while the former is explicit. We used explicit integral formulas in several complex variables to establish a number of basic results in the theory. Then, using an idea of Kerzman-Stein, we were able to relate the explicit formulas to the canonical ones. Finally, we used this connection between explicit and canonical formulas to return to the question of symmetries of domains. We established results which explain how the automorphism group of a domain Ω depends on the geometry of $\partial\Omega$.

There are still many open problems in the study of automorphism groups of domains. One of the most compelling is to decide which domains have non-compact automorphism groups. Another is to relate the dimension of Aut (Ω) as a Lie group to the rank of the Levi form on $\partial\Omega$. I hope that the survey presented here will inspire some new people to consider these questions.

DEPARTMENT OF MATHEMATICS
THE PENNSYLVANIA STATE UNIVERSITY
UNIVERSITY PARK, PA. 16802

BIBLIOGRAPHY

[1] L. Boutet de Monvel and J. Sjöstrand, Sur la Singularité des noyaux de Bergman et Szegö, *Soc. Mat. de France Asterisque* 34-35 (1976), 123-164.

[2] R. Burckel, *An Introduction to Classical Complex Analysis*, Birkhäuser, Basel, 1979.

[3] D. Catlin, Necessary conditions for subellipticity of the $\bar{\partial}$-Neumann problem, *Ann. of Math.* (2) 117 (1983), 147-172.

[4] ________, Boundary invariants of pseudoconvex domains, to appear.

[5] D. Ebin, The Manifold of Riemannian metrics, *Proc. Symp. in Pure Math.*, Vol XV (Global Analysis), AMS (1970), 11-40.

[6] C. Fefferman, The Bergman kernel and biholomorphic mappings of pseudoconvex domains, *Invent. Math.* 26 (1974), 1-65.

[7] G. Folland and J. J. Kohn, *The Neumann Problem for the Cauchy-Riemann Complex*, Princeton Univ. Press, Princeton, 1972.

[8] G. Folland and E. M. Stein, Estimates for the $\bar{\partial}_b$ complex and analysis of the Heisenberg group, *Comm. Pure Appl. Math.* 27 (1974), 429-522.

[9] J. Fornaess, Strictly pseudoconvex domains in convex domains, *Am. Jour. Math.* 98 (1976), 529-569.

[10] B. A. Fuks, *Introduction to the Theory of Analytic Functions of Several Complex Variables*, Translations of Mathematical Monographs, American Mathematical Society, Providence, 1963.

[11] H. Grauert and I. Lieb, Das Ramirezche Integral und die Gleichung $\bar{\partial}f = \alpha$ im Bereich der Beschrankten Formen, *Rice Univ. Studies* 56 (1970), 29-50.

[12] R. E. Greene and S. G. Krantz, Stability of the Bergman kernel and curvature properties of bounded domains, in *Recent Developments in Several Complex Variables*, J. E. Fornaess, ed., Princeton Univ. Press, Princeton, 1981.

[13] ________, Deformations of complex structures, estimates for the $\bar{\partial}$ equation, and stability of the Bergman kernel, *Adv. Math.* 43 (1982), 1-86.

[14] ________, The automorphism groups of strongly pseudoconvex domains, *Math. Ann.*, 261 (1982), 425-446.

[15] ________, Characterization of complex manifolds by the isotropy subgroups of their automorphism groups, preprint.

[16] P. Greiner and E. M. Stein, *Estimates for the $\bar{\partial}$-Neumann Problem*, Princeton Univ. Press, 1977.

[17] G. M. Henkin, A uniform estimate for the solution of the $\bar{\partial}$-problem on a Weil region, *Uspekhi Math. Nauk.* 26 (1971), 221-212 (Russ.).

[18] ________, Integral representation of functions holomorphic in strictly pseudoconvex domains and some applications, *Mat. Sb.* 78 (120) (1969), 611-632; *Math. U.S.S.R. Sbornik* 7 (1969), 597-616.

[19] G. M. Henkin and A. Romanov, Exact Hölder estimates of solutions of the $\bar{\partial}$ equation, *Izvestija Akad. SSSR*, *Ser. Mat.* (1971), 1171-1183, *Math. U.S.S.R. Sb.* 5 (1971), 1180-1192.

[20] L. Hormander, L^2 estimates and existence theorems for the $\bar{\partial}$ operator, *Acta Math.* 113 (1965), 89-152.

[21] T. Iwinski and M. Skwarczynski, The convergence of Bergman functions for a decreasing sequence of domains, in *Approximation Theory*, Reidel, Boston, 1972.

[22] N. Kerzman, Hölder and L^p estimates for solutions of $\bar{\partial}u = f$ on strongly pseudoconvex domains, *Comm. Pure Appl. Math.* XXIV (1971), 301-380.

[23] N. Kerzman and E.M. Stein, The Szegö kernel in terms of Cauchy-Fantappié kernels, *Duke Math. J.* 45 (1978), 85-93.

[24] S. Kobayashi and K. Nomizu, *Foundations of Differential Geometry*, Vol. I, II, Interscience, New York, 1963, 1969.

[25] J. J. Kohn, Harmonic integrals on strongly pseudoconvex manifolds I, *Ann. Math.* 78 (1963), 112-148; II, *ibid* 79 (1964), 450-472.

[26] ________, Global regularity for $\bar{\partial}$ on weakly pseudoconvex manifolds, *Trans. Am. Math. Soc.*, 181 (1973), 273-292.

[27] ________, Sufficient conditions for subellipticity on weakly pseudoconvex domains, *Proc. Nat. Acad. Sci. (USA)* 74 (1977), 2214-2216.

[28] ________, Subellipticity of the $\bar{\partial}$-Neumann problem on pseudo-convex domains: sufficient conditions, *Acta Math.* 142 (1979), 79-122.

[29] ________, Methods of partial differential equations in complex analysis, *Proc. Symp. Pure Math.* 30, Part 2 (1977), 215-237.

[30] S. Krantz, Optimal Lipschitz and L^p estimates for the equation $\bar{\partial}u = f$ on strongly pseudoconvex domains, *Math. Ann.* 219 (1976), 233-260.

[31] ________, *Function Theory of Several Complex Variables*, John Wiley and Sons, New York, 1982.

[32] ________, Characterization of smooth domains in $\mathbb{C}$ by their biholomorphic self-maps, *Am. Math. Monthly* (1983), 555-557.

[33] E. Ligocka, The Hölder continuity of the Bergman projection and proper holomorphic mappings, preprint.

[34] Lu Qi-Keng, On Kähler manifolds with constant curvature, *Acta Math. Sinica* 16 (1966), 269-281 [Chinese]; *Chinese J. Math.* 9 (1966), 283-298.

[35] N. Øvrelid, Integral representation formulas and L^p estimates for the $\bar{\partial}$ equation, *Math. Scand.* 29 (1971), 137-160.

[36] D.H. Phong and E.M. Stein, Estimates for the Bergman and Szegö projections on strongly pseudoconvex domains, *Duke Math. Jour.* 44 (1977).

[37] I. Ramadanov, Sur une propriete de la fonction de Bergman, *C. R. Acad. Bulgare des Sci.* 20 (1967), 759-762.

[38] R.M. Range, The Caratheodory metric and holomorphic maps on a class of weakly pseudoconvex domains, *Pac. Jour. Math.* 78 (1978), 173-189.

[39] N. Sibony, Un exemple de domain pseudoconvexe regulier ou l'equation $u = f$ n'admet pas de solution bornee pur f bournee, *Invent. Math.* 62 (1980), 235-242.

[40] N. Suita and A. Yamada, On the Lu Qi-Keng conjecture, *Proc. Am. Math. Soc.* 59 (1976), 222-224.

[41] B. Wong, Characterizations of the ball in $\mathbb{C}^n$ by its automorphism group, *Invent. Math.* 41 (1977), 253-257.

VECTOR FIELDS AND NONISOTROPIC METRICS

Alexander Nagel*

The main object of this paper is to show how nonisotropic metrics constructed from vector fields play an important role in certain recent developments in partial differential equations and several complex variables. As we shall see, these metrics are useful in describing boundary behavior of holomorphic functions in pseudoconvex domains, in estimating the kernel of the Szegö projection in some of these domains, and in estimating the size of approximate fundamental solutions to certain nonelliptic, hypoelliptic partial differential operators.

The exposition is divided into three parts. In the first, we set the stage by recalling certain classical theorems which are models for and which motivate the more recent results. In the second part, we outline the construction of metrics from a given family of vector fields. In the third part, we show how to apply this construction to some examples from several complex variables and partial differential equations, and obtain in this way analogues of the results sketched in part one.

It is a pleasure to thank Professor M. T. Cheng, and my other hosts at the University of Peking for their invitation to participate in the Summer Symposium in Analysis in China. It was an honor and a privilege to attend the Symposium, and I am grateful for the very warm hospitality I received. It is also a pleasure to thank E. M. Stein for organizing and directing the Symposium. All of his efforts are greatly appreciated.

---

*Research supported in part by an NSF grant at the University of Wisconsin, Madison.

Much of this paper is an exposition of joint work with Eli Stein and Steve Wainger, and I am particularly grateful to them for many years of stimulation, encouragement, and collaboration.

Part I. *Some classical theorems and examples*

In order to motivate our later discussion, we begin by considering three examples of metrics: the standard Euclidean metric; a nonisotropic but translation invariant metric on R^n; and the translation invariant metric on the Heisenberg group. In the Euclidean case, we see how the balls and metric are involved in Fatou's theorem on nontangential limits of Poisson integrals, and in estimates for the Newtonian potential and related singular integral operators. In the other two examples, we see how analogous estimates can be made for kernels related to the heat operator and to the Kohn Laplacian, and how nonisotropic balls on the Heisenberg group are involved in Korányi's extension of Fatou's theorem to existence of admissible limits of holomorphic functions.

In each of these settings, there is a naturally given family of first order linear homogeneous differential operators, or vector fields. In part II of this paper, we shall see that the general construction of metrics applied to these families of vector fields gives back the natural metric in these classical settings.

The discussion of results in these examples will be very brief, but references are given for the complete proofs of all the results.

§1. *The isotropic Euclidean metric and the Laplace operator*

The standard metric on R^n is defined by

$$|x-y| = \left[\sum_{j=1}^{n} |x_j - y_j|^2\right]^{1/2} .$$

In this example, the important first order operators are just the partial derivatives with respect to the n variables $\frac{\partial}{\partial x_1}, \cdots, \frac{\partial}{\partial x_n}$. The Laplace operator

$$\Delta = \sum_{j=1}^{n} \frac{\partial^2}{\partial x_j^2}$$

is of course just the sum of squares of these first order operators.

We first study the role of the Euclidean metric in the solution of the Dirichlet problem for $R_+^{n+1} = \{(x_1, \cdots, x_n, y) = (x,y) | y > 0\}$. We identify the boundary of R_+^{n+1} with $R^n = R^n \times \{0\}$, and, given a function f on R^n, we want a function $u(x,y)$ harmonic in R_+^{n+1} such that $u(x,y) \to f(x_0)$ as $(x,y) \to (x_0, 0)$.

For continuous boundary data, the problem is completely solved by the Poisson integral formula. Thus suppose f is continuous on R^n, and $f \in L^1(R^n) + L^\infty(R^n)$. Set

$$Pf(x,y) = P_y * f(x) = c_n y \int_{R^n} \frac{f(t)}{[|x-t|^2 + y^2]^{\frac{n+1}{2}}} dt$$

where

$$c_n = \Gamma\left(\frac{n+1}{2}\right) / \pi^{\frac{n+1}{2}} .$$

Then:

(a) Pf is harmonic on R_+^{n+1}.

(b) Pf extends continuously to the boundary and takes on the boundary values f.

(c) $\int_{R^n} |Pf(x,y)|^p dx \le \|f\|_p$ for $1 \le p \le \infty$.

(d) Suppose u is harmonic on R_+^{n+1} and

$$\sup_y \int |u(x,y)|^p dx < \infty .$$

Then if $s > 0$ and $f_s(x) = u(x,s)$,

$$P(f_s)(x,y) = u(x,y+s) .$$

Proofs of these assertions, along with many other of the results discussed here, can be found in Stein [16], Chapter III.

Assertions (c) and (d) above suggest a generalization of the Dirichlet problem to certain classes of discontinuous boundary functions. For $1 \leq p \leq \infty$, let h_p denote the space of functions $u(x,y)$ harmonic on R_+^{n+1} which satisfy

$$\sup_{y>0} \int_{R^n} |u(x,y)|^p dx = \|u\|_{h^p}^p < \infty \quad \text{if} \quad p < \infty$$

$$\sup_{R_+^{n+1}} |u(x,y)| = \|u\|_{h^\infty} < \infty \quad \text{if} \quad p = \infty .$$

Now we are interested in the boundary behavior (along $y = 0$) of functions u in h_p, and it is here that the Euclidean metric begins to play an important role for us. To study the boundary behavior, a fundamental concept is that of a nontangential approach region. For $\alpha > 0$ and $x_0 \in R^n$ define:

$$\Gamma_\alpha(x_0) = \{(x,y) \in R_+^{n+1} \mid |x-x_0| < \alpha y\} .$$

Note that if $B(x_0, \delta) = \{x \in R^n \mid |x-x_0| < \delta\}$ are the balls defined by the standard Euclidean metric, then

$$\Gamma_\alpha(x_0) = \{(x,y) \in R_+^{n+1} | x \in B(x_0, \alpha y)\} .$$

Thus the nontangential approach regions in R_+^{n+1} are really defined in terms of the projection $\pi(x,y) = x$ of R_+^{n+1} onto the boundary, the "height function" $h(x,y) = y$, and the family of Euclidean balls on the boundary. We shall later see that in other examples, natural approach regions can be defined in essentially the same way.

We say that a function $u(x,y)$ has a nontangential limit at $x_0 \in R^n$ if and only if for all $\alpha > 0$, $\lim u(x,y)$ exists as (x,y) approaches $(x_0,0)$ and $(x,y) \in \Gamma_\alpha(x_0)$. In 1906 Fatou [4] proved:

THEOREM 1. *For* $1 \leq p \leq \infty$, *if* $u \in h_p$, *then* u *has a nontangential limit at almost every point of* R^n.

A standard modern approach to this theorem involves two main estimates. The first involves the Hardy-Littlewood maximal operator. Let $f \in L^1_{loc}(R^n)$ and set

$$Mf(x_0) = \sup |B|^{-1} \int_B |f(y)|\,dy$$

where the supremum is taken over all Euclidean balls B which contain x_0. The basic estimates for the maximal operator are given in:

THEOREM 2 (Hardy and Littlewood). *For* $1 \leq p < \infty$, *there are constants* $A_p < \infty$ *so that*

(i) $\|Mf\|_p \leq A_p \|f\|_p$ *if* $1 < p \leq \infty$

(ii) $|\{x \in R^n | Mf(x) > \lambda\}| \leq A_1 \lambda^{-1} \|f\|_1$ *if* $p = 1$.

The very definition of the maximal operator involves the family of Euclidean balls, and the proof of the crucial estimate (ii) depends on a covering lemma for these balls.

The second basic estimate needed to prove Fatou's theorem involves the nontangential supremum of a function defined on R^{n+1}_+. Thus for any $\alpha > 0$ and any $v(x,y)$ defined on R^{n+1}_+ set

$$N_\alpha v(x_0) = \sup_{(x,y) \in \Gamma_\alpha(x_0)} |v(x,y)|\,.$$

For Poisson integrals, this non-tangential supremum is point-wise dominated by the Hardy-Littlewood maximal function of the boundary data:

THEOREM 3 (Hardy and Littlewood). *For* $\alpha > 0$ *there exists a constant* $C_\alpha < \infty$ *so that if* $f \in L^1(R^n) + L^\infty(R^n)$ *and if* $u(x,y) = P_y * f(x)$, *then for all* $x \in R^n$

$$N_\alpha u(x) \leq C_\alpha Mf(x) .$$

These are the two quantitative estimates which underlay the qualitative statement of Fatou's theorem. Complete proofs of these results can be found for example in Stein [16], Chapters I and III. However, since we shall appeal to this kind of argument again, we now recall how Fatou's theorem follows from these two theorems.

Let $p < \infty$ and let $u \in h_p$. If $s > 0$ and if we let $f_s(x) = u(x,s)$, then

$$\begin{aligned}\sup_{(x,y)\in\Gamma_\alpha(x_0)} |u(x,y+s)| &= N_\alpha[P(f_s)](x_0)\\ &\leq C_\alpha[M(f_s)](x_0) .\end{aligned}$$

Therefore if $\lambda > 0$

$$\begin{aligned}&|\{x_0 \in R^n| \sup_{(x,y)\in\Gamma_\alpha(x_0)} |u(x,y+s)| > \lambda\}|\\ &\leq |\{x_0 \in R^n| M(f_s)(x_0) > C_\alpha^{-1}\lambda\}|\\ &\leq [C_\alpha \lambda^{-1} \|M(f_s)\|_p]^p\\ &\leq [C_\alpha A_p \lambda^{-1} \|f_s\|_p]^p\\ &\leq [C_\alpha A_p \lambda^{-1} \|u\|_{h_p}]^p .\end{aligned}$$

Since $s > 0$ was arbitrary, we obtain for any $u \in h_p$

$$|\{x_0 \in R^n| N_\alpha u(x_0) > \lambda\}| \leq [C_\alpha A_p \lambda^{-1} \|u\|_{h^p}]^p . \tag{1}$$

Now let $u \in h_p$ be real valued, and set

$$\Omega_\alpha u(x_0) = \limsup u(x,y) - \liminf u(x,y)$$

where the limits are taken as (x,y) approaches $(x_0,0)$ and $(x,y) \in \Gamma_\alpha(x_0)$. Then the following facts are easy to verify:

(a) $\Omega_\alpha u(x) \leq 2N_\alpha u(x)$

(b) $\Omega_\alpha(u+v)(x) \leq \Omega_\alpha u(x) + \Omega_\alpha v(x)$

(c) $\Omega_\alpha u(x_0) = 0$ if and only if u has a limit within $\Gamma_\alpha(x_0)$

(d) $\Omega_\alpha u(x) \equiv 0$ if $u = Pf$ and f is continuous.

Now let $u_n(x,y) = u\left(x, y + \frac{1}{n}\right)$. Then

$$\Omega_\alpha u = \Omega_\alpha(u - u_n + u_n) \leq \Omega_\alpha(u-u_n) + \Omega_\alpha(u_n)$$

$$= \Omega_\alpha(u-u_n) \leq 2N_\alpha(u-u_n)$$

Hence, by inequality (1), we have:

$$|\{x \in R^n | \Omega_\alpha u(x) > \lambda\}| \leq |\{x \in R^n | N_\alpha(u-u_n) > \lambda/2\}|$$

$$\leq [2C_\alpha A_p \lambda^{-1} \|u-u_n\|_{h^p}]^p .$$

Since $p < \infty$, $\|u-u_n\|_{h^p} \to 0$ as $n \to \infty$, and since $\lambda > 0$ is arbitrary, it follows that

$$|\{x \in R^n | \Omega_\alpha u(x) > 0\}| = 0 .$$

By taking a countable sequence of α's which increase to infinity, we obtain a proof of Fatou's theorem.

It is clear that the family of Euclidean balls plays an important role in this theorem, not only in the definition of nontangential approach regions, but also crucially in the definition of the Hardy-Littlewood maximal operator and the proof of its boundedness. We now recall how these balls are also involved in studying the fundamental solution for the Laplace operator.

An important fundamental solution for Δ is given by the Newtonian potential:

$$N(x) = \begin{cases} \dfrac{1}{(2-n)\omega_n} |x|^{-n+2} & \text{if } n \geq 3 \\ \\ \dfrac{1}{2\pi} \log |x| & \text{if } n = 2 \end{cases}$$

where $\omega_n = 2\pi^{n/2}/\Gamma\left(\frac{n}{2}\right)$. Then $\Delta N = \delta$ as distributions. In particular, if $\phi \in C_0^\infty(R^n)$

$$\phi(x) = \int_{R^n} N(x-y)\Delta\phi(y)\,dy \tag{2}$$

and if

$$\psi(x) = \int_{R^n} N(x-y)\phi(y)\,dy \tag{3}$$

then $\Delta\psi = \phi$.

Proofs of these facts can be found in Folland [5], Chapter 2.

A great deal is known about the operator

$$f \to N * f(x) = \int_{R^n} N(x-y) f(y)\,dy \,.$$

Basically, the fundamental idea is that, when measured with appropriate norms, $N * f$ has two more orders of smoothness than f itself. For example, if f satisfies a Hölder continuity condition of order α, $0 < \alpha < 1$, then $f * N$ is of class C^2 and all second derivatives again satisfy a Hölder continuity condition of order α. (See Bers, John, and Schechter [1], page 232.) Proofs of these continuity properties of the operator $f \to N * f$ depend on size estimates of the kernel $N(x,y) = N(x-y)$ which can be written:

$$|N(x,y)| \leq C\,\delta^2 |B(x,\delta)|^{-1}$$
$$|\nabla_x N(x,y)| + |\nabla_y N(x,y)| \leq C\,\delta |B(x,\delta)|^{-1} \tag{4}$$

where $\delta = |x-y|$. Written in this way, these inequalities again make clear the important role played by the Euclidean metric and the Euclidean balls.

This importance can also be seen when we consider certain singular integral operators. We claimed earlier that $N * f$ is two orders smoother than f. Using formula (2), this means that for all i, j, $\frac{\partial^2 \phi}{\partial x_i \partial x_j}$ should be as smooth as $\Delta\phi$. Thus we are led to the study of the operator which carries $\Delta\phi$ to $\frac{\partial^2 \phi}{\partial x_i \partial x_j}$. There are two ways in which we can think of this.

Formally differentiating equation (2) we see that

$$\frac{\partial^2 \phi}{\partial x_i \partial x_j}(x) = \int_{R^n} k_{ij}(x-y)\Delta\phi(y)\,dy \tag{5}$$

where $k_{ij}(y) = c_n \frac{y_i y_j}{|y|^2} |y|^{-n}$ for an appropriate constant $c_n \neq 0$. We note that the kernel k_{ij} is not locally integrable at 0, so we must study the integral in equation (5) in the principal value sense. Now the kernel $k_{ij}(x,y) = k_{ij}(x-y)$ satisfies the following estimates in terms of the Euclidean metric:

$$|k_{ij}(x,y)| \leq C|B(x,\delta)|^{-1}$$

$$|\nabla_x k_{ij}(x,y)| + |\nabla_y k_{ij}(x,y)| \leq C\delta^{-1}|B(x,\delta)|^{-1}$$

where $\delta = |x-y|$. It is well known how to combine estimates of this type with the Calderon-Zygmund decomposition of L^1 functions to prove $L^p(\mathbf{R}^n)$ boundedness of the operator $\Delta\phi \to \frac{\partial^2\phi}{\partial x_i \partial x_j}$ provided that one knows that this operator is bounded on $L^2(\mathbf{R}^n)$ (see for example, Stein [16], Chapter 2). Boundedness on $L^2(\mathbf{R}^n)$ follows from the cancellation property

$$\int_{a<|x|<b} k_{ij}(x)\,dx = 0 ,$$

and can also be established by studying the operator $\Delta\phi \to \frac{\partial^2\phi}{\partial x_i \partial x_j}$ in terms of Fourier transforms. Define:

$$\hat{\phi}(\xi) = \int_{\mathbf{R}^n} e^{-2\pi i x\cdot\xi}\phi(x)\,dx$$

so that

$$\phi(x) = \int_{\mathbf{R}^n} e^{2\pi i x\cdot\xi}\hat{\phi}(\xi)\,d\xi$$

if, for example, $\phi \in C_0^\infty(\mathbf{R}^n)$. Then:

$$(\Delta\phi)\hat{\ }(\xi) = -4\pi^2|\xi|^2\hat{\phi}(\xi)$$

$$\left(\frac{\partial^2\phi}{\partial x_i \partial x_u}\right)\hat{\ }(\xi) = -4\pi^2\xi_i\xi_j\hat{\phi}(\xi)$$

so

$$\left(\frac{\partial^2\phi}{\partial x_i\,\partial x_j}\right)\hat{}(\xi) = \frac{\xi_i\,\xi_j}{|\xi|^2}(\Delta\phi)\hat{}(\xi)\,.$$

Our operator is thus given on the Fourier transform side by multiplication by a bounded function. It follows from Plancherel's theorem that the operator is bounded on $L^2(R^n)$. Thus we have

THEOREM 4. *For* $1 < p < \infty$ *there are constants* $A_p < \infty$ *so that if* $\phi \in C_0^\infty(R^n)$

$$\left\|\frac{\partial^2\phi}{\partial x_i\,\partial x_j}\right\|_{L^p} \leq A_p\|\Delta\phi\|_{L^p}\,.$$

§2. *Spaces of homogeneous type*

In our discussion of the Laplace operator, we emphasized the role of the Hardy-Littlewood maximal operator and the Calderon-Zygmund decomposition of L^1 functions. These basic tools of Euclidean harmonic analysis can be used in much more general settings, and can be applied to a variety of interesting non-Euclidean examples. What is really necessary for the theory to work is a measure space together with an appropriate family of balls, and many people have been involved in these generalizations; see for example Korányi and Vagi [10], Rivière [14], and Coifman and Weiss [3]. Here I want to briefly sketch the approach of Coifman and Weiss to what they call "spaces of homogeneous type."

A locally compact space X is a *space of homogeneous type* if there is a continuous map $\rho : X \times X \to [0,\infty)$ and a non-negative Borel measure $d\mu$ on X so that:

(1) ρ is a pseudometric; i.e., for all $x,y,z \in X$,

(a) $\rho(x,y) = 0$ if and only if $x = y$

(b) $\rho(x,y) = \rho(y,x)$

(c) $\rho(x,z) \leq K[\rho(x,y) + \rho(y,z)]$.

(2) The measure μ has the "doubling property" relative to the family of falls $B(x,\delta) = \{y \epsilon X \,|\, \rho(x,y) < \delta\}$: There is a constant A so that for all $x \epsilon X$, $\delta > 0$

$$\mu(B(x,2\delta)) \leq A\mu(B(x,\delta)) .$$

In order to emphasize the crucial importance of properties (1c) and (2), we now recall the proof of part (ii) of Theorem 2 in this general setting. Thus if $f \epsilon L^1_{loc}(X,d\mu)$, define

$$Mf(x_0) = \sup |B|^{-1} \int_B |f(y)|\, d\mu(y)$$

where the supremum is taken over all "balls" B which contain x_0. We prove:

THEOREM 5. *There is a constant* A_1 *so that if* $f \epsilon L^1(X,d\mu)$,

$$\mu\{x \epsilon X \,|\, Mf(x) > \lambda\} < A_1 \lambda^{-1} \|f\|_1 .$$

Proof. For $\lambda > 0$, let $E = \{Mf > \lambda\}$, and let $\Sigma \subset E$ be any compact subset. If $x \epsilon \Sigma$, since $Mf(x) > \lambda$ there is a ball B_x containing x so that

$$|B_x|^{-1} \int_{B_x} |f(y)|\, d\mu(y) > \lambda .$$

The balls $\{B_x\}_{x \epsilon \Sigma}$ cover Σ, and since Σ is compact we can find a subcover, which we call $B_1, \cdots, B_N$. Suppose $B_j = B(x_j, \delta_j)$ so B_j has "radius" δ_j. Choose B_{i_1} so that $\delta_{i_1} \geq \delta_j$ for all j. We can then inductively choose $B_{i_1}, \cdots, B_{i_k}$ so that

(1) B_{i_k} is disjoint from $B_{i_1}, \cdots, B_{i_{k-1}}$.

(2) $\delta_{i_k} \geq \delta_j$ for all j such that B_j is disjoint from $B_{i_1}, \cdots, B_{i_{k-1}}$.

In this way we get a finite subsequence $B_{i_1}, \cdots, B_{i_m}$ which are disjoint.

Suppose B_j is a ball from our original finite sequence which does not appear in the subsequence. Then there is a first k so that $B_{i_k} \cap B_j \neq \emptyset$. By property (2), $\delta_{i_k} \geq \delta_j$. Let x_{i_k} be the center of B_{i_k} and x_j the center of B_j. Then if $z \in B_{i_k} \cap B_j$,

$$\begin{aligned} \rho(x_{i_k},x_j) &\leq K[\rho(x_{i_k},z) + \rho(z,x_j)] \\ &\leq 2K\delta_{i_k} . \end{aligned}$$

If $y \in B_j$, then $\rho(x_j,y) < \delta_j$, and so

$$\begin{aligned} \rho(x_{i_k},y) &\leq K[\rho(x_{i_k},x_j) + \rho(x_j,y)] \\ &\leq K(2K+1)\delta_{i_k} . \end{aligned}$$

Let $B^*_{i_k} = B(x_{i_k},K(2k+1)\delta_{i_k})$. Then $B_j \subset B^*_{i_k}$, and so

$$\Sigma \subset \bigcup_{j=1}^{N} B_j \subset \bigcup_{k=1}^{m} B^*_{i_k} .$$

Hence $\mu(\Sigma) \leq \sum_{k=1}^{m} \mu(B^*_{i_k})$. But by property (2) of spaces of homogeneous type

$$\begin{aligned} \mu(B^*_{i_h}) &\leq A^{1+\log_2 K(2K+1)} \mu(B_{i_k}) . \\ &= A_1 \mu(B_{i_k}) . \end{aligned}$$

Thus

$$\mu\left(\Sigma\right) \leq A_1 \sum_{k=1}^{m} \mu(B_{i_k}) \leq A_1\lambda^{-1} \sum_{k=1}^{m} \int_{B_{i_k}} |f(y)|\,dy \leq A_1\lambda^{-1} \|f\|_1$$

since the balls $\{B_{i_k}\}$ are disjoint. Since Σ was an arbitrary compact subset of E, we obtain the same estimate for the measure of E.

§3. *The heat operator and a nonisotropic metric*

We turn now to an example which, though elementary, involves truly nonisotropic phenomena. We consider the heat operator

$$L = \frac{\partial}{\partial t} - \sum_{j=1}^{n} \frac{\partial^2}{\partial x_j^2} = \frac{\partial}{\partial t} - \Delta_x$$

on R^{n+1}, where we use coordinates $(x_1,\cdots,x_n,t) = (x,t)$. Unlike the Laplace operator Δ on R^n, L is not elliptic. Nevertheless there is a remarkable fundamental solution for L. Define

$$E(x,t) = \begin{cases} (4\pi t)^{-\frac{n}{2}} e^{-|x|^2/4t} & \text{if } \quad t > 0 \\ 0 & \text{if } \quad t \leq 0 . \end{cases}$$

Then E is C^∞ on $R^{n+1}\setminus\{(0,0)\}$, and $LE = \delta$ in the sense of distributions. (See Folland [5], Chapter 4.) Thus if $\phi \in C_0^\infty(R_+^{n+1})$

$$\phi(x,t) = \int_0^\infty \int_{R^n} (4\pi s)^{-\frac{n}{2}} e^{-|y|^2/4s}\, \tilde{L}\phi(x-y,t-s)\, dy\, ds$$

$$= \int_{-\infty}^{t} \int_{R^n} (4\pi(t-s))^{-\frac{n}{2}} e^{-(x-y)^2/4(t-s)}\, \tilde{L}\phi(y,s)\, dy\, ds$$

where $\tilde{L} = -\frac{\partial}{\partial t} - \Delta_x$ is the formal adjoint of L.

In order to study the operator $f \to E * f$, we would like to obtain size estimates on the kernel E and its derivatives analogous to those for the

Newtonian potential N in equation (4). However, if $B((x,t),\delta)$ denotes the ball in the standard Euclidean metric, the estimate

$$|E(x,t)| \leq C\,\delta^2 |B((x,t),\delta)|^{-1} \approx \delta^{-n+1}$$

with $\delta = (|x|^2+t^2)^{1/2}$ is false. To see this, let $t = |x|^2$ with $|x|$ small, so $E(x,t) = C|x|^{-n}$, while $\delta = (|x|^2+|x|^4)^{1/2} \approx |x|$, so $\delta^{-n+1} \approx |x|^{-n+1}$. Thus we cannot obtain the appropriate kind of estimate with the Euclidean metric.

To remedy this situation, we make the crucial observation that, like the Newtonian potential N, the fundamental solution E possesses a certain homogeneity, though now this homogeneity is nonisotropic. For $\lambda > 0$ define

$$\delta_\lambda(x,t) = (\lambda x, \lambda^2 t)\ .$$

It is easy to check that

$$E(\delta_\lambda(x,t)) = E(\lambda x, \lambda^2 t) = \lambda^{-n} E(x,t)\ .$$

We associate to the family of dilations a pseudometric

$$\rho((x,t),(y,s)) = (|x-y|^4 + (t-s)^2)^{1/4}$$

so that

$$\rho(\delta_\lambda(x,t), \delta_\lambda(y,s)) = \lambda\rho((x,t),(y,s))\ .$$

The corresponding family of balls

$$B_\rho((x,t),\delta) = \{(y,s) \in \mathbf{R}^{n+1} \,|\, \rho((x,t),(y,s)) < \delta\}$$

are now ellipsoids of size δ in the directions of $x_1, \cdots, x_n$, and of size δ^2 in the direction of t. Thus

$$|B_\rho((x,t),\delta)| \approx \delta^{n+2}\ .$$

Moreover, it follows from the homogeneity of E that we now have:

$$\begin{aligned} &\text{(a)}\quad |E(x,t)| \leq C\,\delta^2 |B_\rho((x,t),\delta)|^{-1} \\ &\text{(b)}\quad |\nabla_x E(x,t)| \leq C\delta |B_\rho((x,t),\delta)|^{-1} \\ &\text{(c)}\quad \left|\frac{\partial E}{\partial t}(x,t)\right| \leq C|B_\rho((x,t),\delta)|^{-1} \\ &\text{(d)}\quad \left|\frac{\partial^2 E}{\partial x_i\,\partial x_j}(x,t)\right| \leq C|B_\rho((x,t),\delta)|^{-1} \end{aligned} \tag{7}$$

where $\delta = \rho((0,0),(x,t))$.

Thus we obtain estimates for the fundamental solution $E(x,t)$ which are exactly analogous to those we have for $N(x)$, provided we view the operator $\frac{\partial}{\partial t}$ as acting like a second order operator, so in equation (7c) we loose two powers of δ rather than one.

We can now use the general theory of spaces of homogeneous type to show that L satisfies certain subelliptic estimates analogous to the elliptic estimates for N given in Theorem 4. For example, one can prove:

THEOREM 6. *For* $1 < p < \infty$ *there are constants* $A_p < \infty$ *so that if* $\phi \in C_0(R^{n+1})$ *then*

$$\left\|\frac{\partial \phi}{\partial t}\right\|_p \leq A_p \|L\phi\|_p$$

$$\left\|\frac{\partial^2 \phi}{\partial x_i\,\partial x_j}\right\|_p \leq A_p \|L\phi\|_p\,.$$

The operator L is the sum of squares of the first order operators $\frac{\partial}{\partial x_1}, \cdots, \frac{\partial}{\partial x_n}$ minus the operator $\frac{\partial}{\partial t}$, which we shall count as an operator of order two. We shall later see how the appropriately weighted vector fields $\frac{\partial}{\partial x_1}, \cdots, \frac{\partial}{\partial x_n}$, and $\frac{\partial}{\partial t}$ give back the nonisotropic metric ρ.

§4. *The Siegel upper half space and the Heisenberg group*

Nonisotropic balls and metrics play an important role in the theory of boundary behavior of holomorphic functions in strongly pseudoconvex domains. This is discussed for example in the monograph by Stein [17]. Here we recall what happens in the case of a model strictly pseudoconvex domain, the generalized upper half space, and its boundary, the Heisenberg group.

We let

$$\Omega = \Big\{(z_1,\cdots,z_n,z_{n+1}) = (z,z_{n+1}) \in \mathbf{C}^{n+1} \,|\, \mathrm{Im} z_{n+1} > \sum_{j=1}^{n} |z_j|^2 = |z|^2\Big\} .$$

Recall that Ω is the image of the unit ball $B = \Big\{(w_1,\cdots,w_{n+1}) \,|\, \sum_{j=1}^{n+1} |w_j|^2 < 1\Big\}$ under the biholomorphic map

$$z_k = \frac{w_k}{1+w_{n+1}} \qquad 1 \leq k \leq n$$

$$z_{n+1} = i\,\frac{1-w_{n+1}}{1+w_{n+1}} .$$

The boundary of Ω is the set

$$\partial\Omega = \{(z,z_{n+1}) \in \mathbf{C}^{n+1} \,|\, \mathrm{Im} z_{n+1} = |z|^2\}$$

$$= \{(z,t+i|z|^2) \,|\, z \in \mathbf{C}^n, t \in \mathbf{R}\} .$$

Thus we can identify $\partial\Omega$ with $\mathbf{C}^n \times \mathbf{R}$ by associating to the point $(z,t+i|z|^2) \in \partial\Omega$ the point $(z,t) \in \mathbf{C}^n \times \mathbf{R}$. We can make $\mathbf{C}^n \times \mathbf{R}$ into a group H_n, the Heisenberg group, by defining

$$(z,t) \cdot (w,s) = (z+w,\, t+s+2\mathrm{Im}\langle z,w\rangle)$$

where $\langle z,w\rangle = \sum_{j=1}^{n} z_j \overline{w}_j$. This definition can be motivated and the group properties verified by the following considerations: to each point $h = (w,s) \in \mathbf{C}^n \times \mathbf{R}$ we associate a holomorphic map $T_h : \mathbf{C}^{n+1} \to \mathbf{C}^{n+1}$ by the formula

$$T_h(z,z_{n+1}) = (z+w, z_{n+1} + (s+i|w|^2) + 2i<z,w>) .$$

T_h is clearly holomorphic. Moreover, if $h_j = (w_j,s_j)$, $j = 1,2$, then a simple calculation shows that

$$T_{h_2} \cdot T_{h_1}(z,z_{n+1}) = T_{h_3}(z,z_{n+1})$$

where

$$\begin{aligned} h_3 &= (w_1+w_2, s_1+s_2+2\mathrm{Im}<w_2w_1>) \\ &= h_2 \cdot h_1 . \end{aligned}$$

From this it follows that H_n is a group, and

$$T_{h_2} \cdot T_{h_1} = T_{h_2 \cdot h_1} .$$

Let $\rho(z,z_{n+1}) = \mathrm{Im} z_{n+1} - |z|^2$ be the "height function" for Ω so that $(z,z_{n+1}) \in \Omega$ if and only if $\rho(z,z_{n+1}) > 0$, and $(z,z_{n+1}) \in \partial\Omega$ if and only if $\rho(z,z_{n+1}) = 0$. If $h = (w,s) \in H_n$ we compute

$$\begin{aligned} \rho(T_h(z,z_{n+1})) &= \mathrm{Im}(z_{n+1} + s + i|w|^2 + 2i<z,w>) - |z+w|^2 \\ &= \mathrm{Im} z_{n+1} - |z|^2 \\ &= \rho(z,z_{n+1}) . \end{aligned}$$

Thus T_h carries Ω to itself and $\partial\Omega$ to itself. If $h = (w,s) \in H_n$, we have identified (w,s) with the point $(w, s+i|w|^2) \in \partial\Omega$, and this is the same as $T_h(0,0)$. Thus under the identification, H_n acts on $\partial\Omega$ as follows: if $(z,t) \in \partial\Omega$ and if $h = (w,s) \in H_n$, then

$$\begin{aligned} T_h(z,t) &= T_h T_{(z,t)}(0,0) \\ &= T_{h\cdot(z,t)}(0,0) \\ &= (w,s)\cdot(z,t) \\ &= (z+w,\ t+s-2\mathrm{Im}\langle z,w\rangle) . \end{aligned}$$

We now make H_n (or $\partial\Omega$) into a space of homogeneous type. Define $d : H_n \times H_n \to [0,\infty)$ by

$$d((z,t);(w,s)) = [|z-w|^4 + |t-s+2\mathrm{Im}\langle z,w\rangle|^2]^{1/4} ,$$

or, if we identify H_n with $\partial\Omega$,

$$d((z,z_{n+1});(w,w_{n+1})) = [|z-w|^4 + \mathrm{Re}(z_{n+1}-w_{n+1}-2i\langle z,w\rangle)|^2]^{1/4} .$$

The function d is invariant under the action of H_n. Thus if $h = (u,r) \in H_n$, easy algebraic manipulations show that

$$d(T_h(z,t), T_h(w,s)) = d((z,t),(w,s)) .$$

If we define the balls

$$B((w,s),\delta) = \{(z,t) \mid d((z,t),(w,s)) < \delta\}$$

then this invariance property means that

$$B((w,s),\delta) = T_{(w,s)}(B((0,0),\delta)) .$$

We claim that d is a pseudometric. For if $\tilde{z} = (z,t)$, $\tilde{w} = (w,s)$ and $\tilde{u} = (u,r)$ are in H_n with

$$d(\tilde{z},\tilde{w}) < \delta,\ d(\tilde{w},\tilde{u}) < \delta$$

then

$$\begin{aligned} &|z-w| < \delta \qquad && |t-s+2\mathrm{Im}\langle z,w\rangle| < \delta^2 \\ &|w-u| < \delta \qquad && |s-r+2\mathrm{Im}\langle w,u\rangle| < \delta^2 . \end{aligned}$$

Hence

$$|z-u| \leq |z-w| + |w-u| < 2\delta ,$$

and

$$\begin{aligned} |t-r+2\mathrm{Im}<z,u>| &\leq |t-s+2\mathrm{Im}<z,w>| + |s-r+2\mathrm{Im}<w,u>| \\ &\quad + |2\mathrm{Im}\,(<z,u>-<z,w>-<w,u>)| \\ &\leq 2\delta^2 + 2|\mathrm{Im}<z-w,u-w>| \\ &< 4\delta^2 = (2\delta)^2 . \end{aligned}$$

Therefore

$$\begin{aligned} d(\tilde{z},\tilde{u}) &\leq 2\max\,(d(\tilde{z},\tilde{w}), d(\tilde{w},\tilde{u})) \\ &\leq 2[d(\tilde{z},\tilde{w})+d(\tilde{w},\tilde{u})] . \end{aligned}$$

What do the corresponding balls $B(\tilde{z},\delta)$ look like? We see that $\tilde{w} \in B(\tilde{z},\delta)$ is essentially equivalent to the pair of inequalities:

$$|z-w| < \delta$$

$$|\mathrm{Re}\,(z_{n+1}-w_{n+1}-2i<z,w>)| < \delta^2 .$$

Fix $\tilde{z} = (z,z_{n+1}) \in \partial\Omega$. The complex tangent space to $\partial\Omega$ at $\tilde{z}$ is given by the equation

$$\omega_{n+1} - \tilde{z}_{n+1} - 2i<w,z> = 0 .$$

If $\tilde{w} = (w,w_{n+1}) \in \partial\Omega$, the distance from $\tilde{w}$ to this complex tangent plane is essentially

$$\begin{aligned} &|w_{n+1}-\overline{z}_{n+1}-2i<w,z>| \\ &= |\mathrm{Re}\,(w_{n+1}-z_{n+1})+i(|w|^2+|z|^2-2<w,z>)| \\ &= |\mathrm{Re}\,(w_{n+1}-z_{n+1}-2i<z,w>)+i|w-z|^2| \\ &\approx |w-z|^2+(\mathrm{Re}\,(w_{n+1}-z_{n+1}-2i<z,w>)| \\ &\lesssim \delta^2 . \end{aligned}$$

Thus the balls $B(\tilde{z},\delta)$ are essentially "ellipsoids" of size δ in the directions of the complex part of the tangent space to $\partial\Omega$ at $\tilde{z}$, and of size δ^2 in the orthogonal real direction, and hence in particular

$$|B(\tilde{z},\delta)| \approx \delta^{2n+2} .$$

Thus the doubling property of the balls is verified, and $\partial\Omega$ (or H_n) equipped with the pseudometric d is indeed a space of homogeneous type.

We now want to discuss the analogue of Fatou's theorem for boundary behavior of holomorphic functions in Ω. This problem was first studied by Koranyi [9] for domains like Ω, and was later generalized by Stein [17] to general smoothly bounded domains in $\mathbf{C}^n$. Here we want to emphasize the role of the nonisotropic balls on the boundary, in analogy with the role of Euclidean balls on $\mathbf{R}^n$ in Fatou's theorem.

We begin by defining appropriate *nonisotropic approach regions* in Ω. Let $\pi:\Omega \to \partial\Omega$ be the projection

$$\pi(z,z_{n+1}) = (z,z_{n+1}-i\rho(z,z_{n+1})) .$$

For $\alpha > 0$ and $\tilde{w} = (w,s+i|w|^2) \in \partial\Omega$ let

$$A_\alpha(\tilde{w}) = \{(z,z_{n+1}) \in \Omega \mid \pi(z,z_{n+1}) \in B(\tilde{w},\alpha\rho(z,z_{n+1})^{1/2})\}$$

where of course $B(w,\delta)$ is the nonisotropic ball defined by the pseudometric d. It is clear that this definition is analogous to our earlier definition of nontangential approach regions $\Gamma_\alpha(x_0)$ in $\mathbf{R}^{n+1}_+$.

Now $(z,t+i|z|^2+iy) \in A_\alpha(\tilde{w})$ is essentially equivalent to the following pair of inequalities:

$$|z-w| < \alpha y^{1/2}$$

$$|t-s+2\mathrm{Im}\langle z,w\rangle| < \alpha y .$$

As above, $|t-s+2\mathrm{Im}\langle z,w\rangle|$ is essentially the distance from $(z,t+i|z|^2)$ to the complex part of the tangent space at $\tilde{w}$. Thus $A_\alpha(\tilde{w})$ allows tangential approach to $\tilde{w}$ of order two in the complex directions at $\tilde{w}$, but requires nontangential approach in the complementary real direction, and so the sets $\{A_\alpha(\widetilde{w})\}$ are essentially the admissible approach regions introduced by Korányi [9].

The invariance of the balls $B(\widetilde{w},\delta)$ under the action of H_n, together with the fact that the height function ρ is invariant under the mappings T_h, imply that the approach regions $A_\alpha(\widetilde{w})$ are carried into each other under the action of H_n. Thus it is easy to check that if $h = (w,s) \in H_n$, then $(z,z_{n+1}) \in A_\alpha((0,0))$ if and only if $T_h(z,z_{n+1}) \in A_\alpha(w,s+i|w|^2)$.

Let F be a function defined on Ω, and for $\zeta \in \partial\Omega$, set

$$N_\alpha F(\zeta_0) = \sup_{(z,z_{n+1})\in A_\alpha(\zeta_0)} |F(z,z_{n+1})| .$$

If $f \in L^1_{loc}(\partial\Omega)$, define

$$Mf(\zeta_0) = \sup |B|^{-1} \int_B |f(\zeta)|\, d\sigma(\zeta)$$

where the supremum is taken over all nonisotropic balls B containing ζ_0, and $d\sigma$ is surface area measure on $\partial\Omega$.

We now have the following analogues of Theorems 2 and 3, which were used to prove Fatou's theorem.

THEOREM 7. *For* $1 \le p \le \infty$ *there are constants* $A_p < \infty$ *so that*

(i) $\|Mf\|_p \le A_p \|f\|_p \qquad 1 < p \le \infty$

(ii) $|\{\zeta \in \partial\Omega |\, Mf(\zeta) > \lambda\}| \le A_1 \lambda^{-1} \|f\|, \quad$ *if* $\quad p = 1$.

THEOREM 8. *Suppose* u *is continuous on* $\overline{\Omega}$ *and pleurisubharmonic on* Ω. *For* $\alpha > 0$ *there is a constant* C_α *(independent of* u *) so that for all* $\zeta \in \partial\Omega$

$$N_\alpha u(\zeta) \le C_\alpha Mu(\zeta) .$$

Theorem 7 of course follows from Theorem 5 and the Marcinkiewicz interpolation theorem (see Stein [16], Chapter I), since we already know $\partial\Omega$ is a space of homogeneous type. Before proving Theorem 8, we point out some of the consequences of these results.

For $1 \leq p \leq \infty$, we let $H_p(\Omega)$ denote the space of holomorphic functions on Ω which satisfy

$$\sup_{y>0} \int_{\partial\Omega} |F(z,t+i|z|^2+iy)|^p \, d\sigma(z,t) = \|F\|_{H_p}^p < \infty \quad \text{if} \quad p < \infty$$

$$\sup_{\Omega} |F(z,z_{n+1})| = \|F\|_{H_\infty} < \infty \quad \text{if} \quad p = \infty .$$

COROLLARY. *For* $1 \leq p \leq \infty$ *and* $\alpha > 0$. *There are constants* $A_{p,\alpha} < \infty$ *so that if* $F \in H_p(\Omega)$

(i) $\|N_\alpha F\|_{L^p} \leq A_{p,\alpha} \|F\|_{H_p}$ *for* $1 < p \leq \infty$

(ii) $|\{\zeta \in \partial\Omega | N_\alpha F(\zeta) > \lambda\}| \leq A_{1,\alpha} \lambda^{-1} \|F\|_{H_1}$ *if* $p = 1$.

Proof. Put $F_\varepsilon(z,z_{n+1}) = F(z,z_{n+1}+i\varepsilon)$. Then F_ε is continuous on $\overline{\Omega}$, and pleurisubharmonic on Ω, so by Theorem 8

$$N_\alpha F_\varepsilon(\zeta) \leq C_\alpha M F_\varepsilon(\zeta) .$$

Since $\|F_\varepsilon\|_{L^p} \leq \|F\|_{H_p}$ for $1 \leq p \leq \infty$ we see that

$$|\{N_\alpha F_\varepsilon > \lambda\}| \leq A_1 C_\alpha \lambda^{-1} \|F\|_{H_1}$$

and

$$\|N_\alpha F_\varepsilon\|_{L^p} \leq A_p C_\alpha \|F\|_{H_p} \quad \text{if} \quad p > 1 .$$

But $N_\alpha F_\varepsilon(\zeta) \nearrow N_\alpha F(\zeta)$ as $\varepsilon \to 0$, and

$$\{N_\alpha F > \lambda\} = \bigcup_{\varepsilon < 0} \{N_\alpha F_\varepsilon > \lambda\}$$

so the corollary follows by the monotone convergence theorem and the regularity of the measure $d\sigma$.

We can now apply exactly the same argument we used for Fatou's theorem to prove

THEOREM 9 (Korányi). *If* $F \in H_p(\Omega)$, *then* F *has admissible limits at almost every point of* $\partial\Omega$.

It remains to prove Theorem 8. For simplicity, we deal with the case $n = 1$. Since the approach regions $A_\alpha(\zeta)$ are carried into each other by the action of H_n, it suffices to prove the estimate at the origin $(0,0)$. We first obtain an estimate on the *radial* maximal function:

$$|u(0,iy)| \le \frac{4}{\pi y^2} \iint_{|s|^2+|t-y|^2<\left(\frac{y}{2}\right)^2} |u(0,s+it)|\,ds\,dt$$

$$\le \frac{2}{\pi^2 y^2} \iint_{|s|^2+|t-y|^2<\left(\frac{y}{2}\right)^2} \int_0^{2\pi} |u(\sqrt{t}e^{i\theta},s+it)|\,d\theta\,ds\,dt$$

$$\le \frac{2}{\pi^2 y^2} \int_0^{2\pi} \int_{-\frac{y}{2}}^{\frac{y}{2}} \int_{\frac{y}{2}}^{\frac{3y}{2}} |u(\sqrt{t}e^{i\theta},s+it)|\,dt\,ds\,d\theta\,.$$

Letting $r = \sqrt{t}$, and interchanging the order of integration, we get

$$|u(0,iy)| \leq \frac{4}{\pi^2 y^2} \int_{-\frac{y}{2}}^{\frac{y}{2}} \int_0^{2\pi} \int_{\sqrt{\frac{y}{2}}}^{\sqrt{\frac{3y}{2}}} |u(re^{i\theta}, s+ir^2)| r dr d\theta ds$$

$$\leq \frac{4}{\pi y^2} \int_{B((0,0),2\sqrt{y})} |u(\zeta)| d\sigma(\zeta)$$

$$\approx |B((0,0),2\sqrt{y})|^{-1} \int_{B((0,0),2\sqrt{y})} |u(\zeta)| d\sigma(\zeta) .$$

Thus, by translation invariance, we see there is a constant A so that for $(z,z_{n+1}) \epsilon \Omega$

$$|u(z,z_{n+1})| \leq A|B|^{-1} \int_B |u(\zeta)| d\sigma(\zeta)$$

where $B = B(\pi(z,z_{n+1}), 2\rho(z,z_{n+1})^{1/2})$.

Now let $(z,z_{n+1}) \epsilon A_\alpha((0,0))$. Then $\pi(z,z_{n+1}) \epsilon B((0,0), \alpha\rho(z,z_{n+1})^{1/2})$, so

$$d(\pi(z,z_{n+1}), (0,0)) < \alpha\rho(z,z_{n+1})^{1/2} .$$

If $\zeta \epsilon B(\pi(z,z_{n+1}), 2\rho(z,z_{n+1})^{1/2})$ it follows that

$$d(\zeta, (0,0)) \leq (2\alpha+4)\, \rho(z,z_{n+1})^{1/2} .$$

But now using the doubling property of the balls, it follows that there is a constant A so that if $(z,z_{n+1}) \epsilon A_\alpha(0,0)$

$$|u(z,z_{n+1})| \leq A'|B|^{-1} \int_B |u(\zeta)| d\sigma(\zeta) \tag{8}$$

where $B = B((0,0), (2\alpha+4)\rho(z,z_{n+1})^{1/2})$, and this last average is certainly dominated by $A'Mu(0,0)$. We remark that estimates of type (7) are actually false for Poisson integrals of harmonic functions.

The nonisotropic balls on $\partial\Omega$ are also useful in studying singular integrals and fundamental solutions. For example, the Cauchy-Szegö projection of $L^2(\partial\Omega)$ onto $H_2(\Omega)$ is given by the operator

$$Sf(z,z_{n+1}) = \int_{\partial\Omega} f(w,w_{n+1})S((z,z_{n+1}),(w,w_{n+1}))\,d\sigma$$

where

$$S((z,z_{n+1}),(w,w_{n+1})) = c_n[i(\overline{w}_{n+1}-z_{n+1})-2<z,w>]^{-(n+1)}$$

with $c_n = 2^{n-1}n!/\pi^{n+1}$ (see Nagel and Stein [11], page 23). In particular, for $z = (\widetilde{z},t)$ and $\widetilde{w} = (w,s)$ on $\partial\Omega$ we have,

$$S(z,w) = c_n(i)^{-(n+1)}[(s-t-2\mathrm{Im}<z,w>)-i|z-w|^2]^{-(n+1)}.$$

Thus $|S(\widetilde{z},\widetilde{w})| \approx |B(\widetilde{z},\delta)|^{-1}$ where $\delta = d(\widetilde{z},\widetilde{w})$. It is also easy to check that derivatives of S yield estimates with corresponding negative powers of δ. Thus S behaves like a singular integral kernel relative to this family of nonisotropic balls, and from this follows L^p and Lipschitz estimates for the operator $f \to Sf$ (see Korányi and Vagi [10]).

Finally, we consider fundamental solutions. On H_n let

$$X_j = \frac{\partial}{\partial x_j} + 2y_j\frac{\partial}{\partial t}, \quad Y_j = \frac{\partial}{\partial y_j} - 2x_j\frac{\partial}{\partial t}, \quad T = \frac{\partial}{\partial t}$$

where we write $z_j = x_j + iy_j$. These vector fields form a basis for the left invariant vector fields on H_n. Put

$$Z_j = \frac{1}{2}(X_j - iY_j), \quad \overline{Z}_j = \frac{1}{2}(X_j + iY_j)$$

and consider for $\alpha \in \mathbf{C}$.

$$\mathcal{L}_\alpha = -\frac{1}{2}\sum_{j=1}^{n} (Z_j\overline{Z}_j + \overline{Z}_j Z_j) + i\alpha T \ .$$

This second order operator arises in the following way: if we identify H_n with $\partial\Omega$, the vector fields $\overline{Z}_j$ annihilate the boundary values of holomorphic functions. Thus, in analogy with the operator $\overline{\partial}$, we consider

$$\overline{\partial}_b f = \sum_{j=1}^{n} \overline{Z}_j d\overline{\zeta}_j \, , \qquad \text{on functions,}$$

and we extend this in the usual way to $(0,q)$ forms on $\partial\Omega$. In $L^2(H_n)$ we can define a formal adjoint $(\overline{\partial}_b)^*$, and the Kohn Laplacian is then

$$\Box_b = \overline{\partial}_b\overline{\partial}_b^* + \overline{\partial}_b^*\overline{\partial}_b \ .$$

On q forms, $\Box_b^{(q)}$ acts diagonally, and is given by the operator $\mathcal{L}_\alpha$ where $\alpha = n - 2q$. The operator $\Box_b$ is not elliptic but Kohn's fundamental work [8a] showed that one can obtain subelliptic estimates for $\mathcal{L}_\alpha$. Folland and Stein [6] discovered a fundamental solution for $\mathcal{L}_\alpha$. Define:

$$\phi_\alpha(z,t) = (|z|^2 - it)^{-\left(\frac{n+\alpha}{2}\right)}(|z|^2 + it)^{-\left(\frac{n-\alpha}{2}\right)}$$

THEOREM 10 (Folland and Stein). *$\mathcal{L}_\alpha\phi_\alpha = c_\alpha\delta$ in the sense of distributions, where c_α is a constant, and $c_\alpha \neq 0$ if $\alpha \neq \pm n$, $\pm(n+2)$, $\pm(n+4)$, etc.*

Thus except for the exceptional values of α, $\frac{1}{c_\alpha}\phi_\alpha$ is a fundamental solution for $\mathcal{L}_\alpha$, and it is easy to verify that

$$|\phi_\alpha(z,t)| \leq C\delta^2 |B((0,0),\delta)|^{-1}$$

where $\delta = d((0,0),(z,t))$. One also obtains corresponding estimates for derivatives of ϕ_α, so that again there is a complete analogy with the estimates (4) for the Newtonian potential.

In the case of the Heisenberg group, the basic vector fields are $X_1, \cdots, X_n, Y_1, \cdots, Y_n$, and the vector field T which is given "weight" two. We shall later see how the general construction applied to these vector fields gives the nonisotropic pseudometric d.

Part II. *Metrics defined by vector fields*

Our object in this part of the paper is to outline the construction of metrics from certain families of vector fields. Many details of the arguments will be omitted, and complete proofs can be found in [13]. In [7], Hörmander studied differentiability along noncommuting vector fields, and used the techniques of exponential mappings and the Campbell-Hausdorff formula. The case of vector fields of type 2 was studied in [11]. Balls reflecting commutation properties of vector fields have also been studied by Fefferman and Phong [4a], by Folland and Hung [5a], and by Sánchez-Calle [15a].

Let $\Omega \subset \mathbf{R}^N$ be a connected open set, and let $Y_1, \cdots, Y_q$ be C^∞ real vector fields defined on a neighborhood of $\overline{\Omega}$. We associate to each vector field Y_j an integer $d_j = d(Y_j) \geq 1$ which we call the formal degree, and we make two fundamental assumptions about this collection of vector fields.

(1) For each $x \in \overline{\Omega}$, the vectors $\{Y_1(x), \cdots, Y_q(x)\}$ span $\mathbf{R}^N$.

(2) For all j, k, we can write $[Y_j, Y_k] = \sum_{d_\ell \leq d_j + d_k} c^\ell_{jk}(x) Y_\ell$ where $c^\ell_{jk} \in C^\infty(\overline{\Omega})$. Here $[X,Y] = XY - YX$ is the commutator of the two vector fields.

There are several basic examples to keep in mind.

(A) Let $N = q = n$, let $Y_j = \frac{\partial}{\partial x_j}$ and let $d_j = 1$ for $1 \leq j \leq n$. In this case we shall recover the standard Euclidean metric on $\mathbf{R}^n$.

(B) Let $N = q = n+1$, let $Y_j = \frac{\partial}{\partial x_j}$, and let $d_j = 1$ for $1 \leq j \leq n$ and let $d_{n+1} = 2$. In this case we shall obtain the nonisotropic metric on $\mathbf{R}^{n+1}$ appropriate for the study of the heat operator

$$\frac{\partial}{\partial x_{n+1}} - \sum_{j=1}^{n} \frac{\partial^2}{\partial x_j^2} .$$

(C) Let $N = q = 2n+1$ and let

$$Y_j = \frac{\partial}{\partial x_j} + 2y_j \frac{\partial}{\partial t} \qquad 1 \le j \le n$$

$$\text{with } d_j = 1$$

$$Y_{n+j} = \frac{\partial}{\partial y_j} - 2x_j \frac{\partial}{\partial t} \qquad 1 \le j \le n$$

$$Y_{2n+1} = [Y_j, Y_{n+j}] = -4 \frac{\partial}{\partial t} \qquad \text{with } d_{2n+1} = 2 .$$

In this case of course we are dealing with the Heisenberg group, and we shall recover the invariant metric on H_n defined earlier.

(D) (An example of Grushin type). Let $N = 2$, $q = 3$, and let

$$Y_1 = \frac{\partial}{\partial x_1}; \quad Y_2 = x_1 \frac{\partial}{\partial x_2}; \quad Y_3 = [Y_1, Y_2] = \frac{\partial}{\partial x_2}$$

with $d_1 = d_2 = 1$ and $d_3 = 2$. This example leads to a metric appropriate for studying the hypoelliptic operator $\frac{\partial^2}{\partial x_1^2} + x_1^2 \frac{\partial^2}{\partial x_2^2}$. See Grushin [6a].

(E) (This example generalizes (A), (C), and (D).) Let $X_1, \cdots, X_p$ be C^∞ real vector fields on $\bar{\Omega}$. Let

$$X^{(1)} = \{X_1, \cdots, X_p\}$$

$$X^{(2)} = \{\cdots, [X_i, X_j], \cdots\}$$

$$X^{(3)} = \{\cdots, [X_i, [X_j, X_k]], \cdots\}, \text{ etc.}$$

so that $X^{(k)}$ is a vector whose components are all the commutators of $X_1, \cdots, X_p$, of length k. We say that the vector fields $X_1, \cdots, X_p$ are of Hörmander type m or finite type m if at each $x \in \bar{\Omega}$, the components of $X^{(1)}, \cdots, X^{(m)}$ span R^N.

Let $Y_1, \cdots, Y_q$ be some enumeration of the components of $X^{(1)}, \ldots, X^{(m)}$, and if Y_i is an element of $X^{(j)}$ we set $d(Y_i) = j$. Then property (1) follows from the assumption of finite type, while property (2) follows from the Jacobi identity for commutators.

This example leads to a metric appropriate for studying the hypoelliptic operator $X_1^2 + \cdots + X_p^2$ and its variants.

(F) (This generalizes (B).) Let $X_0, X_1, \cdots, X_p$ be C^∞ real vector fields as in (E) of finite type, but this time let $d(X_j) = 1$ for $1 \le j \le n$ while $d(X_0) = 2$, with appropriate weighting of higher commutators. This example leads to a metric appropriate for studying the generalization of the heat operator $X_1^2 + X_1^2 + \cdots + X_p^2 - X_0$.

Our object now is to construct natural metrics out of the family of vector fields $Y_1, \cdots, Y_q$. I want to distinguish between two general approaches to this problem.

On the one hand, one can define a metric in terms of a "global definition." Here, we let the metric be given as the infimum of some functional over a large class of curves. Then the defining properties of a metric, such as the triangle inequality, are relatively easy to verify, but the local geometry of the corresponding family of balls is hard to understand.

On the other hand, we can define a metric in terms of a "local definition"; here we want the metric to be given in terms of an exponential mapping. Now the local geometry is clearer, but it is much more difficult to verify that one really has a metric.

We now discuss each of these approaches, and then try to sketch why the two definitions are in fact equivalent.

§5. *Global definitions*

DEFINITION 1. Let $x_0, x_1 \in \Omega$, and say $\rho(x_0, x_1) < \delta$ if and only if there is an absolutely continuous map $\phi : [0,1] \to \Omega$ with $\phi(j) = x_j$, $j = 0, 1$, so that for almost all $t \in (0,1)$

$$\phi'(t) = \sum_{j=1}^{q} a_j(t) Y_j(\phi(t))$$

where $|a_j(t)| < \delta^{d(Y_j)}$.

PROPOSITION. *ρ is a metric on Ω. For every compact set $\Sigma \subset\subset \Omega$ there are constants C_1, C_2 so that if $x_0, x_1 \in \Sigma$*

$$C_1 |x_0 - x_1| \leq \rho(x_0, x_1) \leq C_2 |x_0 - x_1|^{\frac{1}{m}} .$$

where $|x_0 - x_1|$ is the standard Euclidean metric.

We sketch the proof:

(1) That $\rho(x,y) = 0$ if and only if $x = y$ is clear.

(2) $\rho(x,y) = \rho(y,x)$ since we can replace $\phi(t)$ by $\phi(1-t)$.

(3) Given x and y, there *is* a smooth curve joining them, so $\rho(x,y) < \infty$.

(4) The triangle inequality: Suppose $x,y,z \in \Omega$. Given $\varepsilon > 0$, there are curves $\phi, \psi : [0,1] \to \Omega$ with $\phi(0) = x$, $\phi(1) = y$, $\psi(0) = y$, $\psi(1) = z$, $\phi'(t) = \sum_1^q a_j(t) Y_j(\phi(t))$, $\psi'(t) = \sum_1^q b_j(t) Y_j(\psi(t))$, with

$$|a_j(t)| \leq (d(x,y) + \varepsilon)^{d(Y_j)}, \quad |b_j(t)| \leq d(y,z) + \varepsilon)^{d(Y_j)} .$$

Define $\theta: [0,1] \to \Omega$ by

$$\theta(t) = \begin{cases} \phi(\alpha t) & 0 \leq t \leq 1/\alpha \\ \psi\left(\dfrac{\alpha t - 1}{\alpha - 1}\right) & \dfrac{1}{\alpha} \leq t \leq 1 \end{cases}$$

where $\alpha = 1 + \dfrac{d(y,z)+\varepsilon}{d(x,y)+\varepsilon}$. Then $\theta(0) = x$, $\theta(1) = z$ and $\theta'(t) = \sum_{j=1}^{q} c_j(t) Y_j(\theta(t))$ where

$$|c_j(t)| \leq (d(x,y) + d(y,z) + 2\varepsilon)^{d(Y_j)} .$$

This shows, since ε is arbitrary, that $d(x,z) \leq d(x,y)+d(y,z)$.

Now let $\Sigma \subset \Omega$ be compact. Then there is a constant $C = C_\Sigma$ so that if $x,y \in \Sigma$, there is a smooth curve ϕ joining x to y with $|\phi'(t)| \leq C|x-y|$. Since $Y_1,\cdots,Y_q$ span R^N at every point, we can write $\phi'(t) = \sum_{j=1}^{q} c_j(t)Y_j$ with $|c_j(t)| \leq C|\phi'(t)|$. But then:

$$|c_j(t)| \leq C|x-y| = C[|x-y|^{\frac{1}{d(Y_j)}}]^{d(Y_j)}$$

$$\leq C[|x-y|^{\frac{1}{m}}]^{d(Y_j)}$$

and so $\rho(x,y) \leq C|x-y|^{\frac{1}{m}}$.

Conversely, if $x,y \in \Sigma$ and $\rho(x,y) = \delta$, there is a curve ϕ joining x to y with $\phi'(t) = \sum_{j=1}^{q} a_j(t)Y_j(\phi(t))$ and $|a_j(t)| \leq (2\delta)^{d(Y_j)}$ almost everywhere. But then

$$|x-y| = \left| \int_0^1 \phi'(t)\,dt \right|$$

$$< \int_0^1 \sum_{j=1}^{q} |a_j(t)|\,|Y_j(\phi(t))|\,dt$$

$$\leq C\delta$$

since the lengths of the vectors $\{Y_j\}$ are uniformly bounded on $\overline{\Omega}$.

We now give two other possible global definitions of metrics.

DEFINITION 2. Let $x_0,x_1 \in \Omega$ and say $\rho_2(x_0,x_1) < \delta$ if and only if there is a C^∞ curve $\phi: [0,1] \to \Omega$ with $\phi(j) = x_j$, $j = 0,1$, and $\phi'(t) = \sum_{j=1}^{q} a_j Y_j(\phi(t))$ where $a_j \in R$ and $|a_j| < \delta^{d(Y_j)}$. Note that in this definition, we have somewhat restricted the class of curves over which we take an infimum. However, in general it is not true that $\rho_2(x,y)$ is finite for any two points of Ω. For example, if Ω is not convex, $N = q = n$, $Y_j = \frac{\partial}{\partial x_j}$, and $d_j = 1$, $1 \leq j \leq n$, then $\rho_2(x,y) < \infty$ if and only if x and y can be joined by a straight line contained in Ω so ρ_2 is not even a pseudometric in this case. On the other hand, it is clear that at least in this example, the metric ρ_2 is locally equivalent to the standard Euclidean metric.

Now suppose we are in the case of vector fields $\{X_1,\cdots,X_p\}$ of finite type m.

DEFINITION 3. Let $x_0,x_1 \in \Omega$ and say $\rho_3(x_0,x_1) < \delta$ if and only if there is an absolutely continuous map $\phi: [0,1] \to \Omega$ with $\phi(j) = x_j$, $j = 0,1$, with $\phi'(t) = \sum_{j=1}^{q} a_j(t) X_j(\phi(t))$ and $|a_j(t)| < \delta$ almost everywhere.

Note that in this definition, we only allow the derivative $\phi'(t)$ to belong to the span of $\{X_1(\phi(t)),\cdots,X_p(\phi(t))\}$, which may not be all of R^N. It is again easy to verify that ρ_3 satisfies the triangle inequality, but it is not *a priori* clear that $\rho(x,y)$ is finite for any two points of Ω. This is in fact a consequence of the finite type hypothesis, and was first proved by Carathéodory in 1909 [2].

To get some feeling for why ρ_3 is finite, and to begin to understand the role of commutators, let us consider the following example of Grushin type in R^2. Let the coordinates be x and y and let $X_1 = \frac{\partial}{\partial x}$, $X_2 = x^k \frac{\partial}{\partial y}$. Note that these vectors fail to span R^2 along $x = 0$, but

$$[X_1,X_2] = kx^{k-1} \frac{\partial}{\partial y}, \quad [X_1[X_1,X_2]] = k(K-1)x^{k-2} \frac{\partial}{\partial y}, \cdots ,$$

and

$$[X_1, [X_1,\cdots, [X_1,X_2]\cdots]] = k! \frac{\partial}{\partial y}$$

where the commutator is of length $k+1$, so $\{X_1,X_2\}$ are of finite type $k+1$.

What points belong to the ball centered at $(x_0,0)$ of radius δ? We shall consider two special cases: $x_0 = 1$ and $x_0 = 0$.

Case 1. If we let

$$\phi_1(t) = (1+\delta t, 0) \text{ so } \phi_1'(t) = \delta \frac{\partial}{\partial x}$$

and

$$\phi_2(t) = (1, \delta t) \text{ so } \phi_2'(t) = \delta \cdot (1)^k \frac{\partial}{\partial y}$$

then this shows that the points $(1+\delta, 0)$ and $(1, \delta)$ belong to the ρ_3 ball centered at $(1,0)$ of radius δ. In fact, this ball is essentially the Euclidean ball of radius δ centered at $(1,0)$.

Case 2. Again, starting at $(0,0)$ we can go to $(\delta, 0)$ by using the vector field $\frac{\partial}{\partial x}$, but it is not immediately clear how to get from $(0,0)$ to other points on the y axis. To do this, we need to use a curve which is only piecewise smooth. Let

$$\phi_1(t) = (\delta t, 0) \qquad 0 \le t \le 1 \quad \text{so} \quad \phi_1' = \delta X_1$$

$$\phi_2(t) = (\delta, \delta^{k+1} t) \qquad 0 \le t \le 1 \quad \text{so} \quad \phi_2' = \delta X_2$$

$$\phi_3(t) = (\delta - \delta t, \delta^{k+1}) \qquad 0 \le t \le 1 \quad \text{so} \quad \phi_3' = -\delta X_1 .$$

This shows that the ρ_3 distance from $(0,0)$ to $(0, \delta^{k+1})$ is at most 3δ. In fact the ρ_3 ball with center $(0,0)$ is an "ellipsoid" of size δ in the x direction and size δ^{k+1} in the y direction.

Thus we have given three possible definitions of a metric or pseudometric in terms of the family of vector fields $Y_1, \cdots, Y_q$. It is also clear from the definitions that we have the following inequalities:

$$\rho(x,y) \le \rho_2(x,y)$$

$$\rho(x,y) \le \rho_3(x,y) \quad \text{(when } \rho_3 \text{ is defined)}.$$

In fact, these three quantities are locally equivalent. One can prove that for every $x_0 \in \Omega$ there is a neighborhood U of x_0 and constants C_1, C_2 so that for all $x,y \in \Omega$, $x \ne y$, and $j = 2,3$.

$$0 < c_1 < \frac{\rho_j(x,y)}{\rho(x,y)} < c_2 < \infty \, .$$

§6. *Local definitions*

In order to begin to see why these pseudometrics are locally equivalent, we need to consider a local definition of metric, and this in turn relies on the notion of an exponential map.

Given a point $p \in \Omega$, we let $T_p\Omega$ denote the tangent space to Ω at p. Suppose we are given C^∞ vector fields $S_1, \cdots, S_N$ defined near p which form a basis for R^N at p. Then we can construct a map from a neighborhood of $0 \in T_p\Omega$ to a neighborhood of p in Ω as follows: every tangent vector $\vec{v}$ at p can be uniquely written as $\sum_{j=1}^{N} a_j S_j(p) = \vec{v}$ with $(a_1, \cdots, a_N) \in R^N$, and so $\sum_{j=1}^{N} a_j S_j$ is a smooth vector field defined near p. We can flow along the integral curve of this vector field for unit time if $\Sigma |a_j|$ is sufficiently small, and the result is by definition $\exp\left(\sum_{j=1}^{N} S_j\right)(p)$, the exponential map of $\vec{v}$.

Given vector fields $S_1, \cdots, S_N$ we can identify $T_p\Omega$ with R^N via $(a_1, \cdots, a_N) \leftrightarrow \sum_{j=1}^{N} a_j S_j(p)$, and then the exponential mapping

$$(a_1, \cdots, a_N) \to \exp\left(\sum a_j S_j\right)(p)$$

introduces a coordinate system centered at p, the so-called canonical coordinates relative to the vector fields $S_1, \cdots, S_N$. The Jacobian of this exponential map at $0 \in R^N$ is just $\det(S_1, \cdots, S_N)$, the volume of the "parallelopiped" spanned by $S_1, \cdots, S_N$. It is important to remember however that this exponential map from $T_\rho\Omega$ to Ω depends on the choice of vector fields $S_1, \cdots, S_N$.

Now we return to the general situation of vector fields $Y_1, \cdots, Y_q$ in $\Omega \subset R^N$. On each tangent space $T_x\Omega$ there is a natural notion of length.

If $\vec{v} \in T_x\Omega$ we say

$$N_x(\vec{v}) < \delta$$

if and only if

$$\vec{v} = \sum_{j=1}^{q} a_j Y_j(x)$$

where $|a_j| < \delta^{d(Y_j)}$. Of course this representation of $\vec{v}$ need not be unique. Nevertheless questions about the set

$$\{\vec{v} \in T_x\Omega \mid N_x(\vec{v}) < \delta\}$$

are presumably just problems in elementary linear algebra. In giving a local definition of metric, we want to transfer these "balls" in the tangent space at x to genuine balls in Ω, and this suggests the use of an exponential map. The question is: how does one choose an appropriate N-tuple of vector fields in order to construct such a map?

To motivate the answer, we first ask a simple question: what is the volume in $T_x\Omega$ of the set $\{\vec{v} \mid N_x(v) < \delta\}$? This amounts to the following problem. Let $Y_1, \cdots, Y_q$ be vectors in $\mathbf{R}^N$ which span, and consider the map

$$\Theta(a_1, \cdots, a_q) = \sum_{j=1}^{q} a_j Y_j .$$

What is the volume of the image under Θ of the box

$$Q_\delta = \{(a_1, \cdots, a_q) \in \mathbf{R}^q \mid |a_j| < \delta^{d(Y_j)}\} ?$$

For any N-tuple $I = (i_1, \cdots, i_N)$ let $d(I) = d(Y_{i_1}) + \cdots + d(Y_{i_N})$, and let $\lambda_I = \det(Y_{i_1}, \cdots, Y_{i_N})$.

LEMMA. *There are universal constants* C_1, C_2 *so that* $0 < C_1 \leq |\Theta(Q_\delta)| / \sum_I |\lambda_I| \delta^{d(I)} \leq C_2 < \infty$. (*Here the sum is taken over all* N-*tuples* I.)

Proof. For each N-tuple I, the image $\Theta(Q_\delta)$ contains all vectors of the form $\sum_{j=1}^{N} a_j Y_{i_j}$ where $|a_j| < \delta^{d(Y_{i_j})}$, and this is just the image under a linear map from R^N to R^N with determinant λ_I. Thus

$$|\lambda_I| \delta^{d(I)} < |\Theta(\Omega_\delta)|$$

$$\sum_I |\lambda_I| \delta^{d(I)} \leq \binom{q}{N} |\Theta(\Omega_\delta)| .$$

We must now prove the reverse inequality. Pick an N-tuple I_0 so that

$$|\lambda_{I_0}| \delta^{d(I_0)} \geq |\lambda_J| \delta^{d(J)}$$

for all N-tuples J. By renumbering, we may assume $I_0 = (1, \cdots, N)$. Since $\lambda_{I_0} \neq 0$, we can write

$$Y_j = \sum_{k=1}^{N} b_{jk} Y_k \qquad 1 \leq j \leq q ,$$

and using Cramer's rule, we see that

$$b_{jk} = \frac{\lambda_{I_{jk}}}{\lambda_{I_0}}$$

where I_{jk} is obtained from I_0 by replacing Y_k by Y_j. From our choice of I_0, it now follows that

$$|b_{jk}| \leq \delta^{d(Y_k) - d(Y_j)} .$$

Now let $v \in \Theta(Q_\delta)$, so $v = \sum_{j=1}^{q} a_j Y_j$ with $|a_j| < \delta^{d(Y_j)}$. Then

$$v = \sum_{k=1}^{N} \left(\sum_{j=1}^{q} a_j b_{jk} \right) Y_k$$

and

$$\left| \sum_{j=1}^{q} a_j b_{jk} \right| \leq \sum_{j=1}^{q} \delta^{d(Y_j)} \delta^{d(Y_k) - d(Y_j)} = q \delta^{d(Y_k)}$$

$$\leq (q\delta)^{d(Y_k)} .$$

Hence

$$|\Theta(Q_\delta)| < q^{d(I_0)} |\lambda_{I_0}| \delta^{d(I_0)}$$

$$\leq q^{mN} \sum_I |\lambda_I| \delta^{d(I)} .$$

This lemma suggests the following further definitions. For each $x \in \Omega$ and each N-tuple $I = (i_1, \cdots, i_N)$ let

$$B_I(x,\delta) = \left\{ y \in \Omega \,\middle|\, y = \exp\left(\sum_{j=1}^{N} a_j Y_{i_j} \right)(x) \text{ with } |a_j| < \delta^{d(Y_{i_j})} \right\} .$$

Clearly for every I

$$B_I(x,\delta) \subset \{ y \in \Omega \,|\, \rho_2(x,y) < \delta \} \equiv B_2(x,\delta) .$$

Hence $\bigcup_I B_I(x,\delta) \subset B_2(x,\delta)$. We also have

$$B_2(x,\delta) \subset B(x,\delta) = \{ y \in \Omega \,|\, \rho(x,y) < \delta \}$$

since $\rho(x,y) \leq \rho_2(x,y)$.

It is now reasonable to conjecture that for each $x \in \Omega$ and each $\delta > 0$, if we choose an N-tuple I_0 so that

$$|\lambda_{I_0}(x)|\,\delta^{d(I_0)} \geq |\lambda_J(x)|\,\delta^{d(J)} \text{ for all } J,$$

then for this particular I_0

$$B(x,\delta) \subset B_{I_0}(x, C\delta)$$

where C is a constant independent of x and δ. This would of course show that ρ and ρ_2 are locally equivalent, and also show that the local and global definitions of metric are equivalent.

We give a very brief sketch of a proof. Suppose $y \in B(x,\delta)$. Then there is $\phi : [0,1] \to \Omega$ with $\phi(0) = x$, $\phi(1) = y$ and

$$\phi'(t) = \sum_{j=1}^{q} b_j(t)\,Y_j(\phi(t))$$

with $|b_j(t)| \leq \delta^{d(Y_j)}$ almost everywhere. Now assume without loss that $I_0 = (1,\cdots,N)$ and let $(u_1,\cdots,u_N)$ denote canonical coordinates near x relative to the exponential map using $\{Y_1,\cdots,Y_N\}$. Then the curve $\phi(t)$ is given in canonical coordinates by $(u_1(t),\cdots,u_N(t))$ and our object is to show that there is a uniform constant C with $|u_j(1)| < (C\delta)^{d(Y_j)}$. But

$$u_j(1) = u_j(1) - u_j(0) = \int_0^1 \frac{d}{dt}[u_j(t)]\,dt$$

$$= \int_0^1 \sum_{k=1}^{q} b_k(t)\,Y_k(\phi(t))(u_j)(t)\,dt .$$

Now since $\{Y_1,\cdots,Y_N\}$ span near x we can write

$$Y_k = \sum_{\ell=1}^{N} a_k^\ell Y_\ell$$

where $a_k^\ell \in C^\infty$. Thus

$$u_j(1) = \int_0^1 \sum_{\ell=1}^{N} \sum_{k=1}^{q} b_k(t) a_k^\ell(\phi(t))(Y_\ell u_j)(t)\,dt\ .$$

We now need to make two kinds of estimates:

(1) $|a_k^\ell(\phi(t))| \le C\delta^{d(Y_\ell)-d(Y_k)}$

(2) $|Y_\ell u_j(t)| \le C\delta^{d(Y_j)-d(Y_\ell)}$.

Combined with the estimate $|b_k(t)| \le \delta^{d(Y_k)}$, this gives $|u_j(1)| \le C\delta^{d(Y_j)}$ which is what we want.

Now estimates (1) are obtained by using Cramer's rule to write $a_k^\ell(y) = \lambda_J(y)/\lambda_{I_0}(y)$ as in our earlier lemma. At x, our choice of I_0 shows that $|\lambda_J(x)|/|\lambda_{I_0}(x)| \le \delta^{d(Y_\ell)-d(Y_k)}$, and in order to obtain a similar estimate for nearby points y, we expand a_k^ℓ in a Taylor series in canonical coordinates about x.

The proof of (2) is more complicated, and involves the Campbell-Hausdorff formula. However, note that *at* x, $Y_\ell u_j = \delta_{j\ell}$, which is the right estimate there. Complete proofs can be found in [13].

As a corollary of our analysis of the metric ρ, we can estimate the volume of the balls $B(x,\delta)$. For every compact $\Sigma \subset\subset \Omega$ there are constants C_1 and C_2 so that

$$C_1\left(\sum_I |\lambda_I(x)|\delta^{d(I)}\right) \le |B(x,\delta)| \le C_2\left(\sum_I |\lambda_I(x)|\delta^{d(I)}\right).$$

In particular, the balls $B(x,\delta)$ satisfy the doubling property, and so Ω with metric ρ and Lebesgue measure is a space of homogeneous type.

Part III. *Applications*

In this part of the paper, we show how the constructions outlined in part II can be applied in partial differential equations and several complex variables.

§7. *Estimates for approximate fundamental solutions*

We can use the metrics constructed from vector fields to obtain estimates for the integral kernels of parametricies of certain hypoelliptic differential operators. We briefly describe the setting. In 1967 Hörmander [7] obtained a far reaching generalization of Kohn's result on the hypoellipticity of $\square_b$. Suppose $X_1,\cdots,X_p$ are C^∞ real vector fields on $\Omega \subset \mathbf{R}^N$ of finite type m. Hörmander showed that the second order operators $X_1^2+\cdots+X_p^2$ or $X_1+X_2^2+\cdots+X_p^2$ are hypoelliptic. As in the work of Folland and Stein [6] on $\square_b$, Rothschild and Stein [15] want to construct parametricies for these more general operators by inverting model operators on appropriate nilpotent groups. There is in general no nilpotent group of dimension N which works, but Rothschild and Stein overcome this difficulty by proving their "lifting theorem." Given $X_1,\cdots,X_p$ on $\Omega \subset \mathbf{R}^N$ (with coordinates $(x_1,\cdots,x_N)$) of type m, they show that one can find additional variables $(t_1,\cdots,t_s) \in \mathbf{R}^s$ and form new vector fields

$$\widetilde{X}_j = X_j + \sum_{\ell=1}^{s} a_{j\ell}(x,t)\frac{\partial}{\partial t_\ell}.$$

These new vector fields will again be of type m on a neighborhood of $\Omega \times \{0\} \subset \Omega \times \mathbf{R}^s$, and in addition they are free up to step m, so that one can model $\widetilde{X}_1,\cdots,\widetilde{X}_p$ by vector fields on a free nilpotent group of step m with p generators.

Rothschild and Stein are able to construct a parametrix for $\widetilde{X}_1^2+\cdots+\widetilde{X}_p^2$ on $\Omega \times \mathbf{R}^s$, given by a kernel $k((x,t),(y,s))$ which comes from a homogeneous kernel on the nilpotent group. In particular, this kernel satisfies

$$|k((x,t),(y,s))| \leq C\delta^2|\widetilde{B}((x,t),\delta)|^{-1}$$

where $\delta = \widetilde{\rho}((x,t),(y,s))$, and $\widetilde{\rho}$ is the metric constructed from the vector fields $\widetilde{X}_1,\cdots,\widetilde{X}_p$. They then define a restriction operator

$$Rk(x,y) = \int_{\mathbf{R}^s} k((x,0),(y,s))\phi(s)\,ds$$

where $\phi \in C^\infty(\mathbf{R}^S)$ is supported near $s = 0$, and if $k((x,t),(y,s))$ is the parametrix for $\ddot{X}_1^2 + \cdots + \widetilde{X}_p^2$, then $D(x,y) = Rk(x,y)$ is a parametrix for $X_1^2 + \cdots + X_p^2$. Using the properties of the metric ρ constructed from $X_1,\cdots,X_p$ one can now prove, for example:

THEOREM 11. *Let* $D(x,y)$ *be the Rothschild-Stein parametrix for* $X_1^2 + \cdots + X_p^2$. *Then, if* $N \geq 3$

$$|D(x,y)| \leq C\delta^2|B(x,\delta)|^{-1}$$

and if $N \geq 2$,

$$|X_{1_1}X_{1_1}\cdots X_{1_j}D(x,y)| \leq C\delta^{2-j}|B(x,\delta)|^{-1}$$

where $\delta = \rho(x,y)$.

Thus we get estimates analogous to those for the Newtonian potential and Kohn Laplacian discussed in part I. These estimates have also been obtained by Sánchez-Calle [15a]. One also gets estimates of this type for the Rothschild-Stein parametrix of the operator $X_1 + X_2^2 + \cdots + X_p^2$, in analogy with the estimates in part I for the heat operator. For details of the argument. see [13].

§8. *Nonisotropic metrics on domains of finite type*

We now apply the theory of metrics to boundaries of domains in $\mathbf{C}^{n+1}$. (Some of these results were announced in [12].) Thus let $\rho \in C^\infty(\mathbf{C}^{n+1})$, with $d\rho \neq 0$ when $\rho = 0$ and let

$$\Omega = \{z \in \mathbf{C}^{n+1} \mid \rho(z) < 0\}$$

so that Ω is a domain with smooth boundary. If we fix a point $\zeta_0 \in \partial\Omega$, then near ζ_0 we can find n linearly independent tangential holomorphic vector fields $L_1, \cdots, L_n$. For example, if $\frac{\partial\rho}{\partial z_{n+1}}(\zeta_0) \neq 0$, we can take

$$L_j = \frac{\partial\rho}{\partial z_{n+1}} \frac{\partial}{\partial z_j} - \frac{\partial\rho}{\partial z_j} \frac{\partial}{\partial z_{n+1}} \qquad 1 \leq j \leq n .$$

We write $L_j = \frac{1}{2}(X_j - iX_{n+j})$, $1 \leq j \leq n$, so that $X_1, \cdots, X_{2n}$ are C^∞ real vector fields defined near ζ_0 on $\partial\Omega$. Since the real dimension of $\partial\Omega$ is $2n+1$, near ζ_0 we can find an additional real tangential vector field T, so that $\{X_1, \cdots, X_{2n}, T\}$ is a basis for the real tangent space to $\partial\Omega$ near ζ_0. Again if $\frac{\partial\rho}{\partial z_{n+1}}(\zeta_0) \neq 0$, we can let

$$T = i\left[\frac{\partial\rho}{\partial \bar{z}_{n+1}} \frac{\partial}{\partial z_{n+1}} - \frac{\partial\rho}{\partial z_{n+1}} \frac{\partial}{\partial \bar{z}_{n+1}}\right].$$

The notion of *type* of a point was introduced by Kohn [8]. If $\zeta \in \partial\Omega$ is near ζ_0, ζ is of *type* $\underline{m}$ if every commutator of $X_1, \cdots, X_{2n}$ of length at most $m-1$ at ζ lies in the span of $\{X_1, \cdots, X_{2n}\}$, while some commutator of length m does not lie in this span, and hence has nonzero T component. One easily checks that the type of a point does not depend on the particular choice of the vector fields $L_1, \cdots, L_n$ and T, but is really a biholomorphic invariant.

We say that the domain Ω is of finite type m if every point $\zeta \in \partial\Omega$ is of type $\leq m$. In this case of course, the vector fields $\{X_1, \cdots, X_{2n}\}$, viewed as being defined on an open set in $\mathbf{R}^{2n+1}$, are of finite type as defined earlier. In particular, we then have a metric defined on $\partial\Omega$, and various equivalent descriptions of the associated family of balls.

Our first object is to show that we can define an equivalent family of balls on $\partial\Omega$ in terms of a single exponential map using the vector fields $\{X_1,\cdots,X_{2n},T\}$. For each finite sequence $i_1,\cdots,i_k$ of integers with $1 \leq i_j \leq 2n$, we can write the commutator

$$X_{i_1,\cdots,i_k} = [X_{i_k},[X_{i_{k-1}},\cdots,[X_{i_2},X_{i_1}]\cdots]]$$

$$\equiv \lambda_{i_1\cdots i_k} T \quad (\mathrm{mod}\ X_1,\cdots,X_{2n})$$

where $\lambda_{i_1,\cdots,i_k} \in C^\infty(\Omega)$ (near ζ_0), since $\{X_1,\cdots,X_{2n},T\}$ is a basis for the tangent spaces near ζ_0. Let $\mathcal{I}_k$ be the ideal in $C^\infty(\Omega)$ generated by the functions $\{\lambda_{i_1,\cdots,i_\ell}\}$ with $\ell \leq k$. It is easy to check that this ideal does not depend on the choice of the vector fields $L_1,\cdots,L_n$ or T. Also

$$X_{i_{k+1}}(\lambda_{i_1,\cdots,i_k}) - \lambda_{i_1,\cdots,i_{k+1}} \in \mathcal{I}_k$$

so in particular, $X_{i_{k+1}}(\lambda_{i_1\cdots i_k}) \in \mathcal{I}_{k+1}$. One can verify this claim by noting that $X_{i_1,\cdots,i_{k+1}} \equiv \lambda_{i_1,\cdots,i_{n+1}} T$, and also $X_{i_1,\cdots,i_{k+1}} = [X_{i_{k+1}}, X_{i_1,\cdots,i_k}]$, and then expanding this commutator.

DEFINITION. For $\zeta \in \partial\Omega$ near ζ_0, set

$$\Lambda_k(\zeta) = \sum |\lambda_{i_1,\cdots,i_\ell}(\zeta)|\ ,$$

and

$$\Lambda(\zeta,\delta) = \sum_{j=2}^{m} \Lambda_j(\zeta)\delta^j$$

where the first sum is over all the generators of $\mathcal{I}_k$. Note that $\Lambda_k(\zeta)$ is a function whose size measures how much T component the commutators of $X_1,\cdots,X_{2n}$ of length $\leq k$ can have. In particular, since $\partial\Omega$ is of type m, then $\Lambda_m(\zeta) \geq \mu_0 > 0$.

We are now in a position to define balls in terms of a single exponential map. For $\zeta \in \partial\Omega$ set

$$\widetilde{B}(\zeta,\delta) = \left\{ \eta \in \partial\Omega \,|\, \eta = \exp\left[\sum_1^{2n} a_j X_j + \gamma T\right](\zeta), \text{ where} \right.$$

$$\left. |a_j| < \delta,\ 1 \le j \le 2n, \text{ and } |\gamma| < \Lambda(\zeta,\delta) \right\}.$$

Thus $\widetilde{B}(\zeta,\delta)$ is the image of a box of size δ in the "complex directions" given by $X_1,\cdots,X_{2n}$, and of size $\Lambda(\zeta,\delta)$ in the complementary real direction. Our first result is then:

THEOREM 12. *The "balls" $\widetilde{B}(\zeta,\delta)$ are equivalent to the balls $B(\zeta,\delta)$ defined in part II in terms of the vector fields $\{X_1,\cdots,X_{2n}\}$; i.e., there is a constant C so that $\widetilde{B}(\zeta,\delta) \subset B(\zeta,C\delta)$ and $B(\zeta,\delta) \subset \widetilde{B}(\zeta,C\delta)$.*

We sketch only part of the proof. Let $\eta \in \widetilde{B}(\zeta,\delta)$, so that

$$\eta = \exp\left[\sum_1^{2n} a_j X_j + \gamma T\right](\zeta)$$

with $|a_j| < \delta$, $|\gamma| < \Lambda(\zeta,\delta) = \sum_2^m \Lambda_j(\zeta)\delta^j$. Since $\Lambda_j(\zeta) = \Sigma |\lambda_{i_1,\cdots,i_\ell}(\zeta)|$, it follows that

$$\gamma = \sum_2^m b_j \lambda_{I_j}(\zeta)$$

where $|b_j| < \delta^j$ and I_j is a j-tuple of integers. Hence

$$\gamma T = \sum_2^m b_j \left(X_{I_j} + \sum_{\ell=1}^{2n} \beta_{I_j,\ell} X_\ell \right)$$

where $\beta_{I_j,\ell} \in \mathbb{R}$ and X_{I_j} is the jth order commutator defined earlier. Hence we can join ζ to η with a curve $\phi(t)$ (the integral curve of $\sum_1^{2n} a_j X_j + \gamma T$) with $\phi'(t) = \sum_{j=1}^q a_j Y_j$ and $|a_j| < C\delta^{d(Y_j)}$. Thus $\eta \in B(\zeta, C\delta)$.

The opposite inclusion is more technical, involving ideas used in the proof of the equivalence of the families of balls introduced in part II and we omit the details. A major ingredient is the use of Taylor series expansions of functions in canonical coordinates. If f is a smooth function defined near ζ, and if

$$g(\alpha_1,\cdots,\alpha_{2n},\gamma) = f\left(\exp\left(\sum \alpha_j x_j + \gamma T\right)(\zeta)\right)$$

then g has a formal Taylor series at the origin, and it is given by:

$$g(\alpha_1,\cdots,\alpha_{2n},\gamma) \sim \sum_{k=0}^{\infty} \frac{1}{k!}\left(\sum \alpha_j X_j + \gamma T\right)^k f|_0$$

where $(\Sigma \alpha_j X_j + \gamma T)^k$ is a kth order differential operator. (See Rothschild-Stein [15] for further details.)

To give an idea of how this formula is used, we can easily see that the function $\eta \to \Lambda(\eta,\delta)$ is essentially constant on $\widetilde{B}(\zeta,\delta)$; i.e. there is a constant C so that for $\eta \in \widetilde{B}(\zeta,\delta)$,

$$|\Lambda(\eta,\delta)| \leq C|\Lambda(\zeta,\delta)| .$$

A typical term in $|\Lambda(\eta,\delta)|$ is $\delta^k|\lambda_{i_1,\cdots,i_k}(\eta)|$. We can expand $\lambda_{i_1,\cdots,i_k}(\eta)$ about ζ in canonical coordinates. Any term involving applying T to this $\lambda_{i_1,\cdots,i_k}$ is certainly of the right size since $|\gamma| < \Lambda(\zeta,\delta)$. On the other hand, if we differentiate $\lambda_{i_1,\cdots,i_k}$ with one of the vector fields $X_1,\cdots,X_{2n}$, this gives us an element of $\mathcal{I}_{k+1}$, so again we get the correct estimate.

§9. *Balls in terms of a polarization*

We now want to find a description of the balls on the boundary of a domain $\Omega \subset \mathbf{C}^{n+1}$ directly in terms of inequalities involving the defining function for Ω. As before, let

$$\Omega = \{z \in C^{n+1} | \rho(z) < 0\}$$

where $\rho : C^{n+1} \to R$ is C^∞ and $d\rho \neq 0$ when $\rho = 0$. A polarization of ρ is a function $R: C^{n+1} \times C^{n+1} \to C$ of class C^∞ satisfying:

(a) $R(z,z) = \rho(z)$

(b) $\bar{\partial}_z R(z,w)$ vanishes to infinite order on the diagonal $z = w$.

(c) $R(z,w) - \overline{R(w,z)}$ vanishes to infinite order on the diagonal $z = w$.

Polarizations always exist, and are unique up to functions vanishing to infinite order on $z = w$. When ρ is a polynomial or is real analytic, it is easy to construct a polarization. If

$$\rho(z) = \sum a_{\alpha\beta} z^\alpha \bar{z}^\beta$$

then

$$R(z,w) = \sum a_{\alpha\beta} z^\alpha \bar{w}^\beta$$

is a polarization which is holomorphic in z and antiholomorphic in w, and in all that follows, we shall be dealing with polynomial ρ.

We now define a new family of balls on the boundary. For $w \in \partial\Omega$ and $\delta > 0$ set

$$B^\#(w,\delta) = \{z \in \partial\Omega \,\| z-w| < \delta \,|R(z,w)| < \Lambda(w,\delta)\} .$$

Note that for $w \in \partial\Omega$, $V_w = \{z \in C^{n+1} | R(z,w) = 0\}$ is a holomorphic hypersurface tangent to $\partial\Omega$ at w. Thus $B^\#(w,\delta)$ is essentially the set of points $z \in \partial\Omega$ within Euclidean distance δ of w, and within Euclidean distance $\Lambda(w,\delta)$ of V_w.

Let us quickly calculate what all this means for the Siegel upper half space discussed in part I, where $\rho(z) = \sum_{j=1}^{n} |z_j|^2 - \operatorname{Im} z_{n+1}$. Then

$$R(z,w) = \sum_{j=1}^{n} z_j \bar{w}_j - \frac{1}{2i}(z_{n+1} - \bar{w}_{n+1}) .$$

Also, on the Heisenberg group $[X_j, X_{n+j}] = -4T$ so $\Lambda(w,\delta) \approx \delta^2$ for all $w \in \partial\Omega$. Thus in this case

$$B^{\#}(w,\delta) = \left\{ z \in \partial\Omega \,|\, |z-w| < \delta, \left| \sum_{j=1}^{n} z_j \overline{w}_j - \frac{1}{2i}(z_{n+1} - \overline{w}_{n+1}) \right| < \delta^2 \right\}$$

and it is easy to check that this defines essentially the same balls as we did earlier in part I.

We now want to show that the balls $B^{\#}(w,\delta)$ are equivalent to the balls $\widetilde{B}(w,\delta)$ defined in terms of the exponential mapping. For simplicity we will do this only in the special case

$$\Omega = \Omega_\phi = \{(z_1,z_2) \in \mathbf{C}^2 \,|\, \operatorname{Im} z_2 > \phi(z_1)\}$$

where $\phi : \mathbf{C} \to \mathbf{R}$ is a polynomial of degree m. Here the defining function is

$$\rho(z_1,z_2) = \phi(z_1) - \frac{1}{2i}(z_2 - \overline{z}_2) .$$

It is easy to check that if

$$\overline{L} = \frac{\partial}{\partial \overline{z}_1} - 2i \frac{\partial \phi}{\partial \overline{z}_1}(z_1) \frac{\partial}{\partial \overline{z}_2}$$

then $\overline{L}$ globally spans the space of tangential antiholomorphic vector fields. We also set

$$T = \frac{\partial}{\partial z_2} + \frac{\partial}{\partial \overline{z}_2} .$$

We want to investigate the geometry of the balls near a fixed boundary point, say $(0, i\phi(0))$. Since all our families of balls are invariant under biholomorphic mappings, we can choose a special coordinate system to study the geometry.

Let

$$\psi(z) = \phi(z) - \phi(0) - 2 \operatorname{Re}\left(\sum_{j=1}^{m} \frac{1}{j!} \frac{\partial^j \phi}{\partial z^j}(0) z^j \right) .$$

If we make the change of variables:

$$z_1' = z_1$$

$$z_2' = z_2 - i\left[\phi(0) + 2\sum_{j=1}^{m} \frac{1}{j!}\frac{\partial^j \phi}{\partial z^j}(0) z^j\right]$$

our original domain now has the form

$$\{(z_1', z_2') \in \mathbf{C}^2 \mid \operatorname{Im} z_2' > \psi(z_1')\}$$

and moreover ψ now satisfies

$$\psi(0) = \frac{\partial^j \psi}{\partial z^j}(0) = \frac{\partial^j \psi}{\partial \bar{z}^j}(0) = 0 \qquad 1 \leq j \leq m .$$

We shall work in this coordinate system near $(0,0) \in \partial\Omega$.

We identify $\partial\Omega$ with $\mathbf{C}\times\mathbf{R}$ so that $(z, t+i\psi(x))$ corresponds to (z,t). Then our basic vector fields are:

$$L = \frac{\partial}{\partial z} + i\frac{\partial\psi}{\partial z}(z)\frac{\partial}{\partial t}, \quad \bar{L} = \frac{\partial}{\partial \bar{z}} - i\frac{\partial\psi}{\partial \bar{z}}(z)\frac{\partial}{\partial t}$$

$$T = \frac{\partial}{\partial t} .$$

We can calculate the various functions $\lambda_{i_1 \cdots i_k}$ using the vector fields L and $\bar{L}$ (instead of $\operatorname{Re} L$, $\operatorname{Im} L$). Thus:

$$[L, \bar{L}] = -2i\frac{\partial^2\psi}{\partial z \partial \bar{z}}\frac{\partial}{\partial t}$$

so the ideal $\mathcal{I}_2$ is generated by $\frac{\partial^2\psi}{\partial z\partial\bar{z}}$ or $\Delta\psi$. Thus

$$\Lambda_2(z,t) = |\Delta\psi(z)|$$

$$\Lambda_k(z,t) = \sum_{\substack{\alpha+\beta\leq k \\ \alpha,\beta>1}} \left| \frac{\partial^k \psi}{\partial z^\alpha \partial \bar{z}^\beta}(z) \right| .$$

To investigate the exponential map $\exp(\alpha L+\bar{\alpha}\bar{L}+\gamma T)(0)$ with $\alpha \in \mathbf{C}$, $\gamma \in \mathbf{R}$, we must find a curve

$$(\phi_1, \phi_2) : [0,1] \to \mathbf{C} \times \mathbf{R}$$

with

$$\phi_1'(t) = \alpha$$

$$\phi_2'(t) = \gamma + i\alpha \frac{\partial \psi}{\partial z}(\phi_1(t)) - i\bar{\alpha} \frac{\partial \psi}{\partial \bar{z}}(\phi_1(t))$$

with $(\phi_1(0), \phi_2(0)) = (0,0)$. Thus $\phi_1(t) = \alpha t$ and

$$\phi_2(t) = \gamma t + i \int_0^t \left(\alpha \frac{\partial \psi}{\partial z}(\alpha s) - \bar{\alpha} \frac{\partial \psi}{\partial \bar{z}}(\alpha s) \right) ds .$$

And so

$$\exp(\alpha L+\alpha\bar{L}+\gamma T)(0) = \left(\alpha, \gamma+i \int_0^1 \left(\alpha \frac{\partial \psi}{\partial z}(\alpha s) - \alpha \frac{\partial \psi}{\partial \bar{z}}(\alpha s) \right) ds \right) .$$

We can estimate the integral by expanding $\frac{\partial \psi}{\partial z}$ and $\frac{\partial \psi}{\partial z}$ about 0. Since $\frac{\partial^j \psi}{\partial_z j}(0) = 0 \quad 1 \leq j \leq m$, all the terms will be dominated by the corresponding $\Lambda_j(0)$, so

$$\left| \int_0^1 \left(\alpha \frac{\partial \psi}{\partial z}(\alpha s) - \bar{\alpha} \frac{\partial \psi}{\partial \bar{z}}(\alpha s) \right) ds \right| \leq |^2\Lambda_2(0) + \cdots + |\alpha|^m \Lambda_m(0)$$

$$= \Lambda(0, |\alpha|) .$$

In particular, we see that the image under the exponential map of the box

$$\{(\alpha,\gamma) \in \mathbf{C}\times\mathbf{R} \mid |\alpha| < \delta, |\gamma| < \Lambda(0,\delta)\}$$

is essentially

$$\{(z,t) \in \mathbf{C}\times\mathbf{R} \mid |z| < \delta, |t| < \Lambda(0,\delta)\} .$$

On the other hand, the polarization of our defining function is

$$R((z_1 z_2)(w_1 w_2)) = \sum \frac{1}{\alpha!\beta!} \frac{\partial^{\alpha+\beta}\psi}{\partial z^\alpha \partial \bar{z}^\beta}(0)\, z_1^\alpha \bar{w}_1^\beta - \frac{1}{2i}(z_2 - \bar{w}_2)$$

and

$$R((z_1 z_2), (0,0)) = -\frac{1}{2i} z_2 = -\frac{1}{2i}(t + i\psi(z_1))$$

if $(z_1, z_2) = (z_1, t + i\psi(z_1)) \in \partial\Omega$. Thus the inequalities

$$|R((z_1 z_2), (0,0))| < \Lambda(0,\delta), \qquad |z_1| < \delta$$

are equivalent to the inequalities

$$|t| < \Lambda(0,\delta), \qquad |z_1| < \delta$$

and so the families of balls are equivalent.

§10. *Estimating pleurisubharmonic functions*

We continue our study of domains of the form

$$\Omega = \{(z_1, z_2) \in \mathbf{C}^2 \mid \operatorname{Im} z_2 > \phi(z_1)\} ,$$

but we now make the additional hypothesis that $\Delta\phi \geq 0$; i.e. ϕ is a subharmonic polynomial of degree m, which is not harmonic. Our object is to obtain an analogue of Theorem 8 in part I. We will prove:

THEOREM. *Suppose* u *is continuous on* $\overline{\Omega}$ *and pleurisubharmonic on* Ω. *Let* $\pi : \Omega \to \partial\Omega$ *be the projection onto the boundary. Then if* $(z_1, z_2) \in \Omega$

$$|u(z_1,z_2)| \leq C|B|^{-1} \int_B |u(\zeta)|d\sigma(\zeta)$$

where $B = B(\pi(z_1,z_2), A\delta)$ *and* $\Lambda(\pi(z_1,z_2),\delta) = \operatorname{Im} z_2 - \phi(z_1)$. *Here* A, C *are constants independent of* u *and* (z_1,z_2).

As before, by making a holomorphic change of variables, we can assume $\phi(0) = \frac{\partial\phi}{\partial z_j}(0) = \frac{\partial\phi}{\partial \bar{z}_j}(0) = 0$, $1 \leq j \leq m$, and it is enough to estimate u at the point $(0, iy)$. We begin as we did in part I:

$$|u(0,iy)| < \frac{4}{\pi y^2} \int_{|s|^2+|t-y|^2<(y/2)^2} |u(0,s+it)|\, ds\, dt \,. \tag{8}$$

Now for each $s + it$ we want to imbed an analytic disc into Ω, whose boundary is mapped to $\partial\Omega$ and whose center is mapped to $(0,s+it)$. To do this, we try to find $\delta > 0$ and a function $G(\zeta)$, continuous for $|\zeta| \leq 1$, holomorphic on $|\zeta| < 1$, so that the holomorphic map

$$\zeta \to (\delta\zeta, G(\zeta))$$

has the required properties. This means we want

$$\operatorname{Im} G(e^{i\theta}) = \phi(\delta e^{i\theta}) \qquad 0 \leq \theta \leq 2\pi$$

and

$$G(0) = s + it \,.$$

Let $\phi_\delta(e^{i\theta}) = \phi(\delta e^{i\theta})$. Let $P[\phi_\delta]$ be the Poisson integral of ϕ_δ, and let $Q[\phi_\delta]$ be the conjugate Poisson integral, with $Q[\phi_\delta](0) = 0$. Then

$$G(\zeta) = G_{s,t}(\zeta) = s + i[P[\phi_\delta](\zeta) + iQ[\phi_\delta](\zeta)]$$

has the required properties provided that δ is chosen so that

$$t = \frac{1}{2\pi}\int_0^{2\pi} \phi(\delta e^{i\theta})\,d\theta . \tag{9}$$

But by Green's theorem

$$d\delta\left[\frac{1}{2\pi}\int_0^{2\pi} \phi(\delta e^{i\theta})\,d\theta\right] = \frac{1}{2\pi\delta}\int_{x^2+y^2<\delta^2} \Delta\phi(x,y)\,dx\,dy > 0$$

so the function $t(\delta) = \frac{1}{2\pi}\int_0^{2\pi} \phi(\delta e^{i\theta})\,d\theta$ is a monotone increasing function of δ. Thus given t, there is a unique $\delta = \delta(t)$ so that equation (9) holds. We thus obtain

$$|u(0),iy)| < \frac{2}{\pi^2 y^2}\int_{-y/2}^{y/2}\int_0^{2\pi}\int_{y/2}^{3y/2} |u(\delta(t)e^{i\theta},$$

$$s + i[P\phi_{\delta(t)}(e^{i\theta}) + iQ\phi_{\delta(t)}(e^{i\theta})])|\,dt\,d\theta\,ds .$$

We want to make the change of variables $r = \delta(t)$. Let

$$A(\delta) = \frac{1}{\pi\delta^2}\iint_{x^2+y^2\leq\delta^2} \Delta\phi(x,y)\,dx\,dy$$

so that $t'(\delta) = \frac{1}{2}\delta A(\delta)$. We get:

$$|u(0,iy)| \leq \frac{1}{\pi y^2} \int_{-y/2}^{y/2} \int_0^{2\pi} \int_{\delta\left(\frac{y}{2}\right)}^{\delta\left(\frac{3y}{2}\right)} |u(re^{i\theta}, s - Q\phi_r(e^{i\theta}) + i\phi(re^{i\theta})| A(r)\, r dr\, d\theta\, ds \;.$$

We now need the following result, which is where the hypothesis $\Delta\phi \geq 0$ is used:

LEMMA. *There are constants* C_1, C_2 *which depend only on* m *so that for* $\delta > 0$,

(i) $0 < C_1 \leq \Lambda(0,\delta)/t(\delta) \leq C_2 < \infty$

(ii) $0 < C_1 \leq \Lambda(0,\delta)/\delta^2 A(\delta) \leq C_2 < \infty$

(iii) $|Q\phi_\delta(e^{i\theta})| \leq C_2 t(\delta)$

We defer the proof for a moment, and return to the estimate for $u(0,iy)$. In our last integral, when r is between $\delta\,\frac{y}{2}$ and $\delta\,\frac{3y}{2}$, $t(r)$ is between $\frac{y}{2}$ and $\frac{3y}{2}$. It follows from the lemma that in this range, $A(r) \approx y/\delta(y)^2$, and it follows from part (iii) of the lemma that the range of s integration in the integral is contained in $\{|s| < Cy\}$ for some constant C. Thus:

$$|u(0,iy)| \leq \frac{C}{y\delta(y)^2} \int_{|s|<Cy} \int_0^{2\pi} \int_0^{C\delta(y)} |u(re^{i\theta}, s + i\phi(re^{i\theta})|\, r dr\, d\theta\, ds \;.$$

But finally, $r dr\, d\theta\, ds$ is essentially surface area measure, and on the surface $\partial\Omega$ we are integrating over a region centered at (0,0) of size y in the "real" s direction, and of size $\delta(y)$ in the "complex" directions. But this is exactly our nonisotropic ball $B(0,\delta)$ where $\Lambda(0,\delta) = y$ and since $|B(0,\delta)| \approx y\delta(y)^2$, we have shown: there are constants A and C so that

$$|u(0,iy)| \le C|B(0,A\delta)|^{-1} \int_{B(0,A\delta)} |u(\zeta)|d\sigma(\zeta)$$

where $\Lambda(0,\delta) = y$. Thus we have managed to estimate u along the "radius" from $(0,0)$. Just as in part I, we now easily extend this result to obtain the theorem.

We finally turn to the proof of the lemma. The inequalities $t(\delta) \le C\Lambda(0,\delta)$ and $\delta^2 A(\delta) \le C\Lambda(0,\delta)$ follow easily by expanding ϕ or $\Delta\phi$ in a Taylor series about 0. The main content of the lemma is contained in the opposite inequalities, and follows from a homogeneity and compactness argument.

We claim for every integer $m \ge 2$ there is a constant A_m so that if $\phi(z)$ is a real polynomial of degree at most m with $\frac{\partial^j \phi}{\partial z^j}(0) = 0$, $0 \le j \le m$, and if $\Delta\phi(z) \ge 0$ for $|z| \le \delta_0$ then for $\delta < \delta_0$.

$$t(\delta) = \frac{1}{2\pi}\int_0^{2\pi} \phi(\delta e^{i\theta})\,d\theta \ge A_m \sum \left|\frac{\partial^{\alpha+\beta}\phi}{\partial z^\alpha \partial \bar{z}^\beta}(0)\right| \delta^{\alpha+\beta} = A_m \Lambda(0,\delta)$$

$$\delta^2 A(\delta) = \frac{1}{\pi}\int_0^{2\pi}\int_0^{\delta} \Delta\phi(re^{i\theta})\,r\,dr\,d\theta \ge A_m \sum \left|\frac{\partial^{\alpha+\beta}\phi}{\partial z^\alpha \partial z^\beta}(0)\right| \delta^{\alpha+\beta} = A_m \Lambda(0,\delta)\,.$$

In fact, if we can prove this for $\delta = \delta_0 = 1$, then given ϕ, we apply the result to $\psi(z) = \phi(\delta z)$, and the result for general δ follows, so it suffices to study $\delta = \delta_0 = 1$. But now we let

$$\Sigma = \Big\{ \phi(z) \text{ real polynomials of degree } \le m \text{ such that } \frac{\partial^j \phi}{\partial z^j}(0) = 0,\ 0 \le j \le m;\ \Delta\phi(z) \ge 0 \text{ if } |z| \le 1;\ \text{and } \sum \left|\frac{\partial^{\alpha+\beta}\phi}{\partial z^\alpha \partial \bar{z}^\beta}(0)\right| = 1 \Big\}\,.$$

It is easy to see that Σ is a compact set of polynomials and that the maps

$$\phi \to \frac{1}{2\pi}\int_0^{2\pi} \phi(e^{i\theta})\,d\theta$$

$$\phi \to \frac{1}{\pi}\int_0^{2\pi}\int_0^1 \phi(re^{i\theta})\,r\,dr\,d\theta$$

are continuous on Σ. Moreover, these functions are strictly positive since $\Delta\phi \geq 0$ and $\Delta\phi \not\equiv 0$, on $|z| \leq 1$, and so the functions are bounded below by a constant $A_m > 0$. The general result now follows by dividing a general polynomial ϕ by $\Sigma \left|\frac{\partial^{\alpha+\beta}\phi}{\partial z^{\alpha}\partial \bar{z}^{\beta}}(0)\right|$.

Finally, in order to show

$$\sup_{\theta} |Q\phi_{\delta}(e^{i\theta})| \leq Ct(\delta)$$

when ϕ is a polynomial of degree $\leq m$, it again suffices to check this for $\delta = 1$. But

$$\sup_{\theta} |Q\phi_{\delta}(e^{i\theta})| \equiv |||\phi|||$$

is a norm on this space of polynomials. Hence

$$|||\phi||| \leq C\Lambda(0,1) \leq Ct(1)\ .$$

As an easy corollary of the theorem we obtain an estimate for the Szegö kernel $S(z,\zeta)$ for the domain Ω on the diagonal $z = \zeta$. Recall that the orthogonal projection S of $L^2(\partial\Omega)$ onto $H^2(\Omega)$ is called the Szegö projection, and formally, this projection is given by integrating against a kernel:

$$Sf(z) = \int_{\partial\Omega} f(\zeta)S(z,\zeta)\,d\sigma(\zeta)\ , \qquad z \in \Omega\ .$$

In fact $S(z,\zeta) = \Sigma\, \phi_j(z)\overline{\phi_j(\zeta)}$ where $\{\phi_j\}$ is a complete orthonormal basis for $H^2(\Omega)$, and this series converges uniformly on compact subsets of $\Omega \times \Omega$. (See Krantz [10a], Chapter 1, for further details.) Now it is easy to check that for $z \in \Omega$

$$S(z,z)^{1/2} = \sup |F(z)|$$

where the supremum is taken over all $F \in H^2(\Omega)$ with $\|F\|_{H^2} \leq 1$. But by our theorem, if $F \in H^2(\Omega)$,

$$|F(z)| \leq C|B|^{-1} \int_B |F(\zeta)|\, d\sigma(\zeta)$$

$$\leq C|B|^{-1/2} \left(\int_{\partial\Omega} |F(\zeta)|^2\, d\sigma(\zeta) \right)^{1/2}$$

$$= C|B|^{-1/2} \|F\|_{H^2}$$

where $B = B(\pi(z),\delta)$ is the ball centered at the projection $\pi(z)$ of z, and $\Lambda(\pi(z),\delta) = -\rho(z)$. Thus we obtain:

COROLLARY. *If* $z \in \Omega$

$$S(z,z) \leq C|B(\pi(z),\delta)|^{-1}$$

where $\Lambda(\pi(z),\delta) = \operatorname{Im} z_2 - \phi(z_1)$.

§11. *Estimates for the Szegö kernel on* $\partial\Omega$

As a final application, we show how one can make estimates of the Szego kernel $S(z,\zeta)$ *on the boundary*, at least for certain very special domains Ω. Thus, let

$$\Omega = \{(z_1,z_2) \in \mathbb{C}^2 \,|\, \operatorname{Im} z_2 > \phi(z_1)\}$$

where ϕ is a subharmonic, non-harmonic polynomial of degree m. We also make the very restrictive assumption that

$$\Delta\phi(z) = \Delta\phi(x+iy)$$

is actually *independent of* y.

Our approach is the following: if

$$L = \frac{\partial}{\partial \bar{z}_1} - 2i \frac{\partial \phi}{\partial \bar{z}_1}(z_1) \frac{\partial}{\partial \bar{z}_2}$$

then L is a global tangential antiholomorphic vector field, and

$$H^2(\partial\Omega) = \{f \in L^2(\partial\Omega) | Lf = 0 \text{ as distribution}\},$$

so we identify H^2 with the kernel of the differential operator. When we identify $\partial\Omega$ with $\mathbf{C} \times \mathbf{R}$ in the usual way, and write $z = x + iy$, this operator becomes:

$$\begin{aligned} L &= \frac{\partial}{\partial \bar{z}} - i \frac{\partial \phi}{\partial \bar{z}} \frac{\partial}{\partial t} \\ &= \frac{1}{2}\left[\frac{\partial}{\partial x} + \frac{\partial \phi}{\partial y}\frac{\partial}{\partial t}\right] + \frac{i}{2}\left[\frac{\partial}{\partial y} - \frac{\partial \phi}{\partial x}\frac{\partial}{\partial t}\right]. \end{aligned}$$

We now make a change of variables on $\mathbf{C} \times \mathbf{R}$

$$\Phi(x,y,t) = (x,y,t-A(x,y))$$

where

$$A(x,y) = -\int_0^x \frac{\partial \phi}{\partial y}(t,y)\,dt\,.$$

Then if we put

$$b'(x) = \int_0^x \Delta\phi(t)\,dt$$

and

$$\widetilde{L} = \frac{\partial}{\partial x} + i\left[\frac{\partial}{\partial y} + b'(x)\frac{\partial}{\partial t}\right]$$

it is easy to check that $\widetilde{L}(f \circ \Phi) = 2(Lf) \circ \Phi$.

Thus our object is to obtain estimates for the orthogonal projection onto the kernel of

$$L = \frac{\partial}{\partial x} + i\left[\frac{\partial}{\partial y} + b'(x)\frac{\partial}{\partial t}\right]$$

where $b''(x) \geq 0$. Our finite type hypothesis is now $\sum_{j=2}^{m} |b^{(j)}(x)| \geq \mu_0 > 0$, and we want to obtain estimates in terms of balls defined by the vector fields $\frac{\partial}{\partial x}$ and $\frac{\partial}{\partial y} + b'(x)\frac{\partial}{\partial t}$ in R^3.

If $u = u(x,y,t)$ is a reasonable function we define partial Fourier transforms by

$$\mathfrak{F}u = \hat{u}(x,\eta,\tau) = \iint_{R^2} e^{-2\pi i(y\eta+t\tau)} u(x,y,t)\, dy\, dt$$

so that

$$u(x,y,t) = \iint_{R^2} e^{2\pi i(y\eta+t\tau)}\, \hat{u}(x,\eta,\tau)\, d\eta\, d\tau\ .$$

$\mathfrak{F}$ is an isometry of $L^2(R^3)$ and

$$Lu = \mathfrak{F}^{-1}\hat{L}\mathfrak{F}u$$

where

$$\hat{L}\hat{u} = e^{2\pi(\eta x+b(x)\tau)}\frac{d}{dx}\left(e^{-2\pi(\eta x+b(x)\tau)}\hat{u}\right).$$

Let

$$\psi(x,\eta,\tau) = e^{-2\pi(\eta x+b(x)\tau)}$$

and let

$$M_\psi g(x,\eta,\tau) = \psi(x,\eta,\tau)\, g(x,\eta,\tau) .$$

Then

$$Lu = \mathcal{F}^{-1} M_{\psi^{-1}} \frac{d}{dx} M_\psi \mathcal{F} u .$$

Now

$$M_\psi : L^2(\mathbf{R}^3, dx\, d\eta\, d\tau) \to L^2(\mathbf{R}^3, e^{4\pi(\eta x + \tau b(x))} dx\, d\eta\, d\tau)$$

and

$$M_{\psi^{-1}} : L^2(\mathbf{R}^3, e^{4\pi(\eta x+\tau b(x))} dx\, d\eta\, d\tau) \to L^2(\mathbf{R}^3, dx\, d\eta\, d\tau)$$

are isometries. Thus L is similar to the operator $\frac{\partial}{\partial x}$, acting on functions which satisfy

$$\int |g(x,\eta,\tau)|^2 e^{4\pi(\eta x+\tau b(x))} dx\, d\eta\, d\tau < +\infty .$$

The kernel of this operator thus consists of functions $g(\eta,\tau)$ so that

$$\iint |g(\eta,\tau)|^2 \left[\int e^{4\pi(\eta x+\tau b(x))} dx \right] d\eta\, d\tau < \infty .$$

Let

$$\Sigma = \left\{ (\eta,\tau) \in \mathbf{R}^2 \,\Big|\, \int e^{4\pi(\eta x+\tau b(x))} dx < \infty \right\} .$$

Then since $b''(x) \geq 0$, $\Sigma = \{(\eta,\tau) | \tau < 0\}$. For $(\eta,\tau) \in \Sigma$, the space

$L^2(e^{4\pi(\eta x+\tau b(x))}dx)$ contains the constants. Let $\hat{P}_{\eta,\tau}$ be the projection of $L^2(e^{4\pi(\eta x+\tau b(x))}dx)$ onto the constants, if $(\eta,\tau)\in\Sigma$, and let $\hat{P}_{\eta,\tau}=0$ otherwise. Let

$$\hat{P}: L^2(e^{4\pi(\eta x+\tau b(x))}dx\,d\eta\,d\tau)\to L^2(e^{4\pi(\eta x+\tau b(x))}dx\,d\eta\,d\tau)$$

be defined by

$$\hat{P}g(x,\eta,\tau)=\hat{P}_{\eta,\tau}(g_{\eta,\tau})(x)$$

where $g_{\eta,\tau}(x)=g(x,\eta,\tau)$. It is then easy to check that $\hat{P}$ is an orthogonal projection whose range is precisely the null space of $\frac{d}{dx}$ on $L^2(e^{4\pi(\eta x+\tau b(x))}dx\,d\eta\,d\tau)$. Thus if we set

$$P=\mathcal{F}^{-1}M_{\psi^{-1}}\hat{P}M_{\psi}\mathcal{F}$$

then P is the orthogonal projection of $L^2(dx\,dy\,dt)$ onto the null space of L.

Now if $(\eta,\tau)\in\Sigma$

$$\hat{P}_{\eta,\tau}g=\frac{\langle g,1\rangle 1}{\langle 1,1\rangle}$$

$$=\frac{\int_{\infty}^{\infty}g(r)\,e^{4\pi(\eta r+\tau b(r))}dr}{\int_{-\infty}^{\infty}e^{4\pi(\eta r+\tau b(r))}dr}.$$

Thus

$$\hat{P}_{\eta,\tau}g(x)=\int_{-\infty}^{\infty}g(y)\,K_{\eta,\tau}(y)\,dy$$

where

$$K_{\eta,\tau}(x,y) = \begin{cases} \dfrac{e^{4\pi(\eta y+\tau b(y))}}{\displaystyle\int_{-\infty}^{\infty} e^{4\pi(\eta r+\tau b(r))}\,dr} & \text{if } (\eta,\tau)\in\Sigma \\[2em] 0 & \text{if } (\eta,\tau)\notin\Sigma . \end{cases}$$

Thus if $f \in L^2(R^3, dx\,dy\,dt)$

$$Pf(x,y,t) = \iiint f(r,s,u)\, S((x,y,t);(r,s,u))\,dr\,ds\,dy$$

where

$$S((x,y,t);(r,s,u)) = \iint e^{2\pi i[(y-s)\eta+(t-u)\tau]}\, e^{2\pi[\eta(x-r)+\tau(b(x)-b(r))]}$$

$$K_{\eta,\tau}(x,r)\,d\eta\,d\tau$$

$$= \int_0^{\infty} e^{-2\pi\tau[(b(x)+b(r))+i(t-u)]} \int_{-\infty}^{\infty} \frac{e^{2\pi\eta((x+r)+i(y-s))}}{\displaystyle\int_{-\infty}^{\infty} e^{4\pi(\eta r-\tau b(r))}\,dr}\,d\eta \quad d\tau .$$

This is the kernel we have to estimate.

We begin by estimating the inner integral. For $\tau > 0$, set

$$F(\lambda+it,\tau) = \int_{-\infty}^{\infty} \frac{e^{2\pi\eta(\lambda+it)}}{\displaystyle\int_{-\infty}^{\infty} e^{4\pi[\eta r-\tau b(r)]}\,dr}\,d\eta .$$

Then replacing r by $r+\frac{\lambda}{2}$, and η by $\eta+\tau b'\left(\frac{\lambda}{2}\right)$ it follows that

$$F(\lambda+it,\tau) = e^{2\pi\tau\left[2b\left(\frac{\lambda}{2}\right)+itb'\left(\frac{\lambda}{2}\right)\right]} \int_{-\infty}^{\infty} \frac{e^{2\pi i\eta t}\,d\eta}{\int_{-\infty}^{\infty} e^{4\pi\left[\eta r+r\tau b'\left(\frac{\lambda}{2}\right)-\tau b\left(r+\frac{\lambda}{2}\right)+\tau b\left(\frac{\lambda}{2}\right)\right]}} .$$

Let $G(r) = r\tau b'\left(\frac{\lambda}{2}\right) - \tau b\left(r+\frac{\lambda}{2}\right) + \tau b\left(\frac{\lambda}{2}\right)$. Then

$$G'(r) = \tau b'\left(\frac{\lambda}{2}\right) - \tau b'\left(r+\frac{\lambda}{2}\right)$$

$$G''(r) = -\tau b''\left(r+\frac{\lambda}{2}\right) .$$

Since $G(0) = G'(0) = 0$, we have

$$G(r) = -\tau \sum_{j=2}^{m} \frac{1}{j!}\, b^{(j)}\left(\frac{\lambda}{2}\right) r^j .$$

Hence

$$F(\lambda+it,\tau) = e^{2\pi\tau\left[2b\left(\frac{\lambda}{2}\right)+itb'\left(\frac{\lambda}{2}\right)\right]} \int_{-\infty}^{\infty} \frac{e^{2\pi i\eta t}d\eta}{\int_{-\infty}^{\infty} e^{4\pi\left[\eta r-\tau \sum_{2}^{m}\frac{1}{j!}\, b^{(j)}\left(\frac{\lambda}{2}\right) r^j\right]} dr} .$$

Now choose $\mu = \mu(\lambda,\tau)$ so that

$$\sum_{j=2}^{m} \left|\frac{b^{(j)}\left(\frac{\lambda}{2}\right)\tau\mu^j}{j!}\right|^2 = 1$$

and in the last integral, make the change of variables $r \to \mu r$, $\eta \to \frac{1}{\mu}\eta$.

Then

$$F(\lambda+it,\tau) = e^{2\pi\tau\left[2b\left(\frac{\lambda}{2}\right)+itb'\left(\frac{\lambda}{2}\right)\right]}\mu^{-2}\int_{-\infty}^{\infty}\frac{e^{2\pi i\eta\left(\frac{t}{\mu}\right)}\,d\eta}{\int_{-\infty}^{\infty}e^{4\pi\left[\eta r-\sum_2^m a_j r^j\right]}dr}$$

where $a_j = \frac{1}{j!}b^{(j)}\left(\frac{\lambda}{2}\right)\tau\mu^j$, and hence $\sum_2^m |a_j|^2 = 1$.

We now make two observations. First, in terms of size,

$$\mu(\lambda,\tau)^{-1} \approx \sum_2^m |b^{(j)}\left(\frac{\lambda}{2}\right)|\,\tau^{1/j}\,.$$

This is clear from the definition of μ. Second, the collection of functions

$$\theta_{\vec{a}}(\eta) = \left[\int_{-\infty}^{\infty} e^{4\pi\left[\eta r-\sum_2^m a_j r^j\right]}dr\right]^{-1}$$

where $\vec{a} = (a_2,\cdots,a_m)$, $\Sigma|a_j|^2 = 1$, and $r \to \sum_2^m a_j r^j$ is convex, is a compact set of functions in the Schwartz class $\mathcal{S}(\mathbf{R})$. Thus

$$F(\lambda+it,\tau) = e^{2\pi\tau\left[2b\left(\frac{\lambda}{2}\right)+itb'\left(\frac{\lambda}{2}\right)\right]}\mu(\lambda,\tau)^{-1}\,\hat{\theta}_{\vec{a}}\left(\frac{t}{\mu}\right).$$

From this, one can make estimates on the size of $F(\lambda+it,\tau)$ and its derivatives.

Finally we have

$$S((x,y,t);(r,s,u)) = \int_0^{\infty} e^{-2\pi\tau[b(x)+b(r)+i(t-u)]}F(x+r+i(y-s),\tau)\,d\tau$$

and we can use the estimates on F to estimate S. A consequence is, for example:

$$|S((x,y,t);(r,s,u))| \leq C|B((x,y,t),\delta)|^{-1}$$

where δ is the nonisotropic distance between (x,y,t) and (r,s,u).

ALEXANDER NAGEL
DEPARTMENT OF MATHEMATICS
UNIVERSITY OF WISCONSIN
MADISON, WISCONSIN

REFERENCES

[1] Bers, L., John, F., and Schechter, M., *Partial Differential Equations*, Interscience Publishers, John Wiley and Sons, Inc., New York 1964.

[2] Carathéodory, C., "Untersuchungen über die Grundlagen der Thermodynamik," Math. Ann. 67 (1909), 355-386.

[3] Coifman, R. R., and Weiss, G., *Analyse harmonique non-commutative sur certains espaces homogènes*, Lecture Notes in Math. #242, Springer-Verlag, 1971.

[4] Fatou, P., "Séries trigonométriques et séries de Taylor," Acta Math. 30 (1906), 335-400.

[4a] Fefferman, C., and Phong, D. H., "Subelliptic eigenvalue problems" in *Proceedings of the Conference on Harmonic Analysis in Honor of Antoni Zygmund*, 590-606, Wadsworth Math. Series, 1981.

[5] Folland, G. B., *Introduction to Partial Differential Equations*, Mathematical Notes Series, #17, Princeton University Press, Princeton, N. J. 1976.

[5a] Folland, G., and Hung, H. T., "Non-isotropic Lipschitz spaces" in *Harmonic Analysis in Euclidean Spaces*, Part 2, 391-394; Amer. Math. Soc., Providence, 1979.

[6] Folland, G. B., and Stein, E. M., "Estimates for the $\bar{\partial}_b$ complex and analysis on the Heisenberg group," Comm. Pure Appl. Math. 27 (1974), 429-522.

[6a] Grushin, V. V., "On a class of hypoelliptic pseudo-differential operators degenerate on a sub-manifold," Math. USSR Sbornik 13 (1971), 155-185.

[7] Hörmander, L., "Hypoelliptic second order differential equations," Acta Math. 119 (1967), 147-171.

[8] Kohn, J. J., "Boundary behavior of $\bar{\partial}$ on weakly pseudoconvex manifolds of dimension two," J. Diff. Geom. 6 (1972), 523-542.

[8a] Kohn, J. J., "Boundaries of Complex Manifolds," in *Proceedings of the Conference on Complex Analysis*, Minneapolis, 1964; 81-94; Springer-Verlag, New York, 1965.

[9] Korányi, A., "Harmonic functions on Hermetian hyperbolic space," Trans. Am. Math. Soc. 135 (1969), 507-516.

[10] Korányi, A., and Vagi, S., "Singular integrals in homogeneous spaces and some problems of classical analysis," Ann. Scuola Norm. Sup. Pisa 25 (1971), 575-648.

[10a] Krantz, S., *Function theory of several complex variables*, John Wiley and Sons, New York, 1982.

[11] Nagel, A., and Stein, E. M., *Lectures on Pseudo-differential Operators*, Mathematical Notes Series, #24, Princeton University Press, Princeton, N. J. 1979.

[12] Nagel, A., Stein, E. M., and Wainger, S., "Boundary behavior of functions holomorphic in domains of finite type," Proc. Natl. Acad. Sci. USA, 78 (1981), 6596-6599.

[13] ——— "Balls and metrics defined by vector fields I: Basic properties" to appear in Acta Math.

[14] Rivière, N., "Singular integrals and multiplier operators," Ark. för Mat. 9 (1971), 243-278.

[15] Rothschild, L. P., and Stein, E. M., "Hypoelliptic differential operators and nilpotent groups," Acta Math. 137 (1976), 247-320.

[15a] Sánchez-Calle, A., "Fundamental solutions and geometry of the sum of squares of vector fields," Inventiones Math. 78, 143-160 (1984).

[16] Stein, E. M., *Singular Integrals and Differentiability Properties of Functions*, Princeton University Press, Princeton, N. J. 1970.

[17] ———, *Boundary behavior of holomorphic functions of several complex variables*, Mathematical Notes Series, #11, Princeton Univ. Press, Princeton, N. J. 1972.

OSCILLATORY INTEGRALS IN FOURIER ANALYSIS

E. M. Stein

Introduction

Oscillatory integrals in one form or another have been an essential part of harmonic analysis from the very beginnings of that subject. Besides the obvious fact that the Fourier transform is itself an oscillatory integral par excellence, one needs only bear in mind the occurrence of Bessel functions in the original work of Fourier (1822), the study of asymptotics related to such functions in the early works of Airy (1838), Stokes (1850), and Lipschitz (1859), Riemann's use in 1854* of the method of "stationary phase" in finding the asymptotics of certain Fourier transforms, and the application of all these ideas to number theory, initiated in the first quarter of our century by Voronoi (1904), Hardy (1915), van der Corput (1922) and others.

Given this long history it is an interesting fact that only relatively recently (1967) did one realize the possibility of restriction theorems for the Fourier transform, and that the relation of the above asymptotics to differentiation theory had to wait another ten years to come to light!

The purpose of these lectures is to survey part of this theory and at the same time to describe some new results. We have found it convenient to divide our discussion into oscillatory integrals of the "first kind," and those of the "second kind." The main difference between the two is that for the first kind we are studying the behavior of only *one* function as the parameter increases to infinity, while for the second kind we are dealing

*In Section XIII of his paper on trigonometric series.

with the boundedness properties of an operator which carries an oscillatory factor in its kernel. However this distinction need not be taken literally since sometimes these different types merge.

We begin by considering the more-or-less standard facts about oscillatory integrals of the first kind, first in one dimension and then in n dimensions. Next as a first application we deal with some estimates of the Fourier transform of smooth surface-carried measures in $\mathbf{R}^n$. This leads us naturally to restriction theorems. (Differentiation theorems, which are another application, are not dealt with here; but these are the subject of Wainger's lectures [3].)

Next we discuss oscillatory integrals (of the first kind) arising in the theory of Hilbert transform along curves and their generalizations. We then turn to oscillatory integrals of the second kind suggested by twisted convolution on the Heisenberg group and the theory of Radon singular integrals. Finally we return to restriction theorems and the oscillatory integrals of the second kind they give rise to, which operators are closely related to Bochner-Riesz summability.*

1. *Oscillatory integrals of the first kind,* $n = 1$

We are interested in the behavior for large positive λ of the integral

$$I(\lambda) = \int_a^b e^{i\lambda\phi(x)}\psi(x)\,dx$$

where ϕ is a real-valued smooth function (the "phase"), and ψ is complex-valued and smooth; often, but not always, one assumes that ψ has compact support in (a,b).

---

*The reader will note that there are several related topics not touched on in this survey. Chief among them is the subject of oscillatory integrals arising in the solution of hyperbolic equations and their generalizations – the class of "Fourier integral operators." For an elegant introduction to that subject see [1], Chapter 4.

The basic facts about $I(\lambda)$ can be presented in terms of three principles.

(a) *Localization*: The asymptotic behavior of $I(\lambda)$ is determined by those points where $\phi'(x) = 0$, (assuming that ψ has compact support in (a,b)). More precisely,

PROPOSITION 1. *Suppose* $\psi \in C_0^\infty(a,b)$, *and* $\phi'(x) \neq 0$ *for* x *in* $[a,b]$. *Then* $I(\lambda) = 0(\lambda^{-N})$, *as* $\lambda \to \infty$ *for every* $N \geq 0$.

The proof is very simple. Let D denote the differential operator $Df = \frac{1}{i\lambda\phi'(x)} \frac{d}{dx}$, and let ${}^tD$ denote its transpose, ${}^tDf = \frac{d}{dx}\left(\frac{f}{i\lambda\phi'}\right)$. Then clearly $D^N(e^{i\lambda\phi}) = e^{i\lambda\phi}$ for every N, and integration by parts shows that

$$\int e^{i\lambda\phi}\psi\, dx = \int D^N(e^{i\lambda\phi})\psi\, dx = (-1)^N \int e^{i\lambda\phi}({}^tD)^N \psi\, dx .$$

Thus clearly $|I(\lambda)| \leq A_N\lambda^{-N}$, and the proposition is proved.

(b) *Scaling*: Suppose we only know that $\left|\frac{d^k\phi(x)}{dx^k}\right| \geq 1$ for some fixed k, and we wish to obtain an estimate for $\int_a^b e^{i\lambda\phi(x)}dx$ which is *independent* of a and b. Then a simple scaling argument shows that the only possible estimate for the integral is $0(\lambda^{-1/k})$. That this is indeed the case goes back to van der Corput.

PROPOSITION 2. *Suppose* ϕ *is real-valued and smooth in* $[a,b]$. *If* $|\phi^{(k)}(x)| \geq 1$, *then*

$$\left|\int_a^b e^{i\lambda\phi(x)}dx\right| \leq c_k\lambda^{-1/k} \tag{1.1}$$

holds when

(i) $k \geq 2$

(ii) *or* $k = 1$, *if in addition it is assumed that* $\phi'(x)$ *is monotonic.*

Proof. Let us show (ii) first. We have

$$\int_a^b e^{i\lambda\phi}\,dx = \int_a^b D(e^{i\lambda\phi})\,dx = -\int_a^b e^{i\lambda\phi}\,{}^tD(1)\,dx + \frac{1}{i\lambda\phi'}e^{i\lambda\phi}\Big]_a^b$$

The boundary terms are majorized by $2/\lambda$, while

$$\left|\int_a^b e^{i\lambda\phi}\,{}^tD(1)dx\right| = \left|\int_a^b e^{i\lambda\phi}\frac{1}{i\lambda}\frac{d}{dx}\left(\frac{1}{\phi'}\right)dx\right|$$

$$\leq \frac{1}{\lambda}\int_a^b \left|\frac{d}{dx}\frac{1}{\phi'}\right|$$

$$= \frac{1}{\lambda}\left|\int_a^b \frac{d}{dx}\left(\frac{1}{\phi'}\right)dx\right|$$

by the montonicity of ϕ'. The last expression equals $\frac{1}{\lambda}\left|\frac{1}{\phi'(b)} - \frac{1}{\phi'(a)}\right|$, which is dominated by $2/\lambda$. This gives the desired conclusion with $c_1 = 4$.

We now prove (i) by induction on k. Let us suppose that the case k is known, and assume (taking complex conjugates if necessary) that $\phi^{(k+1)}(x) \geq 1$. Let $x = c$ be the point in $[a,b]$ where $|\phi^{(k)}(x)|$ takes its minimum value. If $\phi^{(k)}(c) = 0$, then outside the interval $(c-\delta, c+\delta)$, we have that $|\phi^{(k)}(x)| \geq \delta$, (and of course $\phi'(x)$ is monotonic when $k = 1$). Write

$$\int_a^b = \int_a^{c-\delta} + \int_{c-\delta}^{c+\delta} + \int_{c+\delta}^b .$$

By the previous case

$$\left| \int_a^{c-\delta} e^{i\lambda\phi}\, dx \right| \le c_k/(\lambda\delta)^{1/k} .$$

Similarly

$$\left| \int_{c+\delta}^{b} e^{i\lambda\phi}\, dx \right| \le c_k/(\lambda\delta)^{1/k} .$$

Clearly however

$$\left| \int_{c-\delta}^{c+\delta} e^{i\lambda\phi}\, dx \right| \le 2\delta .$$

Thus

$$\left| \int_a^b e^{i\lambda\phi}\, dx \right| \le \frac{2c_k}{(\lambda\delta)^{1/k}} + 2\delta .$$

If $\phi^{(k)}(c) \neq 0$, and so c is one of the end-points of $[a,b]$, a similar argument shows that $c_k/(\lambda\delta)^{1/k} + \delta$ is an upper bound to the integral. In either situation the case $k+1$ follows by taking $\delta = \lambda^{-1/k+1}$, which proves (1.1) with $c_{k+1} = 2c_k + 2$.

COROLLARY. *Under the assumptions on* ϕ *in Proposition 2, we can conclude that*

$$\left| \int_a^b e^{i\lambda\phi(x)} \psi(x)\, dx \right| \le c_k \lambda^{-1/k} \left[|\psi(b)| + \int_a^b |\psi'(x)|\, dx \right]. \tag{1.2}$$

This follows from (1.1) by integrating by parts an estimate of the form

$$\left| \int_a^{\bar{x}} e^{i\lambda\phi(x)} dx \right| \leq c_k \lambda^{-1/k}, \quad \text{for } a \leq \bar{x} \leq b .$$

(c) *Asymptotics*: We already know that the behavior of $\int_a^b e^{i\lambda\phi}\psi dx$ is determined by those points x_0, where $\phi'(x_0) = 0$, (the "critical" points of ϕ). Assuming that the support of ψ is so small that it contains only one critical point of ϕ, the character of the asymptotic expansion then depends on the smallest k so that $\phi^{(k)}(x_0) \neq 0$, and is given in terms of powers of λ in a way which is consistent with Proposition 2.

PROPOSITION 3. *Suppose ϕ is real and smooth, $k \geq 2$, and $\phi(x_0) = \phi'(x_0) \cdots = \phi^{(k-1)}(x_0) = 0$, while $\phi^{(k)}(x_0) \neq 0$. If $\psi \in C_0^\infty$ and the support of ψ is a sufficiently small neighborhood of x_0, then*

$$\text{(1.3)} \qquad I(\lambda) = \int e^{i\lambda\phi(x)}\psi(x) dx \sim \lambda^{-1/k} \sum_{j=0}^{\infty} a_j \lambda^{-j/k} ,$$

in the sense that for any non-negative integers N *and* r

$$\text{(1.3')} \qquad \left(\frac{d}{d\lambda}\right)^r \left[I(\lambda) - \sum_{j=0}^{N} a_j \lambda^{-j/k} \right] = 0(\lambda^{-(N+1)/k-r}) \ \ as \ \lambda \to \infty .$$

REMARK. When k is even, then $a_j = 0$ for j odd.

We shall give a proof of the proposition when $k = 2$. There are three steps.

Step 1. This is the observation that

$$\text{(1.4)} \qquad \int_{-\infty}^{\infty} e^{i\lambda x^2} x^\ell e^{-x^2} dx \sim \lambda^{-1/2-\ell/2} \sum_{j=0}^{\infty} c_j^{(\ell)} \lambda^{-j}$$

where ℓ is a non-negative integer; in the case ℓ is odd the integral vanishes. In fact the left-side of (1.4) is $\int_{-\infty}^{\infty} e^{-(1-i\lambda)x^2} x^\ell dx$ which by a change of variables equals $(1-i\lambda)^{-1/2-\ell/2} \int_{-\infty}^{\infty} e^{-x^2} x^\ell dx$. However, when $\lambda > 0$, $(1-i\lambda)^{-1/2-\ell/2} = \lambda^{-1/2-\ell/2}(1/\lambda - i)^{-1/2-\ell/2}$, where we have fixed the principal branch of $z^{-1/2-\ell/2}$ in the plane slit along the negative half-axis. The power series expansion of $(w-i)^{-1/2-\ell/2}$ (which holds for $|w| < 1$), then gives the desired asymptotic expansion (1.4).

Step 2. Observe next that if $\eta \in C_0^\infty$ and ℓ is a non-negative integer, then

$$\left| \int_{-\infty}^{\infty} e^{i\lambda x^2} x^\ell \eta(x)\, dx \right| \leq A\lambda^{-1/2-\ell/2}. \tag{1.5}$$

To prove this let α be a C^∞ function with the property that $\alpha(x) = 1$ for $|x| \leq 1$, and $\alpha(x) = 0$ when $|x| \geq 2$, and write

$$\int e^{i\lambda x^2} x^\ell \eta(x)\, dx = \int e^{i\lambda x^2} x^\ell \eta(x)\, \alpha(x/\varepsilon)\, dx + \int e^{i\lambda x^2} x^\ell \eta(x)\, (1-\alpha(x/\varepsilon))\, dx .$$

The first integral is dominated by $C\varepsilon^{\ell+1}$. The second integral can be written as

$$\int e^{i\lambda x^2} ({}^tD)^N [x^\ell \eta(x)[1-\alpha(x/\varepsilon)]]\, dx$$

with ${}^tDf = \frac{1}{i\lambda} \frac{d}{dx} \frac{f}{2x}$. A simple computation then shows that this term is majorized by

$$\frac{C_N}{\lambda^N} \int_{|x| \geq \varepsilon} |x|^{\ell - 2N} dx = C_N^1 \lambda^{-N} \varepsilon^{\ell-2N-1}$$

if $\ell - 2N < -1$. Altogether then the integral in (1.5) is bounded by

$$C_N\{\varepsilon^{\ell+1}+\lambda^{-N}\varepsilon^{\ell-2N+1}\}$$ and we need only take $\varepsilon=\lambda^{-1/2}$,

$\left(\text{with } N>\frac{\ell+1}{2}\right)$, to get the conclusion (1.5).

A similar (but simpler) argument of integration by parts also shows that

$$\int e^{i\lambda x^2}\xi(x)\,dx = 0(\lambda^{-N}),\ \text{every } N \geq 0 \tag{1.6}$$

whenever $\xi \in \mathcal{S}$, and ξ vanishes near the origin.

Step 3. We prove the proposition first in the case $\phi(x)=x^2$. To do this write

$$\int e^{i\lambda x^2}\psi(x)\,dx = \int e^{i\lambda x}e^{-x^2}(e^{x^2}\psi(x))\tilde{\psi}(x)\,dx\,,$$

where $\tilde{\psi}$ is a C_0^∞ function which is 1 on the support of ψ. Now for each N, write the taylor expansion

$$e^{x^2}\psi(x) = \sum_{j=0}^{N} b_jx^j + x^{N+1}R_N(x) = P(x)+x^{N+1}R_N(x)\,.$$

Substituting in the above gives three terms

(a) $$\sum_{j=0}^{N} b_j \int_{-\infty}^{\infty} e^{i\lambda x^2}e^{-x^2}x^j\,dx$$

(b) $$\int_{-\infty}^{\infty} e^{i\lambda x^2}x^{N+1}R_N(x)e^{-x^2}\tilde{\psi}(x)\,dx$$

(c) $$\int e^{i\lambda x^2}P(x)e^{-x^2}(1-\tilde{\psi}(x))\,dx\,.$$

For (a) we use (1.4); for (b) we use (1.5); and for (c) we use (1.6). It is then easy to see that their combination gives the desired asymptotic expansion for $\int e^{i\lambda x^2}\psi(x)\,dx$.

Let us now consider the general case when $k = 2$. We can then write $\phi(x) = c(x-x_0)^2 + 0(x-x_0)^3$ with $c \neq 0$ and set $\phi(x) = c(x-x_0)^2[1+\varepsilon(x)]$, where ε is a smooth function which is $0(x-x_0)$, and hence $|\varepsilon(x)| < 1$ when x is sufficiently close to x_0. Moreover, $\phi'(x) \neq 0$, when $x \neq x_0$, but x lies sufficiently close to x_0. Let us now fix such a neighborhood of x_0, and let $y = (x-x_0)(1+\varepsilon(x))^{1/2}$. Then the mapping $x \to y$ is a diffeomorphism of that neighborhood of x_0 to a neighborhood of $y = 0$, and of course $cy^2 = \phi(x)$. Thus

$$\int e^{i\lambda\phi(x)}\psi(x)\,dx = \int e^{i\lambda cy^2}\tilde{\psi}(y)\,dy$$

with $\tilde{\psi} \in C_0^\infty$ if the support of ψ lies in our fixed neighborhood of x_0. The expansion (1.3) (for $k=2$), is then proved as a consequence of the special case treated before.

REMARKS:

(1) The proof for higher k is similar and is based on the fact that

$$\int_0^\infty e^{i\lambda x^k} e^{-x^k} x^\ell\,dx = c_{k,\ell}(1-i\lambda)^{-(\ell+1)/k}.$$

(2) Each constant a_j that appears in the asymptotic expansion (1.3) depends on only finitely many derivatives of ϕ and ψ at x_0. Note e.g. that when $k = 2$, we have $a_0 = \sqrt{\pi}(-i\phi''(x_0))^{-1/2}\psi(x_0)$. Similarly the bounds occurring in (1.3′) depend on upper bounds of finitely many derivatives of ϕ and ψ in the support of ψ, the size of the support of ψ, and a lower bound for $\phi^{(k)}(x_0)$.

References: The reader may consult Erdélyi [8], Chapter II, where further citations of the classical literature may be found.

2. *Oscillatory integrals of the first kind*, $n \geq 2$

Only some of the above results have analogues when $n \geq 2$, but the extension of Proposition 1 is simple. Continuing a terminology used above we say that a phase function ϕ defined in a neighborhood of a point x_0 in $\mathbf{R}^n$ has x_0 as a critical point, if $(\nabla\phi)(x_0) = 0$.

PROPOSITION 4. *Suppose* $\psi \in C_0^\infty(\mathbf{R}^n)$, *and* ϕ *is a smooth real-valued function which has no critical points in the support of* ψ. *Then*

$$I(\lambda) = \int_{\mathbf{R}^n} e^{i\lambda\phi(x)}\psi(x)\,dx = 0(\lambda^{-N}), \text{ as } \lambda \to \infty \text{ for every } N \geq 0\,.$$

Proof. For each x_0 in the support of ψ, there is a unit vector ξ and a small ball $B(x_0)$, centered at x_0, so that $(\xi, \nabla_x)\phi(x) \geq c > 0$, for $x \in B(x_0)$. Decompose the integral $\int e^{i\lambda\phi(x)}\psi(x)\,dx$ as a finite sum

$$\sum_n \int e^{i\lambda\phi(x)}\psi_k(x)\,dx\,,$$

where each ψ_k is C^∞ and has compact support in one of these balls. It then suffices to prove the corresponding estimate for each of these integrals. Now choose a coordinate system $x_1, x_2, \cdots, x_n$ so that x_1 lies along ξ. Then

$$\int e^{i\lambda\phi(x)}\psi_k(x)\,dx = \int\left(\int e^{i\lambda\phi(x_1,\cdots,x_n)}\psi_k(x_1,\cdots,x_n)\,dx_1\right)dx_2,\cdots,dx_n\,.$$

But the inner integral is $0(\lambda^{-N})$ by Proposition 1, and so our desired conclusion follows.

We can only state a weak analogue for the scaling principle, Proposition 2; it, however, will be useful in what follows.

PROPOSITION 5. *Suppose* $\psi \in C_0^\infty$, ϕ *is real-valued, and for some multi-index* a, $|a| > 0$,

$$\left|\left(\frac{\partial}{\partial x}\right)^a \phi(x)\right| \geq 1$$

throughout the support of ψ. *Then*

$$(2.1) \qquad \left|\int_{R^n} e^{i\lambda\phi(x)}\psi(x)\,dx\right| \leq c_k(\phi)\cdot\lambda^{-1/k}(\|\psi\|_{L^\infty} + \|\nabla\psi\|_{L^1})$$

with $k = |a|$, *and the constant* $c_k(\phi)$ *is independent of* λ *and* ψ *and remains bounded as long as the* C^{k+1} *norm of* ϕ *remains bounded.*

Proof. Consider the real linear space of homogeneous polynomials of degree k in R^n. Let $d(k,n)$ denote its dimension. Of course $\{x^a\}_{|a|=k}$ is a basis for this space. However it is not difficult to see that there are $d(k,n)$ unit vectors $\xi^{(1)}, \xi^{(2)}, \cdots, \xi^{(d(k,n))}$, so that the homogeneous polynomials $(\xi^{(j)}\cdot x)^k$, $j = 1,\cdots,d(k,n)$, give another basis. This means that if

$$\left|\frac{\partial^a\phi(x^0)}{\partial x^a}\right| \geq 1 ,$$

for some $|a| = k$, there is a unit vector $\xi = \xi(x^0)$, so that

$$|(\xi, \nabla_x)^k \phi(x^0)| \geq a_k, \text{ with } a_k > 0 .$$

Moreover since we can assume that the C^{k+1} norm of ϕ is bounded we can also conclude that $|(\xi, \nabla_x^k)\phi(x)| \geq a_k/2$ whenever $x \in B(x_0)$, where B is the ball centered at x of fixed radius. (The radius of B can be taken to be a small multiple of the C^{k+1} norm of ϕ.) Next choose an appropriate covering of R^n by such balls of fixed radius, and a corresponding partition of unity, $1 = \Sigma\, \eta_j(x)$, with $0 \le \eta_j \le 1$, $\sum_k |\nabla \eta_j| \le b_k$, and each η_j supported in one of our balls. So

$$\int e^{i\lambda\phi}\psi dx = \sum_j \int e^{i\lambda\phi}\psi\eta_j\, dx = \sum \int e^{i\lambda\phi}\psi_j\, dx\,.$$

To estimate $\int e^{i\lambda\phi}\psi_j\, dx$, with ξ determined as above, choose a coordinate system so that x_1 lies along ξ. Then

$$\int e^{i\lambda\phi}\psi_j\, dx = \int\left(\int e^{i\lambda\phi(x_1,\cdots,x_n)}\tilde{\psi}_j(x_1,\cdots,x_n)dx_1\right) dx_2,\cdots,dx_n\,.$$

For the inner integral we invoke (1.2) giving us an estimate of the form

$$c_k a_k^{-1/k}\lambda^{-1/k}\left\{\left(\int\left|\frac{\partial\eta_j}{\partial x_1}(x_1,\cdots,x_n)\right| dx_1\right)\|\psi\|_{L^\infty} + \int\left|\frac{\partial\psi}{\partial x_1}(x_1,\cdots,x_n)\right| dx_1\right\}.$$

A final integration in the other variables then leads to (2.1).

REMARK. Let us note that in R^2 if $\phi(x) = x_1x_2$, the above proposition gives no better than a decrease of order $\lambda^{-1/2}$, while the asymptotics of the proposition below shows that the true order is λ^{-1}.

Let us go back for the moment to the case of one dimension. If ϕ has a critical point at x_0, and ϕ' does not vanish of infinite order at x_0, then after a smooth change of variables ϕ can be transformed to a simple canonical form $\tilde{\phi}$, with $\tilde{\phi}(x) = \pm x^k$ (for x near 0). There is no analogue of this in higher dimensions, except for $k = 1$ and in a special case corresponding to $k = 2$. To the asymptotics of the latter situation we now turn.

Suppose ϕ has a critical point at x^0. If the symmetric $n \times n$ matrix $\left\{\frac{\partial^2\phi(x^0)}{\partial x_i \partial x_j}\right\}$ is invertible, then the critical point is said to be *non-degenerate*. It is an easy matter to see by the use of Taylor's expansion that if x^0 is a non-degenerate critical point, then in fact it is an isolated critical point.

PROPOSITION 6. *Suppose $\phi(x^0) = 0$, and ϕ has a non-generate critical point at x^0. If $\psi \in C_0^\infty$ and the support of ψ is a sufficiently small neighborhood of x^0, then*

$$(2.2) \qquad \int_{R^n} e^{i\lambda\phi(x)}\psi(x)\,dx \sim \lambda^{-n/2} \sum_{j=0}^{\infty} a_j \lambda^{-j}, \text{ as } \lambda \to \infty$$

where the asymptotics hold in the same sense as (1.3), (1.3′).

Note. Again each of the constants a_j appearing in the asymptotic expansion depends on only finite many values of derivatives of ϕ and ψ at x^0. Thus e.g. $a_0 = \left(\pi^{n/2} \prod_{j=1}^{n} (-i\mu_j)^{-1/2}\right) \cdot \psi(x_0)$, where $\mu_1, \mu_2, \cdots, \mu_n$ are the eigenvalues of the matrix $\left\{\frac{1}{2} \frac{\partial^2\phi(x^0)}{\partial x_j \partial x_k}\right\}$. Similarly each of the bounds occurring in the error terms depend only on upper bounds for finitely many derivatives of ϕ and ψ in the support of ψ, the size of the support of ψ, and a lower bound for $\det\left\{\frac{\partial^2\phi(x^0)}{\partial x_j \partial x_k}\right\}$.

The proof of the proposition follows closely the same pattern as that of Proposition 3. First, let $Q(x)$ denote the unit quadratic form given by $Q(x) = x_1^2 + x_2^2 + \cdots + x_m^2 - x_{m+1}^2 - \cdots - x_n^2$, where $0 \le m \le n$, with m fixed. The analogue of (1.4) is

$$(2.3) \qquad \int_{R^n} e^{i\lambda Q(x)} e^{-|x|^2} x^\ell \, dx \sim \lambda^{-n/2-|\ell|/2} \sum_{j=0}^{\infty} c_j(m,\ell) \lambda^{-j}$$

with $\ell = (\ell_1, \cdots, \ell_m)$ a multi-index, $|\ell| = \ell_1 + \ell_2 + \cdots + \ell_n$, and $x^\ell = x_1^{\ell_1}, \cdots, x_n^{\ell_n}$; also note that if one ℓ_j is odd then (2.3) is identically zero. To prove (2.3) write it as a product

$$\prod_{j=1}^n \left(\int_{-\infty}^{\infty} e^{\pm i\lambda x^2} e^{-x^2} x^{\ell_j} dx \right) = \prod_{j=1}^n \left(\int_{-\infty}^{\infty} e^{-x^2} x^{\ell_j} dx \right) (1 \mp i\lambda)^{-1/2-\ell_j/2} ,$$

and expand the function $\prod_{j=1}^n (1/\lambda \mp i)^{-\frac{1}{2}-\ell_j/2}$, (for large λ) in a power series in $1/\lambda$.

The analogue of (1.5) is the statement that

$$(2.4) \qquad \left| \int_{R^n} e^{i\lambda Q(x)} x^\ell \eta(x) \right| dx \le A\lambda^{-n/2-|\ell|/2} ; \text{ if } \eta \in C_0^\infty(R^n) .$$

To prove it we consider the two-sided cones Γ_j defined by $\Gamma_j = \left\{ x \mid |x_j|^2 \ge \frac{1}{2n} |x|^2 \right\}$, and the smaller cases $\Gamma_j^0 = \left\{ x \mid |x_j|^2 \ge \frac{1}{n} |x|^2 \right\}$. Then since $\bigcup_{j=1}^n \Gamma_j^0 = R^n$, we can find functions Ω_j, $j = 1, \cdots, n$, each homogeneous of degree 0, and C^∞ away from the origin, so that $1 = \sum_{j=1}^n \Omega_j(x)$, $x = 0$, with Ω_j supported in Γ_j. Then we can write

$$\int e^{i\lambda Q(x)} x^\ell \eta(x)\, dx = \sum_j \int e^{i\lambda Q(x)} x^\ell \eta(x) \Omega_j(x)\, dx .$$

In the cone Γ_j one uses integration by parts via

$$D_j e^{i\lambda Q(x)} = e^{i\lambda Q(x)} \quad \text{with} \quad D_j(f) = \frac{1}{\pm i\lambda x_j} \frac{\partial f}{\partial x_j} .$$

This, together with the fact $|x_j| \ge \frac{1}{\sqrt{2n}} |x|$ in Γ_j, and $|\, ({}^tD_j)^N \Omega_j(x)| \le$

$\frac{C_N}{\lambda^N}|x|^{-2N}$, allows one to conclude the proof of (2.4) in analogy with that of (1.5).

A similar argument also show that whenever $\xi \in \mathcal{S}$ and ξ vanishes near the origin, then

$$\int e^{i\lambda Q(x)}\xi(x)\,dx = 0(\lambda^{-N}),\ \text{for every}\ N \geq 0\,. \tag{2.5}$$

We then combine (2.3), (2.4) and (2.5) as before to obtain the asymptotic formula (2.2) in the special case when $\phi(x) = Q(x)$.

To pass to the general case one is then fortunate to be able to appeal to the change of variables guaranteed by Morse's lemma: Since $\phi(x^0) = 0$, $(\nabla\phi)(x^0) = 0$, and the critical point is assumed to be non-degenerate, there exists a diffeomorphism of a small neighborhood of x^0 in the x-space to the y-space, under which ϕ is transformed $y_1^2 + y_2^2 + \cdots + y_m^2 - y_{m+1}^2 - \cdots - y_n^2$. Observe that the index m is the same as that of the form corresponding to $\left\{\frac{1}{2}\frac{\partial^2\phi(x^0)}{\partial x_j \partial x_k}\right\}$.

References. For a proof of Morse's lemma see Milnor [20], §2.

§3. *Fourier transforms of surface-carried measures*

Let S denote a smooth m-dimensional sub-manifold of R^n (not necessarily closed). We let $d\sigma$ denote the measure on S induced by the Lebesgue measure on R^n, and fix a function ψ in $C_0^\infty(R^n)$. Consider now the finite Borel measure $d\mu = \psi(x)\,d\sigma$ on R^n, which is of course carried on S. The problem we wish to deal with is that of finding estimates at infinity of the Fourier transform of μ, i.e. $d\hat{\mu}(\xi)$. We shall consider two cases of this problem.

(1) Suppose first $\dim S = n - 1$, and S has *non-zero Gaussian curvature* at each point. By this we mean the following: Let x^0 be any point of S,

and consider a rotation and translation of the underlying $\mathbf{R}^n$ so that the point x^0 is moved to the origin, and the tangent plane of S at x^0 becomes the hyperplane $x_n = 0$. Then near the origin (i.e. near x^0) the surface S can be given as a graph $x_n = \phi(x_1, \cdots, x_{n-1})$ with $\phi \in C^\infty$, and $\phi(0) = 0$, $(\nabla\phi)(0) = 0$. Now consider the $(n-1) \times (n-1)$ matrix $\left\{\frac{1}{2} \frac{\partial^2\phi(0)}{\partial x_j \partial x_k}\right\}_{1 \le j,k \le n-1}$. Its eigenvalues $\mu_1, \cdots, \mu_{n-1}$ are called the principal curvatures of S at x^0, and their product $\left(= \det \frac{1}{2} \frac{\partial^2\phi(0)}{\partial x_j \partial x_k}\right)$ is the Gaussian curvature at x^0.

THEOREM 1. *Suppose* S *is a smooth hypersurface in* $\mathbf{R}^n$, *with non-vanishing Gaussian curvature at each point, and let* $d\mu = \psi d\sigma$ *as above. Then*

$$|(d\mu)\hat{\ }(\xi)| \le A|\xi|^{-(n-1)/2} . \tag{3.1}$$

Proof. It would be convenient (in applying Proposition 6 above) to change notation momentarily by taking n to be $n+1$. Now by the compactness of the support of ψ (and since when the surface is given as a graph $x_{n+1} = \phi(x_1, \cdots, x_n)$, then $d\sigma = \sqrt{1 + |\nabla\phi|^2}\, dx_1, \cdots, dx_n$) we can reduce matters to showing that

$$\left| \int_{\mathbf{R}^n} e^{i\lambda\Phi(x,\eta)} \tilde{\psi}(x)\, dx \right| \le A\lambda^{-n/2} \tag{3.2}$$

where $\Phi(x,\eta) = x_1\eta_1 + x_2\eta_2 + \cdots + x_n\eta_n + \phi(x_1, \cdots, x_n)\eta_{n+1}$, with $\xi = \lambda(\eta_1, \cdots, \eta_{n+1})$, $\sum_{j=1}^{n+1} \eta_j^2 = 1$, and $\lambda > 0$; also $\phi(0) = (\nabla\phi)(0) = 0$, $\det\left(\frac{\partial^2\phi(0)}{\partial x_j \partial x_j}\right) \ne 0$, and the support of $\tilde{\psi}$ is a sufficiently small neighborhood of the origin.

We divide consideration of the unit vector η into three cases.

1° when η is sufficiently close to the "north pole" $\eta_N = (0,\cdots,0,1)$;

2° when η is sufficiently close to the "south pole" $\eta_N = (0,\cdots,0,-1)$;

3° the complementary set on the unit sphere.

Let us consider 1° first. The function $\phi(x,\eta)$ has the property that $(\nabla_x\phi)(0,\eta_N) = 0$. We want to see that for each η sufficiently close to η_N there is a (unique) $x = x(\eta)$, so that $(\nabla_x)\phi(x(\eta),\eta) = 0$. The latter is a series of n equations and one can find the desired solution by the implicit function theorem, which requires that we check that the Jacobian determinant $\det((\nabla_x\nabla_x\phi)(0,\eta_N)) \neq 0$; but this of course is our assumption of non-vanishing curvature. Notice that if the η-neighborhood of η_N is sufficiently small, then $\det\left(\frac{\partial^2\phi(x(\eta),\eta)}{\partial x_j \partial x_k}\right) \neq 0$, so we can invoke Proposition 6 (with $x^0 = x(\eta)$), as long as the support of $\tilde{\psi}$ is sufficiently small. This proves (3.2) when η is in region 1°. The proof when η is in 2° is the same. So we now come to the region 3°. Since $\phi(x,\eta) = \sum_{j=1}^{n} x_j\eta_j + \phi(x)\eta_{n+1}$, then $\nabla_x\phi(x,\eta) = (\eta_1,\cdots,\eta_n) + \eta_{n+1}(\nabla\phi)(x)$. However $(\eta_1^2+\cdots+\eta_n^2)^{1/2} \geq c > 1$, since η is in 3° and $\nabla\phi(x) = 0(x)$ as $x \to 0$; thus $|\nabla_x\phi(x,\eta)| \geq c' > 0$, if the support of $\tilde{\psi}$ is a a sufficiently small neighborhood of the origin. Hence for the η in region 3° we may use Proposition 4 to conclude that the left-side of (3.2) is actually $0(\lambda^{-N})$ for every N. The proof of Theorem 1 is therefore concluded.

REMARK. We have used only a special consequence of the asymptotic formula (2.2), namely the "remainder estimate" analogous to (1.3′) when $N = r = 0$. Had we used the full formula we can get an asymptotic expansion for $d\hat{\mu}(\xi)$; its main term is explicitly expressible in terms of the Gaussian curvature at those points $x \in S$, for which the normal is in the direction ξ or $-\xi$.

(2) We shall now consider the problem in a wider setting. Here S will be a smooth m-dimensional sub-manifold, with $1 \leq m \leq n-1$, and our assumptions on the non-vanishing curvature will be replaced by the more

general assumption that at each point S has at most a finite order contact with any hyperplane. We shall call such sub-manifolds of *finite type*. (These have some analogy with the finite-type domains in several complex variables, which are also discussed in Nagel's lectures [21].) The precise definitions required for our considerations are as follows. We shall assume that we are considering S in a sufficiently small neighborhood of a given point, and then write S as the image of mapping $\phi : R^m \to R^n$, defined in a neighborhood U of the origin in R^n. (To get a smoothly embedded S we should also suppose that the vectors $\frac{\partial\phi}{\partial x_1}, \frac{\partial\phi}{\partial x_2}, \cdots, \frac{\partial\phi}{\partial x_m}$ are linearly independent for *each* x, but we shall not need that assumption.) Now fix any point $x^0 \in U \subset R^m$, and any unit vector η in R^n. We shall assume that the function $(\phi(x)-\phi(x^0))\cdot\eta$ does not vanish of infinite order as $x \to x^0$. Put another way, for each $x^0 \in U$ and each unit vector η, there is a multi-index α, with $1 \le |\alpha|$, so that $\left(\frac{\partial}{\partial x}\right)^\alpha (\phi(x)\cdot\eta)\big|_{x=x^0} \neq 0$. Notice that if (x', η') are sufficiently close to (x^0, η), then also $\left(\frac{\partial}{\partial x}\right)^\alpha \phi(x')\cdot\eta'\big|_{x=x'} \neq 0$. The smallest k so that for each unit vector η then $\exists\alpha$, $|\alpha| \le k$, with $\frac{\partial^\alpha}{\partial x^\alpha}(\phi(x)\cdot\eta)\big|_{x=x^0} \neq 0$ will be called the *type* of ϕ at x^0. Also if U_1 is a compact set in U, the type of ϕ in U_1 will be the least upper bound of the types for x^0 in U_1.

THEOREM 2. *Suppose* S *is a smooth* m*-dimensional manifold in* R^n *of finite type. Let* $d\mu = \psi d\sigma$, *with* $\psi \in C_0^\infty(R^m)$. *Then*

$$|\widehat{d\mu}(\xi)| \le A|\xi|^{-\varepsilon}, \quad \textit{for some } \varepsilon > 0, \tag{3.3}$$

and in fact we can take $\varepsilon = 1/k$, *where* k *is the type of* S *inside the support of* ψ.

Proof. By a suitable partition of unity we can reduce the problem to showing that

$$\int_{R^m} e^{i\phi(x)\cdot\xi}\tilde{\psi}(x)\,dx = 0(|\xi|^{-1/k})$$

with ϕ as described above, and the support of $\tilde{\psi}$ sufficiently small. Now we can write $\xi = \lambda\eta$, with $|\eta| = 1$, and $\lambda > 0$. Then we know that there is an α, with $|\alpha| \leq k$, so that $\left(\frac{\partial}{\partial x}\right)^{\alpha} \phi(x)\cdot\eta \neq 0$, whenever x is in the support of $\tilde{\psi}$ (once the size of the support has been chosen small enough). Thus the conclusion (3.3) follows from (2.1) of Proposition 5.

References. Theorem 1 in its more precise form alluded to in the remark goes back to Hlawka [14]. See also Herz [13], Littman [18], Randol [25], and Hörmander [16]. When S is a real-analytic sub-manifold not contained in any affine hyper-plane, then it is of finite type as defined above. For such real-analytic S estimates of the type (3.3) were proved by Björk [2].

4. *Restriction theorems for the Fourier transform*

The Fourier transform of a function in $L^p(R^n)$, $1 < p \leq 2$ is most naturally thought of as an $L^{p'}$ function (via the Hausdorff-Young Theorem) and so at first sight it is viewed as defined only almost-everywhere. This impression is further supported by the case $p = 2$, when clearly the Fourier transform can be completely arbitrary on any given set of zero Lebesgue measure. It is therefore a noteworthy fact that whenever $n \geq 2$ and S is a sub-manifold of R^n (with some appropriate "curvature") then there exists a $p_0 = p(S)$, $p_0 > 1$, so that every function in L^p, $1 \leq p \leq p_0$ has a Fourier transform restricting to S (i.e. with respect to the induced measure on S). Let us make this precise.

Suppose that S is a given smooth sub-manifold in R^n, with $d\sigma$ its induced Lebesgue measure. We shall say that the L^p restriction property holds for S, if there exists a $q = q(p)$, so that the inequality

$$(4.1) \qquad \left(\int_{S_0} |\hat{f}(\xi)|^q d\sigma(\xi) \right)^{1/q} \leq A_{p,q}(S_0) \|f\|_p$$

holds for each $f \in \mathcal{S}$, whenever S_0 is an open subset of S with compact closure in S.

THEOREM 3. *Suppose* S *is a smooth hypersurface in* $\mathbf{R}^n$ *with non-zero Gaussian curvature. Then the restriction property (4.1) holds for* $1 \leq p \leq \frac{2n+2}{n+3}$, (*with* $q = 2$).

Proof. Suppose $\psi \geq 0$ and $\psi \in C_0^\infty$. It will suffice to prove the inequality

$$(4.2) \qquad \left(\int_{\mathbf{R}^n} |\hat{f}(\xi)|^2 \psi(\xi) d\sigma(\xi) \right)^{1/2} \leq A \|f\|_{L^{p_0}}$$

for $p_0 = \frac{2n+2}{n+3}$, and $f \in \mathcal{S}$; the case $1 \leq p \leq p_0$ will then follow by interpolation.* By covering the support of ψ by sufficiently many small open sets, it will be enough to prove (4.2) when (after a suitable rotation and translation of coordinates) the surface S can be represented (in the support of ψ) as a graph: $\xi_n = \phi(\xi_1, \cdots, \xi_{n-1})$. Now with $d\mu = \psi d\sigma$ we have that

$$\int |\hat{f}(\xi)|^2 d\mu = \int \hat{f}(\xi) \overline{\hat{f}(\xi)} d\mu = \int T(f)(x) \overline{f(x)} dx$$

where $(Tf)(x) = (f * K)(x)$, with

$$K(x) = \int e^{-2\pi i x \cdot \xi} d\mu(\xi) .$$

*In fact the interpolation argument shows that we can take q so that (4.1) holds with $q = \left(\frac{n-1}{n+1}\right) p'$, which is the optimal relation between p and q.

Thus (4.2) follows from Hölder's inequality if we can show that

$$\|T(f)\|_{p_0'} \le A\|f\|_{p_0} , \tag{4.3}$$

where p_0' is the dual exponent to p_0.

To prove (4.3) we consider the function K_s (initially defined for $\text{Re}(s) > 0$) by

$$K_s(x) = \frac{e^{s^2}}{\Gamma(s/2)} \int_{R^n} e^{-2\pi i x\cdot\xi} |\xi_n - \phi(\xi')|^{-1+s} \eta(\xi_n - \phi(\xi'))\tilde{\psi}(\xi') d\xi . \tag{4.4}$$

Here we have abbreviated $(\xi_1, \cdots, \xi_{n-1})$ by ξ'; we have set $\tilde{\psi}(\xi') = \psi(\xi')(1 + |\nabla\phi(\xi')|^2)^{1/2}$, so that $\tilde{\psi}(\xi')d\xi' = d\mu$; also η is a $C_0^\infty(R)$ function which equals 1 near the origin.

Now the change of variables $\xi_n \to \xi_n + \phi(\xi')$ in the above integral shows that it equals

$$\zeta_s(x_n) \int_{R^{n-1}} e^{-2\pi i(x'\cdot\xi' + x_n\phi(\xi'))} \tilde{\psi}(\xi')d\xi' = \zeta_s(x_n)K(x) ,$$

with

$$\zeta_s(x_n) = \frac{e^{s^2}}{\Gamma(s/2)} \int_{-\infty}^{\infty} e^{-2\pi i x_n \xi_n} |\xi_n|^{-1+s} \eta(\xi_n) d\xi_n .$$

Now it is well known that ζ_s has an analytic continuation in s which is an entire function; also $\zeta_0 \equiv 1$; and $|\zeta_s(x_n)| \le c|x_n|^{-\text{Re}(s)}$, where $|x_n| \ge 1$, and the real part of s remains bounded. From these facts it follows that K_s has an analytic continuation to an entire function s (whose values are smooth functions of $x_1, \cdots, x_n$ of at most polynomial growth). One can conclude as well that

(a) $K_0(x) = K(x)$,

(b) $|K_{-n/2+it}(x)| \leq A$, all $x \in \mathbf{R}^n$, all real t ,

(c) $|\hat{K}_{1+it}(\xi)| \leq A$, all $\xi \in \mathbf{R}^n$, all real t .

In fact (c) is immediate from our initial definition (4.4), and (b) follows from Theorem 1.

Now consider the analytic family T_s of operators defined by $T_s(f) = f * K_s$. From (b) one has

(4.5) $\|T_{-n/2+it}(f)\|_{L^\infty} \leq A\|f\|_{L^1}$, all real t ,

and from (c) and Plancherel's theorem one gets

(4.6) $\|T_{1+it}(f)\|_{L^2} \leq A\|f\|_{L^2}$, all real t ,

An application of a known convexity property of operators (see [28]) then shows that $\|T_0(f)\|_{L^{p_0'}} \leq A\|f\|_{L^{p_0}}$, with $p_0 = \dfrac{2n+2}{n+3}$, and the proof of Theorem 3 is complete.

REMARKS:

(i)   For hypersurfaces with non-zero Gaussian curvature this theorem is the best possible, only insofar as it is of the form (4.1) with $q \geq 2$. If q is not required to be 2 or greater, then it may be conjectured that a restriction theorem holds for such hypersurfaces in the wider range $1 \leq p < 2n/(n+1)$. This is known to be true when $n = 2$ (see also §7 below).

(ii)   For hypersurfaces for which only k principal curvatures are non-vanishing, Greenleaf [12] has shown that then the corresponding results hold with $1 \leq p \leq \dfrac{2(k+2)}{k+4}$, giving an extension of Theorem 3.

(iii) In the case of $\dim(S) = 1$ (i.e. in the case of a curve) there are a series of results extending our knowledge of the case $n = 2$ alluded to above. For further details one should consult the references cited below.

It would of course be of interest to know what are the exponents p and q (if any) for which the restriction holds if we are dealing with a given sub-manifold S. This problem is highlighted by the fact quite general sub-manifolds S (those which are of finite type in the sense described in §3) have the restriction property:

THEOREM 4. *Suppose* S *is a smooth* m*-dimensional sub-manifold of* $\mathbf{R}^n$ *of finite type. Then there exists a* $p_0 = p_0(S)$, $1 < p_0$, *so that* S *has the* L^p *restriction property (4.1) with* $q = 2$, *and* $1 \le p \le p_0$. *(In fact if the type of* S *is* k, *we can take* $p_0 = 2nk/(2nk-1$.)

COROLLARY. *Suppose* S *is real-analytic and does not lie in any affine hyperplane. Then* S *has the* L^p *restriction property for* $1 \le p \le p_0$, *for some* $p_0 > 1$.

Proof. As we saw above, it suffices to prove (4.3). However $Tf = f * K$, and $K(x) = d\hat{\mu}(-x)$, Theorem 2 tells us that $|K(x)| \le A|x|^{-1/k}$, So according to the theorem of fractional integration, (see [26], Chapter V), we therefore get (4.3) with $\frac{1}{p_0'} = \frac{1}{p_0} - \frac{\alpha}{n}$, where $\alpha = n - 1/k$, and this relation among exponents is the same as $p_0 = \frac{2nk}{2nk-1}$. Q.E.D.

Further bibliographic remarks. The initial restriction theorem dates from 1967 but was unpublished. The sharp result for $n = 2$ was observed by C. Fefferman and the author and can be found essentially in [9]; see also Zygmund [33]. Further results are in Thomas [30], [31], Strichartz [29], Prestini [24], Christ [4], and Drury [7].

5. *Oscillatory integrals of the first kind related to singular integrals*

A key oscillatory integral used in the theory of Hilbert transforms along curves is the following:

$$\text{p.v.} \int_{-\infty}^{\infty} e^{iP_a(t)} \frac{dt}{t}, \tag{5.1}$$

where $P_a(t)$ is a real polynomial in t of degree d, $P_a(t) = \sum_{j=0}^{d} a_j t^j$. It
was proved by Wainger and the author in [27], that the integral is bounded
with a bound depending only on the degree d and independent of the
coefficients $a_0, a_1, \cdots, a_d$. The relevance of such integrals can be better
understood by consulting Wainger's lectures [32]. We shall be interested
here in giving an n-dimensional generalization of this result. We formulate
it as follows. Let $K(x)$ be a homogeneous function of degree $-n$;
suppose also that $|K(x)| \leq A|x|^{-n}$ (i.e. K is bounded on the unit sphere);
moreover, we assume the usual cancellation property: $\int_{|x'|=1} K(x') d\sigma(x') = 0$.
We let $P(x) = \sum_{|\alpha| \leq d} a_\alpha x^\alpha$ be any real polynomial of degree d.

THEOREM 5:

$$(5.2) \qquad \text{p.v.} \left| \int_{\mathbf{R}^n} e^{iP(x)} K(x) dx \right| \leq A_d$$

with the bound A_d *that depends only on* K *and* d, *and not on the*
coefficients a_α.

Nagel and Wainger observed that if K were odd, one could prove (5.2)
from the one-dimensional form (5.1) by the method of rotations (passage to
polar coordinates). To deal with the general case we need two lemmas.

Let $P_a(t) = \sum_{j=1}^{d} a_j t^j$ denote a real polynomial on $\mathbf{R}^1$, and write also
$P_b(t) = \sum_{j=1}^{d} b_j t^j$.

LEMMA 1:

$$(5.3)\qquad \left|\int_{\varepsilon}^{\infty} \left(e^{iP_a(t)} - e^{iP_b(t)}\right) \frac{dt}{t}\right| \le A'_d\left(1 + \sum_{j=1}^{d} \left|\log\left|\frac{a_j}{b_j}\right|\right|\right),$$

with A'_d *independent of* $\{a_j\}$, $\{b_j\}$ *and* $\varepsilon > 0$.

LEMMA 2. *Let* $P(x) = \sum_{|\alpha|=d} a_\alpha x^\alpha$ *be a homogeneous polynomial of degree* d *on* R^n. *Write*

$$m_P = \int_{|x'|=1} |P(x')|\, d\sigma(x')\,.$$

Then,

$$(5.4)\qquad \int_{|x'|=1} \left|\log\left(\left|\frac{P(x')}{m_P}\right|\right)\right| d\sigma(x') \le B_d\,,$$

with B_d *independent of* P.*

One can if one wishes give an elementary (but complicated) proof of Lemma 2. It may however be more interesting to obtain it as a consequence of a general property of polynomials in R^n related to the class of functions of bounded mean oscillation. This property can be stated as follows. It is very well known that the function $\log|x|$ is in B.M.O., and this is usually the first example discussed in that theory. It is surprising therefore that the following natural generalization seems to be been overlooked.

---

*Of course in writing (5.4) we assume that P is not identically zero, i.e. $m_P > 0$.

THEOREM 6. *Let* $P(x)$ *be any polynomial of degree* $\leq d$ *in* $\mathbf{R}^n$. *Then* $\log|P(x)|$ *is in B.M.O. and in addition*

$$\|\log|P(x)|\|_{BMO} \leq B'_d ,$$

where B'_d *depends only on* d, *and not otherwise on* P.

The proofs of Theorem 6 and Lemma 2 will be given in an appendix. We now pass to the proof of Lemma 1. We prove it by induction on d. The case $d = 1$, i.e. the estimate

$$\left|\int_\varepsilon^\infty (e^{iat} - e^{ibt}) \frac{dt}{t}\right| \leq A\left(1 + \left|\log\left|\frac{a}{b}\right|\right|\right)$$

is classical. Let us now assume (5.3) for polynomials of degree $d-1$, and observe that the estimates (5.3) we wish to prove are unchanged if we replace t by δt, $\delta \neq 0$. Thus we may assume that $b_d = 1$ and $|a_d| \leq 1$. Now write

$$\int_{\varepsilon'}^\infty (e^{iP_a(t)} - e^{iP_b(t)}) \frac{dt}{t} = \int_{\varepsilon'} + \int_1^\infty ,$$

and we treat these two integrals separately. (If $\varepsilon' > 1$, we have only $\int_{\varepsilon'}^\infty$, and that integral is estimate like $\int_1^\infty$.) Let us consider the second integral. It equals

$$\int_1^\infty e^{iP_a(t)} \frac{dt}{t} - \int_1^\infty e^{iP_b(t)} \frac{dt}{t} .$$

Now since $b_d = 1$, we see that $(d/dt)^d P_b(t) = d!$, and hence

$$\left| \int_1^{\infty} e^{iP_b(t)} \frac{dt}{t} \right| \leq c_d ,$$

by the corollary of Proposition 2 in §1. Next, by a change of variables $t \to |a_d|^{-1/d} t$, the integral

$$\int_1^{\infty} e^{iP_a(t)} \frac{dt}{t}$$

becomes

$$\int_{|a_d|^{1/d}}^{\infty} e^{i\tilde{P}_a(t)} dt = \int_{|a_d|^{1/d}}^{1} + \int_1^{\infty}$$

where $\tilde{P}_a(t)$ is a polynomial of degree d, with t^d having coefficient one. Again

$$\left| \int_1^{\infty} e^{i\tilde{P}_a(t)} \frac{dt}{t} \right| \leq c_d ,$$

while

$$\left| \int_{|a_d|^{1/d}}^{1} \right| \leq \int_{|a_d|^{1/d}}^{1} \frac{dt}{t} = \frac{1}{d} \log \left| \frac{1}{d} \right| = \frac{1}{d} \left| \log \left| \frac{b_d}{a_d} \right| \right|$$

since $b_d = 1$. Next

$$\int_{\varepsilon'}^{1} (e^{iP_a(t)} - e^{iP_b(t)}) \frac{dt}{t} = \int_{\varepsilon'}^{1} (e^{iQ_a(t)} - e^{iQ_b(t)}) \frac{dt}{t} + 0\left(\int_{\varepsilon'}^{1} \frac{dt}{t} \right),$$

with

$$Q_a(t) = \sum_{j=1}^{d-1} a_j t^j \,, \quad Q_b(t) = \sum_{j=1}^{d-1} b_j t^j \,,$$

since $|P_a(t) - Q_a(t)| \leq |t|$ and $|P_b(t) - Q_b(t)| \leq |t|$. However, by induction hypothesis (using (5.3) for $\varepsilon' = \varepsilon$, and $\varepsilon = 1$, and $d-1$),

$$\left| \int_{\varepsilon'}^{1} (e^{iQ_a(t)} - e^{iQ_b(t)}) \frac{dt}{t} \right| \leq A'_{d-1} \left(1 + \sum_{j=1}^{d-1} \left| \log \left| \frac{a_j}{b_j} \right| \right| \right) .$$

Gathering all these terms together then proves (5.3).

Armed with Lemmas 1 and 2 we can now prove Theorem 5. We may assume that $P(x) = \sum_{|\alpha| \leq d} a_\alpha x^\alpha$ has no constant term, and using polar coordinates $x = tx'$, $t > 0$, $|x'| = 1$, we write $P(x) = \sum_{j=1}^{d} P_j(x') t^j$, where $P_j(x')$ are restrictions to the unit sphere of homogeneous polynomials of degree j. Let us also set $m_j = \int_{|x'|=1} |P_j(x')| \, d\sigma(x)$, and write $K(x) = t^{-n} \Omega(x')$, with Ω bounded and $\int_{|x'|=1} \Omega(x') \, d\sigma(x') = 0$. Then to prove (5.2) it suffices to show that

$$\left| \int_{\varepsilon_1 \leq |x| \leq \varepsilon_2} e^{iP(x)} K(x) \, dx \right| \leq A_d \,,$$

with A_d independent of ε_1, ε_2 and P. The above integral can be written as

$$\int_{|x'|=1} \left(\int_{\varepsilon_1}^{\varepsilon_2} e^{i\Sigma P_j(x') t^j} \frac{dt}{t} \right) \Omega(x') \, d\sigma(x') \,.$$

Since Ω has vanishing mean-value this integral may be rewritten as

$$\int_{|x'|=1}\left(\int_{\varepsilon_1}^{\varepsilon_2}\left[\exp\left(i\sum P_j(x')t^j\right)-\exp\left(i\sum m_j t^j\right)\right]\frac{dt}{t}\right)\Omega(x')\,d\sigma(x')\,.$$

However by Lemma 1 the inner integral is bounded by

$$A'_d\left(1+\sum_{j=1}^{d}\left|\log\left|\frac{P_j(x')}{m_j}\right|\right|\right)$$

and so an appeal to Lemma 2 shows that

$$\int_{|x'|=1}\left|\int_{\varepsilon_1}^{\varepsilon_2}\exp\left(i\sum P_j(x')t^j\right)-\exp\left(i\sum m_j t^j\right)\frac{dt}{t}\right|(\sup|\Omega(x')|)\,d\sigma(x')\leq A_d$$

proving (5.2) and the theorem.

6. *Oscillatory integrals of the second kind: an example related to the Heisenberg group*

To motivate the interest in this example we recall the definition of the Heisenberg group H^m. The underlying space of H^m is $\mathbf{C}^m\times\mathbf{R}$, i.e. $H^m=\{(z,t)\}$, with $z\in\mathbf{C}^m$, $t\in\mathbf{R}$; the multiplication here is $(z,t)(w,s)=(z+w,t+s+\langle z,w\rangle)$, where $\langle z,w\rangle=2\,\mathrm{Im}\sum_{j=1}^{m}z_j\bar{w}_j$, with $z=(z_j)$, $w=(w_j)$. Now on the Heisenberg group one can consider two types of dilations and their corresponding quasi-distances. The first are the usual dilations $(z,t)\to(\rho z,\rho t)$, $\rho>0$, and the metric could be defined in terms of the usual distance. The second are the dilations $(z,t)\to(\rho z,\rho^2 t)$, and the appropriate quasi-distance (from the origin) is then $(|z|^4+t^2)^{1/4}$. The latter dilations and metric are closely tied with the realization of the Heisenberg group as the boundary of the generalized upper half-space holomorphically equivalent with the unit ball in $\mathbf{C}^{n+1}$. This point of view, as well as related generalizations, is elaborated in Nagel's lectures [21].

In the present context the first type of dilations and corresponding metric would be appropriate if one considered expressions related to ordinary potential theory in H^m viewed as R^{2m+1}. However the two conflicting types of dilations (and related metrics) occur in e.g. the solutions of $\bar{\partial} u = f$. (One sees this for example in Krantz's lectures [17], where in the formula of Henkin we have a kernel made of products of functions each belonging to one of the two above homogeneities.)* Other expressions of this type occur in the explicit formulae for the solutions of the $\bar{\partial}$-Neumann problem (see [1], Chapter 7).

Let us now consider the simplest operator on the Heisenberg group displaying simultaneously these two homogeneities. The prime example is given by

$$Tf = f * K \tag{6.1}$$

where convolution is with respect to the Heisenberg group, and the kernel K is a distribution of the form

$$K(z,t) = L(z)\delta(t) . \tag{6.2}$$

$L(z)$ is a standard Calderón-Zygmund kernel in $C^m = R^{2m}$, i.e. $L(\rho z) = \rho^{-2m} L(z)$, L is smooth away from the origin, and L has vanishing mean-value on the unit sphere. Here $\delta(t)$ is the Dirac delta function in the t-variable, and in an obvious sense is homogeneous $\delta(\rho t) = \rho^{-1}\delta(t)$. Thus K is homogeneous at degree $-2m-1$ with respect to the standard dilations, and at the same time homogeneous of degree $-2m-2$ with respect to the other dilations; in both instances the degrees are the critical ones.

We turn next to the question of proving that the operator (6.1) is bounded on $L^2(H^m)$. The most efficient way is to proceed via the Fourier transform in the t-variable. This leads to the problem of showing that the family of operators T_λ defined by

---

*In particular the terms A_1 and A_2 that appear in §6 of [17].

(6.3) $$(T_\lambda)(F)(z) = \int_{C^m} L(z-w)e^{i\lambda<z,w>}F(w)dw ,$$

(with $<z,w>$ the anti-symmetric form which occurs in the multiplication law for the Heisenberg group) is bounded on $L^2(C^m)$ to itself, uniformly in λ, $-\infty < \lambda < \infty$.*

We now change our notation and call $2m = n$, $L = K$, $-\lambda<\cdot,\cdot> = B(\cdot,\cdot)$, and $F = f$. Then the operators T_λ have the form

(6.4) $$(Tf)(x) = \int_{R^n} K(x-y)e^{iB(x,y)}f(y)dy .$$

We shall suppose B is a real bilinear form, but we shall not suppose that B is necessarily anti-symmetric nor that K is homogeneous of degree $-n$.

THEOREM 7. *Suppose* K *is homogeneous of degree* $-\mu$, $0 \leq \mu \leq n$, *smooth away from the origin, and with vanishing mean-value when* $\mu = n$.

(a) *If* B *is non-degenerate, then the operator* T *given by (6.4) is bounded on* $L^2(R^n)$ *to itself, for* $0 \leq \mu \leq n$; *when* $1 < p < \infty$, *the operator is bounded on* $L^p(R^n)$ *to itself if* $\left|\frac{1}{2} - \frac{1}{p}\right| \leq \frac{\mu}{2n}$.

(b) *If we drop the assumption that* B *is non-degenerate but require that* $\mu = n$, *then* T *is bounded on* $L^p(R^n)$ *to itself for* $1 < p < \infty$. *The bound of* T *can then be taken to be independent of* B.

We shall give only the highlights of the proof, leaving the details, further variants, and applications to the papers cited below. Let us consider first the L^2 part of assertion (a) when $n/2 < \mu \leq n$. Suppose η

*For further details see Mauceri, Picardello and Ricci [19] and Geller and Stein [10].

is a C_0^∞ function, with $\eta(x) = 1$ for $|x| \leq 1/2$, and $\eta(x) = 0$, for $|x| \geq 1$. We write $T = T_0 + T_\infty$, where T_0 is defined as in (6.4), but with K replaced by $K_0 = \eta K$, and T_∞ with K replaced by $K = (1-\eta)K$.

Observe first that since $K_0(x-y)$ is supported where $|x-y| \leq 1$, estimating $T_0(f)(x)$ in the ball $|x| \leq 1$ involves only $f(y)$ in the ball $|y| \leq 2$. We claim

$$(6.5) \qquad \int_{|x|\leq 1} |T_0(f)(x)|^2 \, dx \leq A \int_{|y|\leq 2} |f(y)|^2 \, dy \, .$$

In fact when $|x| \leq 1$,

$$\left| \int K_0(x-y) e^{iB(x,y)} f(y) \, dy - \int K_0(x-y) e^{iB(y,y)} f(y) \, dy \right|$$

$$\leq c \int_{|x-y|\leq 1} |x-y|^{-\mu+1} |f(y)| \, dy \, ,$$

and thus the L^2 theory for $f \to f * K_0$ (which is non-trivial only when $\mu = n$) proves (6.5).

While operators of the type (6.4) are not translation invariant they do satisfy

$$(6.6) \qquad \tau_{-h} T \tau_h f = e^{iB(h,h)} e^{iB(x,h)} T(e^{iB(h,\cdot)} f(\cdot))$$

with $\tau_h(f)(x) = f(x-h)$. Applying this to T_0 gives the following generalization of (6.5)

$$\int_{|x-h|\leq 1} |T_0(f)(x)|^2 \, dx \leq A \int_{|y-h|\leq 2} |f(y)|^2 \, dy \, ,$$

and an integration in h shows that as a consequence

$$\int_{R^n} |T_0(f)(x)|^2 dx \leq A2^n \int_{R^n} |f(y)|^2 dy .$$

We now turn to the proof of

$$\int_{R^n} |T_\infty f(x)|^2 dx \leq A \int_{R^n} |f(x)|^2 dx .$$

This will be done by proving the corresponding result for the operator $T_\infty^* T_\infty$. The kernel L of this operator is given by

$$L(x,y) = \int e^{-iB(z,x-y)} \overline{K}_\infty(z-x) K_\infty(z-y) dz .$$

Now since K_∞ is in $L^2(R^n)$ (here the assumption $n/2 < \mu$ is used), Schwarz's inequality implies

$$|L(x,y)| \leq A .$$

We next integrate by parts in the definition of $L(x,y)$, using the fact that $(D_z)^N e^{-iB(z,x-y)} = e^{-iB(z,x-y)}$, where $D_z = i(a,\nabla_z)/|x-y|$, with $a = B^{-1}\left(\frac{x-y}{|x-y|}\right)$, and B denotes the matrix so that $B(x,y) = (Bx,y)$. The result is $|L(x,y)| \leq A_N|x-y|^{-N}$, for every $N \geq 0$, and hence

$$(6.7) \qquad |L(x,y)| \leq A_N(1+|x-y|)^{-N}, \quad N \geq 0 .$$

This shows that L is the kernel of a bounded operator on L^2 proving the boundedness of $T_\infty^* T_\infty$ and thus of T_∞. The proofs of the L^2 boundedness when $0 \leq \mu \leq n/2$ (in part (a) of the theorem), and the L^2 boundedness when $\mu = n$ but when B is not assumed to be non-degenerate, are refinements of the above argument.

Let us now describe the main idea in proving the L^p inequalities stated in (a) and (b) above. We shall need a generalization of BMO (and of H^1) which may be of interest in its own right. Suppose $E = \{e_Q\}$ is a mapping from the collection of cubes Q in R^n to complex-valued functions on R^n so that

$$|e_Q(x)| = \chi_Q(x), \quad \text{all } x ,$$

where χ_Q denotes the characteristic function of the cube Q. Let us define on "E-atom" to be a function a so that for some cube Q

(i) a is supported in Q

(ii) $|a(x)| \le 1/|Q|$

(iii) $\int a(x)\overline{e_Q}(x)\,dx = 0$.

The space H^1_E is then given by $\{f|f = \Sigma\,\lambda_j a_j$, with each a_j an E atom, and $\Sigma\,|\lambda_j| < \infty\}$. In a similar vein the function $f^\#_E$ will be defined as

$$f^\#_E(x) = \sup_{x\in Q} \frac{1}{|Q|}\int_Q |f - f^E_Q|\,dx , \tag{6.8}$$

where

$$f^E_Q = \frac{e_Q}{|Q|}\int_Q f\,\overline{e_Q}\,dy ,$$

and we take $BMO_E = \{f|f^\#_E \in L^\infty\}$.

Some of the basic facts about the standard H^1 and BMO spaces* go through for H^1_E and BMO_E, and sometimes these come free of charge. One such case is the following assertion: Suppose $f \in L^{p_0}$, $1 \le p \le p_0 < \infty$, and $f^\#_E \in L^p$. Then $f \in L^p$ and

*The standard situation arises of course when $e_Q = \chi_Q$, all Q.

(6.9) $$\|f\|_{L^p} \leq A_p \|f_E^\#\|_{L^p} .$$

To prove this we need only observe that $(|f|)^\# \leq 2f_E^\#$, and use the result (see [10]) for the standard # function.

The point of all of this is that for operators of the form (6.4), there is a naturally associated H_E^1 and BMO_E theory, and it is given by choosing

(6.10) $$e_Q(x) = e^{-iB(x, c_Q)} ,$$

where c_Q is the center of the cube Q. The basic step in the L^p theory (besides an appropriate interpolation which goes via (6.9)), is the proof that when $\mu = n$ our operator T maps L^∞ to BMO_E. Let us give the proof in the case (a). We may assume that $\|f\|_{L^\infty} \leq 1$, and suppose first that Q is a cube centered at the origin. Then we have to show that there exists a constant γ_Q, so that

(6.11) $$\frac{1}{|Q|} \int_Q |Tf - \gamma_Q| \, dx \leq A .$$

The corresponding inequality for a cube centered at another point, say c_Q, then follows from the translation formula (6.6), (and this is the reason for defining e_Q as we do). Turning to (6.11), the argument is not exactly the same as in the standard case (see e.g. Coifman's lectures [5] or [9]), since we must split f into *three* parts to take into account the oscillations of $e^{iB(x,y)}$. Suppose $Q = Q_\delta$, has side-lenghts δ, then write $f = f_1 + f_2 + f_3$, where

$$f_1 = f \text{ in } Q_{2\delta}, \quad f_1 = 0 \text{ elsewhere},$$

$$f_2 = f \text{ in } {}^cQ_{2\delta} \cap Q_{\delta^{-1}}, \quad f_2 = 0 \text{ elsewhere},$$

$$f_3 = f \text{ in } {}^cQ_{2\delta} \cap {}^cQ_{\delta^{-1}}, \quad f_3 = 0 \text{ elsewhere}.$$

(Note that f_2 occurs only when $\delta \leq \sqrt{2}/2$.) We have $F = T(f) = F_1 + F_2 + F_3$, where $F_j = T(f_j)$. For F_1 we make the usual estimate, using the fact that T is bounded on L^2. Next observe that

$$|K(x-y)e^{iB(x,y)} - K(-y)| \leq c\delta\left[\frac{1}{|y|^{n+1}} + \frac{1}{|y|^{n-1}}\right],$$

if $x \in Q_\delta$ and $y \in {}^cQ_{2\delta}$. Thus if $\gamma_Q = \int K(-y) f_2(y)\,dy$, we get that for $x \in Q_\delta$

$$|F_2(x) - \gamma_Q| \leq A\left\{\delta \int_{{}^cQ_{2\delta}} \frac{dy}{|y|^{n+1}} + \delta \int_{Q_{\delta^{-1}}} \frac{dy}{|y|^{n-1}}\right\} \leq A' .$$

Finally

$$F_3(x) = \int K(x-y)e^{iB(x,y)} f_3(y)\,dy$$

$$= \int (K(x-y) - K(-y))e^{iB(x,y)} f_3(y)\,dy + \int K(-y) f_3(y) e^{iB(x,y)}\,dy$$

$$= F_3^1 + F_3^2 .$$

For F_3^1 we again make the standard estimates, and for F_3^2 we use Plancherel's formula (which we may since we have assumed that $B(x,y)$ is non-degenerate). The result is

$$\int_Q |F_3^2(x)|^2\,dx \leq \int_{R^n} |F_3^2(x)|^2\,dx = A \int_{R^n} |K(-y) f_3(y)|^2 dy \leq A^1 \int_{{}^cQ_{\delta^{-1}}} \frac{dy}{|y|^{2n}} =$$

Combining these estimates proves (6.11), and hence the fact that T takes L^∞ to BMO_E.

We shall now state a generalization of this theorem which also includes the oscillatory integral result given in Theorem 5 (in §5). Suppose $P(x,y)$ is a real polynomial on $\mathbf{R}^n \times \mathbf{R}^n$ of total degree d. Consider the operator

$$(6.12) \qquad (Tf)(x) = \text{p.v.} \int_{\mathbf{R}^n} e^{iP(x,y)} K(x-y) f(y)\, dy$$

where K is homogeneous of degree $-n$, smooth away from the origin and with vanishing mean-value.

THEOREM 8. *The operator* T *given by (6.12) is bounded on* $L^2(\mathbf{R}^n)$ *to itself, with a bound that can be taken to depend only on* K *and the degree* d *of* P, *and is otherwise independent of* P.

This is a recent result obtained jointly with F. Ricci. The proof is based in part on a combination of ideas used in the proof Theorems 5 and 7. This result has also many variants, and we now state some of these:

(i) One may also show that the operators (6.12) are bounded on L^p, $1 < p < \infty$.

(ii) Given p, with $1 < p < \infty$, then there is an $\varepsilon = \varepsilon(p,d)$, so that if K is homogeneous of degree $-\mu$, $n - \varepsilon \le \mu \le n$, T is still bounded on L^p. However now the bounds may depend on P, and in addition one must assume that $P(x,y)$ is not of the term $P(x,y) = P_0(x) + P_1(y)$.

(iii) One can replace $K(x-y)$ in (6.12) by a more general "Calderón-Zygmund kernel" $K(x,y)$, a distribution for which the operator when $P \equiv 0$ is bounded in L^2, and which in addition is a function (when $x \neq y$) which satisfies $|K(x,y)| \le A|x-y|^{-n}$, $|\nabla_x K(x,y)| + |\nabla_y K(x,y)| \le A|x-y|^{-n-1}$.

References. For the detailed proof of Theorem 7, other variants, and applications to the $\bar{\partial}$ Neumann problem see the papers of Phong and the author [22], [23].

7. *Further oscillatory integrals related to restriction theorems and Bochner-Riesz summability*

We have seen that if S is a hypersurface in R^n with non-vanishing Gaussian curvature, then

$$\int_S e^{ix\cdot\xi}\psi(x)\,d\sigma(x) = 0(|\xi|^{-(n-1)/2}), \quad \text{as } \xi \quad \infty$$

whenever $\psi \in C_0^\infty$, and this is a typical oscillatory integral of the first kind. We may pass to an oscillatory integral of the second kind when we replace the C_0^∞ function ψ by an L^r function f, and consider the resulting linear operator on f. The resulting operator, as is easy to observe, is in fact the dual to the restriction operator considered in §4. Hence by Theorem 3 we can state that operator is in fact bounded from $L^2(d\sigma)$ to $L^{p'}(R^n)$, where p' is the dual exponent to $\frac{2n+2}{n+3}$. We shall now describe the sharper result in this setting that can be obtained for $n = 2$.

We fix a curve $t \to \gamma(t) = (t, \gamma_2(t))$, $0 \le t \le 1$, lying in R^2, with $\gamma(t) \in C^2$, and having non-vanishing curvature, i.e. $|\gamma_2''(t)| \ge c > 0$. Consider the transformation T, which maps function on the interval $[0,1]$ to functions on R^2, given by

$$(Tf)(\xi) = \int_0^1 e^{i\xi\cdot\gamma(t)} f(t)\,dt\,. \tag{7.1}$$

THEOREM 9. *Under the assumption above* T *is bounded from* $L^p[0,1]$ *to* $L^q(R^2)$, *whenever* $3/q + 1/p = 1$ *and* $1 \le p < 4$.

(Note that when $p \to 4$, then $q \to 4$ in the above relation between p and q.)

Proof. Write

$$(7.2)\qquad F(\xi) = ((Tf)(\xi))^2 = \int_0^1 \int_0^1 e^{i\xi\cdot(\gamma(s)+\gamma(t))} f(s)f(t)\,ds\,dt\ ,$$

and we shall try to apply Plancherel's theorem (more precisely, the Hausdorff-Young inequality) to F. To do this break the above integral into two essentially equal parts according to $t > s$ or $t \leq s$, which divides $[0,1] \times [0,1]$ into the union of two regions R_1 and R_2. We then consider the mapping of $R_1 \to \mathbf{R}^2$ given by $x = \gamma(s) + \gamma(t)$, i.e. $x_1 = s + t$, $x_2 = \gamma_2(s) + \gamma_2(t)$. It is easy to verify on the basis of our assumptions that this mapping is one-one, and its Jacobian J satisfies $|J| = |\gamma_2'(s) - \gamma_2'(t)| \geq c|s-t|$. Therefore

$$(7.3)\qquad \int_{R_1} e^{i\xi\cdot(\gamma(s)+\gamma(t))} f(s)f(t)\,ds\,dt = \int_{\mathbf{R}^2} e^{i\xi\cdot x} f(x_1,x_2)\,dx_1 dx_2$$

with $f(x_1,x_2) = f(s)f(t)|J|^{-1}$.

So if we denote by $F_1(\xi)$ the quantity appearing in (7.3) then whenever $1 \leq r \leq 2$, and $1/r' + 1/r = 1$, we know that

$$(7.4)\qquad \|F_1\|_{L^{r'}(\mathbf{R}^2)} \leq c\|f\|_{L^r(\mathbf{R}^2)}\ .$$

However

$$\|f\|^r_{L^r(\mathbf{R}^2)} = \iint |f(x_1,x_2)|^r\,dx_1 dx_2$$

$$= \int |f(s)|^r\,|f(t)|^r |J|^{1-r}\,ds\,dt$$

$$\leq c \int |f(s)|^r\,|f(t)|^r\,|s-t|^{1-r} ds\,dt\ .$$

To estimate the last integral we need to invoke the theorem of fractional integration in one dimension in the form

$$\int g(s)\,g(t)\,(s-t)^{-1+\alpha}\,ds\,dt \leq A\|g\|_u\,,\quad \frac{2}{u} - 1 = \alpha\,,\quad 0 < \alpha \leq 1\,.$$

So we take $g(t) = |f(t)|^r$, then $\|g\|_u = \|f\|_p^r$ when $p = ur$. Then if we fix α so that $-1+\alpha = 1-r$, then $\frac{3-r}{2} = \frac{1}{u}$, and $p = \frac{2r}{3-r}$. The limitation $0 < \alpha$ becomes $r < 2$, and with $q = 2r'$ we obtain from (7.4) that

$$\|F_1\|_{L^{r'}(R^2)} \leq c'\left(\int_0^1 |f(t)|^p\,dt\right)^{2/p}\,,$$

with a similar estimate for $F_2(\xi)$ which is the analogue of (7.4), but taken over R_2. Since $F = F_1 + F_2$ and $F = (Tf)^2$ we obtain

$$\|T(f)\|_{L^q(R^2)} \leq A\|f\|_{L^p[0,1]}\,.$$

Note that $\frac{3}{q} = \frac{3}{2r'} = \frac{3r-3}{2r} = 1 - \frac{1}{p}$, so $\frac{3}{q} + \frac{1}{p} = 1$, and the limitation $1 \leq r < 2$ is equivalent with $1 \leq p = \frac{2r}{3-r} < 4$. Theorem 9 is therefore proved.

It is clear that inequalities for the Fourier transform play a key role in the above argument. If we want to generalize Theorem 9 it is natural to look for a corresponding extension of the L^2 boundedness of the Fourier transform and the Hausdorff-Young theorem. One result along these lines is as follows.

Suppose we consider the family of operators T, depending on the parameter λ, $\lambda > 0$, defined by

$$T_\lambda(f)(\xi) = \int_{R^n} e^{i\lambda\Phi(x,\xi)}\,\psi(x,\xi)\,f(x)\,dx\,,$$

where ψ is a fixed $C_0^\infty(R^n \times R^n)$ cut-off function; Φ is a real-valued C^∞ phase function which we assume satisfies the assumption that its Hessian is non-vanishing, i.e.

(7.6)
$$\det\left(\frac{\partial^2\Phi(x,\xi)}{\partial x_i \partial \xi_j}\right) \neq 0 .$$

PROPOSITION:

(7.7)
$$\|T_\lambda(f)\|_{L^2(R^n)} \leq A\lambda^{-n/2}\|f\|_{L^2(R^n)} .$$

COROLLARY:

(7.8)
$$\|T_\lambda(f)\|_{L^{p'}(R^n)} \leq A\lambda^{-n/p'}\|f\|_{L^p(R^n)} ,$$

where $1 \leq p \leq 2$, *and* $1/p + 1/p' = 1$.

REMARK. The boundedness of T_λ for any fixed λ is trivial, but what is of interest is the decrease in the norm as $\lambda \to \infty$. This decrease is consistent with the special case when $\Phi(x,\xi)$ is bilinear (and non-degenerate); when we take $\lambda \to \infty$ in that case we recover the usual $(L^p, L^{p'})$ inequalities for the Fourier transform. Notice also that the corollary follows from the proposition by the use of the M. Riesz convexity theorem.

To prove (7.7) we argue as in the proof of Theorem 7; as in the treatment of the operator T_∞ it suffices to show that the operator norm of $T_\lambda^* T_\lambda$ is bounded by $A\lambda^{-n}$. Now this operator has as its kernel the function $K_\lambda(\xi,n)$ given by

(7.9)
$$K_\lambda(\xi,\eta) = \int_{R^n} e^{i\lambda(\Phi(x,\eta)-\Phi(x,\xi))}\psi(x,\eta)\overline{\psi(x,\xi)}\,dx .$$

Now since

$$\Delta(x,\xi,\eta) = (a,\nabla_x)[\Phi(x,\eta)-\Phi(x,\xi)] = \left(\frac{\partial^2\Phi}{\partial x\partial\xi}a,\eta-\xi\right) + 0|\eta-\xi|^2$$

we can find $a = (a_1,\cdots,a_n)$, so that the a_j depend smoothly on x and $|\Delta(x,\xi,\eta) \geq c|\xi-\eta|$ on the support of $K_\lambda(\xi,\eta)$. Set $D_x = \frac{1}{i\lambda\Delta}(a,\nabla_x)$. Then since $(D_x)^N e^{i\lambda(\Phi(x,\eta)-\Phi(x,\xi))} = e^{i\lambda(\Phi(x,\eta)-\Phi(x,\xi))}$, we can integrate by parts N times in (7.9) and obtain

(7.10) $$|K_\lambda(\xi,\eta)| \leq A_N(1+\lambda|\xi-\eta|)^{-N}, \quad N \geq 0 .$$

It follows from (7.10) with $N = n+1$, that the operator $T_\lambda^* T_\lambda$ which has kernel K_λ has a norm bounded by $A\lambda^{-n}$ and the proposition is proved.

We shall now formulate some theorems for oscillatory integrals of the form

(7.11) $$(T_\lambda f)(\xi) = \int_{R^{n-1}} e^{i\lambda\Phi(t,\xi)}\psi(t,\xi)f(t)dt, \quad \xi \in R^n ,$$

which will generalize the restriction theorems (Theorem 9 above, as well as Theorem 3 in §4) and also give results for Bochner-Riesz summability.

Notice that (7.11) are mappings from functions on R^{n-1} to functions on R^n. The basic assumptions on the real phase function Φ are as follows: We consider for each fixed (t^0,ξ^0) the associated bilinear form $B(u,v)$ defined by $B(u,v) = (v,\nabla_t)(u,\nabla_\xi)(\Phi)(t^0,\xi^0)$, with $u \in R^{n-1}$, $v \in R^n$. Our first assumption is that

(7.12^a) B is of rank $n-1$.

Thus there exists (an essentially unique), $\bar{u} \in R^n$, $|\bar{u}| = 1$, so that the scalar function $t \to (\bar{u},\nabla_\xi\Phi(t,\xi))$ has a critical point at (t^0,ξ^0). We shall also assume that this critical point is non-degenerate, i.e. we suppose the non-vanishing of the $(n-1)\times(n-1)$ determinant:

$$(7.12^b) \qquad \det\left(\nabla_t^2(\bar{u},\nabla_\xi \Phi)\right)(t^0,\xi^0) \neq 0 .$$

These assumptions will be supposed to hold at all (t^0,ξ^0) in the support of $\psi(t,\xi)$, where ψ is a fixed function in $C_0^\infty(R^{n-1}\times R^n)$.

THEOREM 10. *Under the assumptions above the operator (7.11) satisfies*

$$(7.13) \qquad \|T_\lambda(f)\|_q \leq A\lambda^{-n/q}\|f\|_p$$

with $q = \left(\frac{n-1}{n+1}\right)p'$, $\frac{1}{p}+\frac{1}{p'}=1$,

(a) *when* $n=2$, *if* $1 \leq p < 4$

(b) *when* $n \geq 3$, *if* $1 \leq p \leq 2$.

REMARKS:

(1) When $\Phi(t,\xi) = t_1\xi_1+t_2\xi_2+\cdots+t_{n-1}\xi_{n-1} + \phi(t_1,\cdots,t_{n-1})\cdot\xi_n$, and $(\nabla\phi)(0)=0$, then the conditions (7.12) are near the origin equivalent with the non-vanishing Gaussian curvature of the graph $t_n = \phi(t_1,\cdots,t_{n-1})$. If we apply the result (7.13), letting $\lambda\to\infty$, it is not difficult to recover Theorem 9 from part (a), and Theorem 3 from part (b).

(2) The proof of part (a) follows the same lines as the proof given for Theorem 9, once we use (7.8) as the substitute for the Hausdorff-Young theorem; further details as well as relations with Bochner-Riesz summability may be found in the papers of Carleson and Sjölin [3] and Hörmander [15]. Since part (b) has not appeared before, we will outline its proof. This will also serve as a good review of many of the notions we have discussed here.

Proof of part (b). It suffices to prove the case $p=2$, since the case $p=1$ is trivial and the rest follows by interpolation. Now the case $p=2$ is equivalent by duality to the statement

$$(7.14) \qquad \|T_\lambda^*(F)\|_{L^2(R^{n-1})} \leq A\lambda^{-n/r'}\|F\|_{L^r(R^n)}$$

with $r = \frac{2(n+1)}{n+3}$, where

$$(T_\lambda^*)(F)(t) = \int_{R^n} e^{-i\lambda\Phi(t,\xi)}\overline{\psi(t,\xi)}F(\xi)\,d\xi\,, \quad t \in R^{n-1}\,.$$

We can calculate

$$\int_{R^{n-1}} T_\lambda^*(F)\overline{T_\lambda^*}(F)\,dt$$

and write as

$$\int_{R^n \times R^n} K_\lambda(\xi,\eta)F(\xi)\overline{F}(\eta)\,d\xi\,d\eta\,,$$

with

$$\text{(7.15)} \qquad K_\lambda(\xi,\eta) = \int_{R^{n-1}} e^{i\lambda(\Phi(t,\eta)-\Phi(t,\xi))}\psi(t,\eta)\overline{\psi}(t,\xi)\,dt\,.$$

It suffices therefore to see that K_λ is the kernel of a bounded operator from $L^r(R^n)$ to $L^{r'}(R^n)$, with norm not exceeding $A\lambda^{-2n/r'}$.

Because of our assumptions on Φ we can construct a phase function $\tilde{\Phi}$ on $R^n \times R^n$ so that the following holds: we will write $x \in R^n$, as (t,x_n) with $t = (x_1,\cdots,x_{n-1}) \in R^{n-1}$. The $\tilde{\Phi}$ we can construct will satisfy:

(i) $\tilde{\Phi}(x,\xi) = \Phi(t,\xi) + \Phi_0(\xi)x_n$

(ii) the determinant of the $n\times n$ matrix $\nabla_x\nabla_\xi\tilde{\Phi}$ is non-vanishing.

In fact $\nabla_t \nabla_\xi \Phi$ already has rank $n-1$ by assumption (7.12[a]), and so we need only choose $\Phi_0(\xi)$ so that $(u,\nabla_\xi)\Phi_0(\xi) \neq 0$ to increase the rank of $\nabla_x \nabla_\xi \tilde{\Phi}$ to n.

Now, as in the proof of Theorem 3 in §4, we form K_λ^s defined by

$$K_\lambda^s(\xi,\eta) = \frac{e^{s^2}}{\Gamma(s/2)} \int_{R^n} e^{i\lambda(\tilde{\Phi}(x,\eta)-\tilde{\Phi}(x,\xi))} \psi(t,\eta)\bar{\psi}(t,\xi)|x_n|^{-1+s}\nu(x_n)\,dx\,,$$

with $dx = dt\,dx_n$, and where ν is a C_0^∞ function which equals 1 near the origin. We easily verify

$$K_\lambda^0(\xi,\eta) = K_\lambda(\xi,\eta) \tag{7.16}$$

since

$$\tilde{\Phi}(x,\xi) = \Phi(t,\xi)\,, \text{ when } x = (t,0)\,.$$

Next

(7.17) K_λ^{1+it} is the kernel of a bounded operator from $L^2(R^n)$ to itself with norm $\leq A\lambda^{-n/2}$.

This follows by applying the estimate (7.7) of the proposition above and using the non-degeneracy of the Hessian of $\tilde{\Phi}(x,\xi)$. Finally we claim that

$$|K_\lambda^{-n/2+1/2+it}(\xi,\eta)| \leq A\,. \tag{7.18}$$

To see this write $K_\lambda^s(\xi,\eta)$ as $K_\lambda(\xi,\eta)\hat{\nu}(\lambda(\Phi_0(\eta)-\Phi_0(\xi)))$ where

$$\hat{\nu}_s(u) = \frac{e^{s^2}}{\Gamma(s/2)} \int_{-\infty}^{\infty} e^{ix_n u}\nu(x_n)|x_n|^{-1+s}\,dx_n\,.$$

Then since $|\hat{\nu}_{-n/2+1/2+it}(u)| \leq c|u|^{n/2-1/2}$, as u ∞ we see that to prove (7.18) it suffices to show that

(7.19) $$|K_\lambda(\xi,\eta)| \leq A(\lambda|\eta-\xi|)^{-n/2+1/2}\ .$$

In proving this estimate for the integral K_λ given by (7.15) we may suppose that the integrand is supported in a sufficiently small neighborhood of a given point $t = t_0$, (for otherwise we can write it as the sum of finitely many such terms). When we write $\Phi(t,\eta) = \Phi(t,\xi) = (\nabla_\xi \Phi)(t,\eta)\cdot(\eta-\xi) + 0(\eta-\xi)^2$ we see that these are two cases to consider as in the proof of Theorem 1 in §3: 1° when the directions $\eta-\xi$ or $\xi-\eta$ are close to the critical direction u arising in condition (7.12^b); or 2° in the opposite case. In the first case we use stationary phase (i.e. Proposition 6 in §2) to obtain (7.19). In the second case, we actually get $0(\lambda|\eta-\xi|)^{-N}$, for every $N \geq 0$ as an estimate, by Proposition 4. This completes the proof of (7.18), and shows that $K_\lambda^{-n/2+1/2+it}$ is the kernel of a bounded operator from $L^1(R^n)$ to $L^\infty(R^n)$, with bounds uniform in λ. The proof of the theorem is then concluded by applying the interpolation theorem, as in the proof of Theorem 3.

8. *Appendix*

Here we shall prove Lemma 2 and Theorem 6 which were stated in §5.

First let $\mathcal{P}_d$ denote the linear space of polynomials in R^n of degree $\leq d$. We claim that there is a constant A_d, so that

(8.1) $$\left(\frac{1}{|Q|}\int_Q |P(x)|^2\,dx\right)^{1/2} \leq A_d \frac{1}{|Q|}\int_Q |P(x)|\,dx$$

holds for all $P \in \mathcal{P}_d$, and all cubes Q.

The space $\mathcal{P}_d$ is invariant under translations and dilations, and so a moment's reflection shows that to prove (8.1) for all $P \in \mathcal{P}_d$, it suffices to prove it for $Q = Q_0$, the unit cube centered at the origin. However

$(\int_{Q_0} |P(x)|^2 dx)^{1/2}$ and $\int_{Q_0} |P(x)| dx$ are two (equivalent) norms on the finite-dimensional space $\mathcal{P}_d$, so (8.1) holds for $Q = Q_0$, and then for general Q.

Now it is well known (see e.g. [6]) that a function which satisfies a "reverse Hölder" inequality belongs to the weight space A_∞. Examining the proof of this fact one obtains an $r = r(d)$, $0 < r < \infty$, and a constant C_d, so that

$$\left(\frac{1}{|Q|}\int_Q |P(x)|\,dx\right)\cdot\left(\frac{1}{|Q|}\int_Q |P(x)|^{-r}dx\right)^{1/r} \le C_d \tag{8.2}$$

for all cubes Q.

From (8.2) and Jensen's inequality, Theorem 6 follows easily.

Let us now assume that P is homogeneous of degree d. Observe also that since (8.2) holds, if we normalize P by the condition that $m_P = \int_{|x|=1} |P(x)|\,d\sigma(x) = 1$, we can conclude that

$$\int_{|x|=1} |P(x)|^{-r}\,d\sigma(x) \le c'_d\,. \tag{8.3}$$

However, when $u > 0$, $|\log u| = \log^+ u + \log^+ \frac{1}{u} \le u + \frac{1}{r} u^{-r}$. Therefore (8.3) implies (5.4) whenever $m_P = 1$, and so that result also holds in general.

E. M. STEIN
DEPARTMENT OF MATHEMATICS
PRINCETON UNIVERSITY
PRINCETON, NEW JERSEY 08544

REFERENCES

[1] M. Beals, C. Fefferman, and R. Grossman, "Strictly pseudo-convex domains," Bull. A.M.S. *8* (1983), 125-322.

[2] J. E. Björck, "On Fourier transforms of smooth measures carried by real-analytic submanifolds of R^n," preprint 1973.

[3] L. Carleson and P. Sjölin, "Oscillatory integrals and a multiplier problem for the disc," Studia Math. *44* (1972), 287-299.

[4] M. Christ, "On the restriction of the Fourier transform to curves," Trans. Amer. Math. Soc., *287* (1985), 223-238.

[5] R. Coifman and Y. Meyer, in these proceedings.

[6] R. Coifman and C. Fefferman, "Weighted norm inequalities for maximal functions and singular integrals," Studia Math. *51* (1979), 241-250.

[7] S. Drury, "Restrictions of Fourier transforms to curves," preprint.

[8] A. Erdélyi, "Asymptotics Expansions," 1956, Dover Publication.

[9] C. Fefferman, "Inequalities for strongly singular convolution operators," Acta Math. *124* (1970), 9-36.

[10] C. Fefferman and E. M. Stein, "H^p spaces of several variables," Acta Math. *129* (1972), 137-193.

[11] D. Geller and E. M. Stein, "Estimates for singular convolution operators on the Heisenberg group," Math. Ann. *267* (1984), 1-15.

[12] A. Greenleaf, "Principal curvature in harmonic analysis," Ind. Univer. Math. J. *30* (1981), 519-537.

[13] C. S. Herz, "Fourier transforms related to convex sets," Ann. of Math. *75* (1962), 81-92.

[14] E. Hlawka, "Über Integrale auf konvexen Körper. I," Monatsh. Math. *54* (1950), 1-36.

[15] L. Hörmander, "Oscillatory integrals and multipliers on FL^p," Ark. Mat. *11* (1973), 1-11.

[16] ————, "The analysis of linear partial differential operators. I," 1983, Springer Verlag.

[17] S. Krantz, "Integral formulas in complex analysis," in these proceedings.

[18] W. Littman, "Fourier transforms of surface-carried measures and differentiability of surface averages," Bull. A.M.S. *69* (1963), 766-770.

[19] G. Mauceri, M. A. Picardello, and F. Ricci, "Twisted convolutions, Hardy spaces, and Hörmander multipliers," Supp. Rend. Cir. Mat-Palermo *1* (1981), 191-202.

[20] J. Milnor, "Morse Theory," Annals of Math. Study #51, 1963, Princeton University Press.

[21] A. Nagel, "Vector fields and nonisotropic metrics," in these proceedings

[22] D.H. Phong and E.M. Stein, "Singular integrals related to the Radon transform and boundary value problems," Proc. Nat. Acad. Sci. USA *80*(1983), 7697-7701.

[23] ———, "Hilbert integrals, singular integrals, and Radon transforms," preprint.

[24] E. Prestini, "Restriction theorems for the Fourier transform to some manifolds in R^n in Harmonic analysis in Euclidean spaces," Proc. Symp. in Pure Math. *35*, part 1 (1979), 101-109.

[25] B. Randol, "On the asymptotic behaviour of the Fourier transform of the indicator function of a convex set," Trans. Amer. Math. Soc. *139*(1969), 279-285.

[26] E.M. Stein, "Singular integrals and differentiability properties of functions," 1970, Princeton University Press.

[27] E.M. Stein and S. Wainger, "The estimation of an integral arising in multiplier transformations," Studia Math. *35*(1970), 101-104.

[28] E.M. Stein and G. Weiss, "Introduction to Fourier analysis on Euclidean spaces," 1971, Princeton University Press.

[29] R.S. Strichartz, "Restrictions of Fourier transforms to quadratic surfaces and decay of solutions of wave equations," Duke Math. J. *44*(1977), 705-713.

[30] P.A. Tomas, A restriction theorem for the Fourier transform," Bull. A.M.S. *81* (1975), 477-478.

[31] ———, "Restriction theorems for the Fourier transform in Harmonic Analysis in Euclidean spaces," Proc. Symp. in Pure Math. *35*, part 1 (1979), 111-114.

[32] S. Wainger, "Averages and singular integrals over lower dimensional sets," in these proceedings.

[33] A. Zygmund, "On Fourier coefficients and transforms of functions of two variables," Studia Math. *50*(1974), 189-201.

AVERAGES AND SINGULAR INTEGRALS OVER LOWER DIMENSIONAL SETS

Stephen Wainger[1]

I. *Introduction*

These lectures deal with work primarily due to Alex Nagel, Nestor Riviere, Eli Stein, and myself dealing with certain averages of and singular integral operators on functions, f, of n variables, $n \geq 2$. These averages and singular integrals differ in character from the classical theory in that the integration is over a manifold of dimension less than n.

Let us begin with an example of the type of problem we have in mind. The classical differentiation theorem of Lebesgue asserts for any locally integrable function f

$$f(x) = \lim_{r\to 0} \frac{1}{|Q_r|} \int_{Q_r} f(x-y)dy \qquad \text{a.e.}$$

(where Q_r is the square, $Q_r = \{x \in R^n | \sup|x_i| \leq r\}$, and $|Q_r|$ denotes the Lebesgue measure of Q_r), and

$$f(x) = \lim_{r\to 0} \frac{1}{|B_r|} \int_{B_r} f(x-y)dt \qquad \text{a.e.,}$$

(where B_r is the ball, $B_r = \{x | |x| \leq r\}$).

*Supported in part by a grant from the National Science Foundation.

Our first problems are the following:

Problem IA:

Does

1) $$\lim_{r\to 0} \int_{\partial Q_r} f(x-y)\, d\sigma_r(y) = f(x) \qquad \text{a.e.?}$$

Here ∂Q_r denotes the boundary of Q_r and $d\sigma_r$ is n–1 dimensional Lebesque measure on ∂Q_r normalized so that $d\sigma_r(\partial Q_r) = 1$.

Problem IB:

Does

2) $$\lim_{r\to 0} \int_{\partial B_r} f(x-y)\, d\mu_r = f(x) \qquad \text{a.e.?}$$

Here $d\mu_r$ is the unit rotationally invariant mass on ∂B_r.

1) and 2) trivially hold if f is continuous, and the questions only become interesting when we consider functions in a class like L^∞, L^2, or L^1.

In questions IA and IB, we are considering certain averages

3) $$M_{Q_r} f(x) = \int_{\partial Q_r} f(x-y)\, d\sigma_r(y)$$

and

4) $$M_{B_r} f(x) = \int_{\partial B_r} f(x-y)\, d\mu_r(y)\,.$$

We are asking if

5) $$M_{Q_r} f(x) \to f(x) \qquad \text{a.e.}$$

and

6) $$M_{B_r} f(x) \to f(x) \qquad \text{a.e.}$$

The standard approach to this type of problem involves consideration of appropriate maximal functions.

We define the maximal functions

7) $$\mathfrak{M}_Q f(x) = \sup_{r>0} M_{Q_r}(|f|)(x) ,$$

and

8) $$\mathfrak{M}_B f(x) = \sup_{r>0} M_{B_r}(|f|)(x) .$$

$\mathfrak{M}_B$ is called the spherical maximal function.

Since 1) and 2) hold for f which are continuous 1) would follow for every f in L^p, $1 \le p < \infty$, if we could show

9) $$|\{x \mid \mathfrak{M}_Q f(x) > \lambda\}| \le \frac{C}{\lambda^p} \|f\|_p^p$$

for every f in L^p, and 2) would follow if we could show

10) $$|\{x \mid \mathfrak{M}_B f(x) > x\}| \le \frac{c}{\lambda p} \|f\|_p^p .$$

The argument showing that 9) and 10) imply 1) and 2) is the same as the argument showing that Lebesgue's differentiation theorem follows from the weak type inequality for the Hardy-Littlewood maximal function given in chapter 1 of [S]. While it is not quite as well known, there are appropriate estimates on maximal functions that guarantee 1) and 2) hold for all L^∞ functions. In our case this means the following:

Let E be a measurable set and χ_E its characteristic function. Then if

(11) $$|\{x \mid \mathfrak{M}_Q \chi_E(x) > \lambda\}| \le C(\lambda)\,|E|$$

where $C(\lambda)$ may depend on λ but not on E, then 1) holds for every f in L^∞. If

$$12) \qquad |\{x|\mathfrak{M}_B\chi_E(x)>\lambda\}| \leq C(\lambda)|E| ,$$

then 2) holds for every f in L^∞. A discussion of this can be found in [BF].

Let us try to see if 9) or 10) could be true in some simple cases. We consider for example the one-dimensional case. Here $B_r = Q_r = \{x|-r<x<r\}$, and

$$M_{Q_r}f(x) = M_{B_r}f(x) = \frac{1}{2}[f(x+r)+f(x-r)] .$$

So if we take $f(x)$ to be $\log \frac{1}{|x|}$ near $x = 0$, have compact support, and be in C^∞ away from the origin, we would have a function f in every L^p class such that $\mathfrak{M}_Q f(x) = \mathfrak{M}_B f(x) = \infty$ for every x. We can also see that 11) and 12) are false in one dimension. We just take $E_\varepsilon = \{x|0\leq x\leq \varepsilon\}$. Then $|E_\varepsilon| \to 0$ but

$$\mathfrak{M}_Q f(x) = \mathfrak{M}_B f(x) \geq \frac{1}{2}$$

for all x.

We could still ask if 1) and 2) hold in some interesting class even though 9), 19), 11), and 12) fail. However an important idea of Stein shows that the failure [SI] of 9), 10), 11), and 12) implies that 1) and 2) fail even in the class of locally bounded functions. The statement of the main theorem of [SI] requires that the underlying space be compact. But if 1) or 2) were true for an L^p class on R^n, it would also hold for the corresponding L^p class on the torus. Furthermore, the theorem of Stein requires the hypothesis that $1 \leq p \leq 2$. However due to the positive nature of the averages under consideration, his ideas can be modified to show that 1) fails for at least some L^∞ functions. See [SW]. Thus we obtain negative results in one dimension. Similar reasoning gives the same negative conclusion for question IA in any number of dimensions.

One need only consider $F(x_1,\cdots,x_n) = f(x_1)h(x_2,\cdots,x_n)$ where f is as above and h is a nice function.

So there are no interesting positive results in problem IA. However as we shall see later there are positive results for problem IB in 3 or more dimensions. One might ask if there is a simple geometric reason why there should be positive answers for the sphere and only negative answers for the boundaries of squares. It turns out that the underlying basic reason that we have positive results for the boundary of balls and negative results for the boundary of squares is that spheres are round and boundaries of squares are flat. In other words an important word for us will be

CURVATURE.

We will come back to the role of curvature in our problem in a little while, but first we shall discuss the other problems that we will consider.

Problem II: Let $\gamma(t)$ be a curve passing through the origin in R^n. Is it true that $\lim \frac{1}{h} \int_0^h f(x-\gamma(t))\,dt = f(x)$ a.e., for f in L^∞ or L^2 or L^1 ?

Problem III: Let $v(x)$ be a smooth vector field in R^n. Does

$$\lim_{h\to 0} \frac{1}{h} \int_0^h f(x-tv(x))\,dt = f(x) \qquad \text{a.e.}$$

for f in L^∞ or L^2 or L^1 ?

Corresponding to problems II and III there are interesting singular integrals. We let $\gamma(t)$ be a curve and $v(x)$ be a smooth vector field as in problems II and III. We set

$$\text{13)} \qquad H_\gamma f(x) = \int_{-a}^{a} f(x-\gamma(t))\,\frac{dt}{t}\,,$$

(where sometimes we wish to think of a as finite and sometimes as ∞), and

14) $$H_v f(x) = \int_{-1}^{1} f(x-tv(x)) \frac{dt}{t} .$$

We call H_γ the Hilbert transform along the curve γ and H_v the Hilbert transform along the vector field $v(x)$. We then have the following two problems:

Problem II′: Can we have an estimate

15) $$\|H_\gamma f\|_{L^p} \leq C_p \|f\|_{L^p}$$

for some p's ?

Problem III′: Can we have an estimate

16) $$\|H_v f\|_{L^p} \leq C_p \|f\|_{L^p}$$

for some values of p ?

The classical development of singular integrals and maximal functions suggests that problems II′ and III′ should be related to problems II and III. In fact the progress on problems I, II, III, II′, and III′ is all interrelated.

We have presented our problems as variants of Lebesgue's Theorem on the differentiation of the integral. These particular variants arose from other considerations.

Riviere was led to problem II′ from the consideration of a problem of singular integrals, namely from trying to generalize the method of Rotations of Calderon and Zygmund. Calderon and Zygmund developed the method of rotations to reduce the study of operators

$$Tf = K * f$$

where K is a kernel having "standard homogeneity" that is

17) $$K(\lambda x) = \lambda^{-n} K(x) \qquad \lambda > 0$$

($K(x)$ is a function on R^n) to the one-dimensional Hilbert transform

$$Hf(x) = \int f(x-t)\frac{dt}{t}$$

(f a function on R^1).

We will explain how the method of rotations can lead to problem II′. Let $K(x,y)$ be a function of two variables x and y which is odd,

$$K(-x,-y) = -K(x,y)$$

and which has a "parabolic homogeneity," that is

18) $$K(\lambda x, \lambda^2 y) = \frac{1}{\lambda^3} K(x,y) .$$

We wish to consider the L^p boundedness of the transformation

19) $$Tf(u,v) = \int_{-\infty}^{\infty}\int_{-\infty}^{\infty} f(u-x,v-y)\, K(x,y)\, dx\, dy .$$

We now introduce parabolic polar coordinates into 19)

$$x = r\cos\theta$$

$$y = r^2\sin\theta$$

and find

20) $$Tf(u,v) = \int_0^{\infty}\int_0^{2\pi} f(u-r\cos\theta, v-r^2\sin\theta)$$

$$K(r\cos\theta, r^2\sin\theta)\, r^2 N(\theta)\, dr\, d\theta$$

where $r^2N(\theta)$ is the Jacobian factor in the change of variables. $N(\theta)$ is smooth and $N(\theta+\pi) = N(\theta)$. By 18 we see that

$$\text{21)}\quad Tf(u,v) = \int_0^{2\pi} N(\theta)K(\cos\theta,\sin\theta)\,d\theta \int_0^{\infty} \frac{1}{r} f(u-r\cos\theta, v-r\sin\theta)\,dr$$

$$= -\int_0^{2\pi} N(\theta)K(\cos(\theta+\pi),\sin(\theta+\pi))\,d\theta \int_0^{\infty} \frac{1}{r} f(u-r\cos\theta, v-r\sin(\theta))\,dr$$

since K is odd. Thus

$$Tf(u,v) = -\int_0^{2\pi} N(\theta)K(\cos\theta,\sin\theta)\,d\theta \int_0^{\infty} \frac{1}{r} f(u+r\cos\theta, v+r\sin\theta)\,dr$$

since

$$N(\theta+\pi) = N(\theta)\ .$$

Finally

$$\text{21A)}\quad Tf(u,v) = \int_0^{2\pi} N(\theta)K(\cos t,\sin\theta)\,d\theta \int_{-\infty}^{\infty} \frac{1}{r} f(u-r\cos\theta; v-r^2\sin\theta)\,dr\ .$$

Now adding 21) and 21A) we find that

$$Tf(u,v) = \frac{1}{2}\int_0^{2\pi} N(\theta)K(\cos\theta,\sin\theta)\,d\theta \int_{-\infty}^{\infty} \frac{1}{r} f(u-r\cos\theta, v-r^2\sin\theta)\,dr\ .$$

If $K(\cos\theta,\sin\theta)$ is in L^1 of $[0,2\pi]$, we can apply Minkowski's inequality

$$\|Tf\|_{L^p} \le C\int_0^{\pi} d\theta\ \|H_\theta f\|_{L^p}$$

where

$$H_\theta f = \int_{-\infty}^{\infty} f(x-\gamma_\theta(r))\,\frac{dr}{t}\ ,$$

with

$$\gamma_\theta(r) = (r\cos\theta, r^2\sin\theta)\ .$$

Now we prove

$$\|Tf\|_{L^p} \le c\|f\|_{L^p}$$

by showing

$$\|H_\theta f\|_{L^p} \le c_p\|f\|_{L^p}\ .$$

This is a problem of the type II′.

Stein was led to consider problem II by his study of Poisson integrals on symmetric spaces. We are not going to launch into a discussion of symmetric spaces, but instead we consider an example. Let

$$M_\varepsilon f(x,y) = \frac{1}{\varepsilon^2}\iint f(x-r,y-s)\,\frac{1}{\left(1+\frac{r^2}{\varepsilon^2}\right)\left(1+\frac{s^2}{\varepsilon^2}\right)}\,dr\,ds$$

$$= f * \frac{1}{\varepsilon^2}K\left(\frac{x}{\varepsilon},\frac{y}{\varepsilon}\right)\ .$$

If $K(x,y)$ were dominated by a decreasing, radial, L^1 function, the classical theory would imply

$$\lim_{\varepsilon\to 0} M_\varepsilon f(x,y) = f(x,y) \qquad \text{a.e.}$$

see [SWE]. However

$$K(x,y) = \frac{1}{(1+x^2)(1+y^2)}\ ,$$

so the smallest radial majorant of K is $\frac{1}{1+x^2+y^2}$ which is not integrable. In effect K has too much of its mass along the coordinate axis. The extreme case of this phenomena would be to have a kernel with all of its mass on the coordinate axis.

In other examples, kernels have too much of their mass along curves, and the extreme case of difficulties arising in problems of Poisson Integrals on symmetric spaces lead to Problem II.

Appropriate positive results to problem III would have implications for the boundary behavior of functions holomorphic in pseudoconvex domains in C^n. The natural balls in these problems are long, thin and twisting. The idealized situation is that of a vector field. In the case of a strictly pseudo convex domain, the balls satisfy the standard properties that ensure that the usual covering arguments apply. See [SBC]. For progress in the case of pseudoconvex domains see [NSW] and [NSWB].

Now that we have seen some of the roots of our problems, let us consider why these problems don't fit into the framework of the standard theory of Maximal functions and singular integrals as presented for example in [S].

In the standard treatment of averages over Euclidean balls an important geometric property of the Euclidean balls is used. If two balls B_1 and B_2 of the same radius, r, intersect, then B_2^*, the ball having the same center as B_2 but having radius 3r contains B_1. To see how badly this property fails for our problems let us suppose we were considering averages

$$\frac{1}{h\varepsilon}\int_0^h \int_{t^2}^{t^2+\varepsilon} f(x-r,y-s)\,dr\,ds$$

over slightly thickened parabolas or balls

$$B_{\varepsilon,h} = \{(r,s)\,|\,0\leq r\leq h,\ t^2\leq s\leq t^2+\varepsilon\}\ .$$

We now consider the intersection of two of these balls of the same size

Clearly one of these "balls" is not contained in a fixed multiple of the other (uniform in ε).

We can also see the difficulty of using the Calderón-Zygmund theory to study Problem II′.

Suppose $\gamma(t)$ is the parabola (t,t^2) in R^2 and

$$Tf(x,y) = \int f(x-t, y-t^2)\frac{dt}{t} .$$

Then we may formally write

$$Tf(x,y) = \iint f(x-t, y-st^2)\frac{\delta(1-s)}{t}\, ds\, dt$$

$$= \iint f(x-t, y-v)\,\frac{1}{t^3}\,\delta\left(1-\frac{v}{t^2}\right) dv\, dt .$$

Or

22) $$Tf = K * f ,$$

where

22A) $$K(x,y) = \frac{1}{x^3}\,\delta\left(1-\frac{y}{x^2}\right),$$

The Calderón-Zygmund theory deals with convolution operators with kernels $K(x,y)$, but in their theory $K(x+h,y+k)-K(x,y)$ should be much less than $K(x,y)$ if h and k are much smaller than x or y. However for our K if (x,y) is a point on the curve $y=x^2$ and $(x+h,y+k)$ is not on the curve, no cancellation in the difference $K(r+h,y+k)-K(x,y)$ can occur, no matter how small h and k are.

The Calderón-Zygmund Theory is based on 4) tools

a) The Fourier transform (The Fourier transform is even used in the L^1 theory)

b) Interpolation

c) Covering lemmas

d) Calderón-Zygmund decomposition.

Perhaps the natural attack on our problems would be to find appropriate covering lemmas and suitable variants of the Calderón-Zygmund decomposition. Some progress in finding covering lemmas for related problems was made by Stromberg [Str] and [STRO] and Cordoba [COR1], [COR2], Cordoba and Fefferman [CF1], [CF2], [CF3], and Fefferman [FEf].

Our approach will however be different. We shall try to use the Fourier transform or other orthogonality methods and interpolation to reduce our problems on averages and singular integrals to the more standard averages and singular integrals. In retrospect we see that some of these ideas occurred in [SPL], [CS], and in [KS].

We have said earlier that *curvature* and *Fourier Transform* would be important for us. Actually they go together. If one has a nice measure on a curved surface, the Fourier transform of that measure decays at infinity even though the measure is singular. Let us consider some examples. Define, for a test function ϕ,

$$\mu(\phi) = \int_0^1 \phi(t,0)\,dt\ .$$

μ is supported on a straight line, namely the x-axis, and

$$\hat{\mu}(\xi,\eta) = \mu(e^{i\xi x} e^{i\eta y}) = \int_0^1 e^{i\xi t}\, dt$$

which is independent of η and hence cannot decay at infinity along the η-axis. Now let us consider a measure supported on a parabola,

$$\nu(\phi) = \int_{-\infty}^{\infty} e^{-t^2} \phi(t,t^2)\, dt \,. \tag{23}$$

Then

$$\hat{\nu}(\xi,\eta) = \nu(e^{i\xi x} e^{i\eta y})$$

$$= \int_{-\infty}^{\infty} e^{-t^2} e^{i\xi t} e^{i\eta t^2}\, dt \,.$$

This integral may be computed exactly by completing the square, and it is easy to see that

$$|\hat{\nu}(\xi,\eta)| \le C \, \frac{e^{-\frac{\xi^2}{\eta}}}{\sqrt{1+|\eta|}} \,.$$

Thus $\hat{\nu}(\xi,\eta)$ tends to zero at infinity.

Another example is afforded by rotationally invariant Lebesgue measure on the $n-1$ dimensional sphere $|x| = 1$ in R^n. If we denote this measure by $d\mu$, we have for $\xi \in R^n$

$$\hat{d\mu}(\xi) = C_n J_{\frac{n-2}{2}}(|\xi|)\, |\xi|^{-\left(\frac{n-2}{2}\right)}.$$

See [SWE]. Thus

$$|\widehat{d\mu}(\xi)| \leq C_n(1+|\xi|)^{-\left(\frac{n-1}{2}\right)}.$$

Of course we want to have a tool to estimate the Fourier transform of measures in general, not in just a few specific cases.

This tool is a lemma of Van Der Corput.

VAN DER CORPUT'S LEMMA. *Let* $h(t)$ *be a real function. For some* j, *assume* $|h^{(j)}(t)| \geq \lambda$ *in an interval* $a \leq t \leq b$. *If* $j = 1$, *assume also that* $h'(t)$ *is monotone, then*

$$\left|\int_a^b \exp(ih(t))\,dt\right| \leq \frac{C_j}{\lambda^{1/j}}.$$

For the proof of Van Der Corput's lemma for $j = 1$ and 2 see [Z]. The proof for higher j is similar.

Let us consider the measure

24) $$d\mu(\phi) = \int_1^2 \phi(t,t^2)\,dt.$$

Then

$$\widehat{d\mu}(\xi,\eta) = \int_1^2 e^{i\xi t}e^{i\eta t^2}\,dt.$$

This integral cannot be evaluated explicitly, but we wish to see that Van Der Corput's lemma may be applied. We take $h(t) = \xi t + \eta t^2$. First we use the fact that $h''(t) = \eta$. Thus by Van Der Corput's lemma with $j = 2$, we see

25) $$|\widehat{d\mu}(\xi,\eta)| \leq C \frac{1}{(1+|\eta|)^{1/2}} .$$

Now if $|\eta| < \frac{|\xi|}{8}$,

$$|h'(t)| = |\xi + 2\eta t| > \xi/8 .$$

So by Van Der Corput's lemma with $j = 1$,

26) $$|\widehat{d\mu}(\xi,\eta)| \leq \frac{C}{|\xi|}$$

if

$$|\xi| > 8|\eta| .$$

Putting 25) and 26) together we have

27) $$|\widehat{d\mu}(\xi,\eta)| \leq \frac{C}{(1+\xi^2+|\eta|)^\delta}$$

for some $\delta > 0$.

Stein pointed out in retrospect that we can already see from an estimate like 27) that $d\mu$ has interesting properties from the point of harmonic analysis – namely even though $d\mu$ is singular,

$$Tf = d\mu * f$$

maps L^p into L^2 continuously for some $p < 2$. For

$$\int (Tf)^2 = \int |\widehat{Tf}(\xi,\eta)|^2 \, d\xi \, d\eta$$

$$= \int |\widehat{d\mu}(\xi,\eta)|^2 |\hat{f}(\xi,\eta)|^2 \, d\xi \, d\eta$$

$$= \int |\widehat{d\mu}(\xi,\eta)|^2 (1+\xi^2+|\eta|)^{2\delta} \frac{|\hat{f}(\xi,\eta)|^2}{(1+\xi^2+|\eta|)^{2\delta}}$$

$$\leq \frac{|\hat{f}(\xi,\eta)|^2}{(1+\xi^2+|\eta|^{2\delta}}$$

$$\leq \left(\int |\hat{f}(\xi,\eta)|^{2q}\right)^{1/q} \left(\int \frac{1}{(1+\xi^2+|\eta|)^{2\delta q'}}\right)^{1/q'} .$$

The second integral is bounded if q' is sufficiently large which means for some $q > 1$. But then the first integral is bounded for $f \in L^p$ where $\frac{1}{p} + \frac{1}{2q} = 1$.

II. *The Hilbert transform along curves*

The first progress in our series of problems was made on the Hilbert transform along curves. The Hilbert transform along a curve can be thought of as a multiplier transformation

$$\text{28)} \qquad \widehat{H_\gamma f}(\xi) = m_\gamma(\xi)\hat{f}(\xi)$$

where

$$\text{29)} \qquad m_\gamma(\xi) = \int_{-\infty}^{\infty} e^{i\xi\cdot\gamma(t)} \frac{dt}{t} .$$

To see that 29) is true we may either substitute the formula

$$f(x) = \int e^{-i\xi \cdot x} \hat{f}(\xi)\, d\xi$$

into 13) or recognize the fact that

$$Hf = D * f$$

where D is a distribution

$$D\phi = \int \phi(\gamma(t))\, dt \, .$$

So

$$\widehat{Hf} = \hat{D}\hat{f} \, ,$$

and $\hat{D}$ may be computed by evaluating D on an exponential.

Thus to prove that H_γ is bounded on L^2 one needs to show that m_γ is bounded. The first result of this type was obtained by Fabes [F]. Fabes showed H_γ is bounded on L^2 in 2-dimensions for the curve

$$\gamma(t) = (t, |t|^\alpha \operatorname{sgn} t) \, , \qquad \alpha > 0 .$$

So Fabes' proof consisted in showing that the integral

$$m(\xi,\eta) = \int_{-\infty}^{\infty} \exp(it\xi + i|t|^\alpha (\operatorname{sgn} t)\eta) \frac{dt}{t}$$

is uniformly bounded in ξ and η. To this end Fabes employed the method of steepest descents. The method of steepest descents is a method of obtaining very precise asymptotic information for large λ about integrals of the form

$$\int \exp(i\lambda h(t))\, dt$$

by contour integration. However to employ the method one has to have very precise information on where the real part of h(z) is positive and negative in the complex plane. Thus already to employ the method of steepest descents for the curve (t,t^2,t^3), one would have to understand the zero set of

$$\text{Real Part }\{\xi_1 z+\xi_2 z^2+\xi_3 z^3\}$$

uniformly in ξ_1, ξ_2 and ξ_3. So it is hard to imagine using the method of steepest descents, and for the curve t,t^2,t^3,t^4,t^5 it would seem close to impossible. Fabes' result was very important in that it gave the first clue that problems such as II and II$'$ could have positive answers. However a better method would have to be found – a method that needed less precise information about h(t). The next step was to show that if $\gamma(t)=(t,t^{a_1},t^{a_2},\cdots,t^{a_{n-1}})$, $1<a_1<a_2<\cdots<a_{n-1}$ (here t^{a_j} can mean either $|t|^{a_j}$ or $|t|^{a_j}\operatorname{sgn}t$) then H_γ was bounded on $L^2(R^n)$ [SWA]. Here we had to prove the boundedness of the integral

$$\int_{-\infty}^{\infty}\exp[i(\xi_1 t+\cdots+\xi_n t^{a_{n-1}})]\,\frac{dt}{t}\,.$$

The proof was by way of the Van Der Corput lemma but was unnecessarily complicated because at that time we only knew the lemma for $j=1,2$. Let us see how Van Der Corputs lemma works in the case $\gamma(t)=(t,t^2,\cdots,t^n)$. We then have to show that

$$\text{30)}\qquad \left|\int_{\varepsilon\le|t|\le R}\exp(i\xi_1 t+\cdots+i\xi_n t^n)\,\frac{dt}{t}\right|\le C(n)\,.$$

where C(n) does not depend on $\varepsilon,R,\xi_1,\cdots,\xi_n$. We shall prove 30) by induction on n. By changing variables, replacing t by $\dfrac{t}{|\xi_n|^{1/n}}$ we may assume $\xi_n = \pm 1$ in 30). Then by using Van Der Corput's lemma with $j = n$, we find

31) $$\left| \int_1^t \exp(i\xi_1 s + \cdots + i\xi_{n-1} s^{n-1} \pm i s^n) \right| \le C(n) .$$

An integration by parts together with 31) shows that

32) $$\left| \int_{1 \le |t| \le R} \exp(i\xi_1 t + \cdots + i\xi_{n-1} t^{n-1} \pm i t^n) \frac{dt}{t} \right| < C(n) .$$

Now

$$\left| \int_{\varepsilon \le |t| \le 1} \exp(i\xi_1 t + \cdots + i\xi_{n-1} t^{n-1} \pm i t^n) \frac{dt}{t} \right.$$

$$\left. - \int_{\varepsilon \le |t| \le 1} \exp(i\xi_1 t + \cdots + i\xi_{n-1} t^{n-1}) \frac{dt}{t} \right| \le \int_{\varepsilon \le t \le 1} t^n \frac{dt}{t} \le C(n) .$$

Hence we have reduced the proof 30) for n to proving 30) for $n-1$. Also the case of $n = 1$ is easy. So we are done by induction.

We now turn to the L^p theory. We wish to emphasize how *curvature* and *Fourier Transform* are joining together to help us. So we shall compare the case of the parabola (t,t^2) to the straight line (t,t). In the case of the parabola we are studying

34) $$H_p f = D_p * f$$

where D_p is a distribution. For a test function ϕ

35) $$D_p\phi = \int \phi(t,t^2)\,\frac{dt}{t}\,.$$

In the case of a straight line we are studying

36) $$H_L f = D_L * f\,,$$

where

37) $$D_L\phi = \int \phi(t,t)\,\frac{dt}{t}\,.$$

38) $$\hat{D}_p(\xi,\eta) = D_p(e^{i\xi x}e^{i\eta y}) = \int_{-\infty}^{\infty} e^{i\xi t}e^{i\eta t^2}\,\frac{dt}{t}\,,$$

and

39) $$\hat{D}_L(\xi,\eta) = D_L(e^{i\xi x}e^{i\eta y})$$

$$= \int_{-\infty}^{\infty} e^{i\xi t}e^{i\eta t}\,\frac{dt}{t}\,.$$

We can calculate $\hat{D}_L$ explicitly, and we find

40) $$\hat{D}_L(\xi,\eta) = c\ \mathrm{sgn}\,(\xi + n)\,.$$

Notice that $\hat{D}_L$ is discontinuous along the line $\xi = -\eta$. We shall show in contrast that $\hat{D}_p(\xi,\eta)$ is continuous away from the origin. It is very easy to see that $D_p(\xi,\eta)$ is C^∞ away from the line $\eta = 0$ by complex integration. If for example $\eta > 0$, we think of t as a complex variable and integrate along the line $\mathrm{lm}\,t = \mathrm{Re}\,t$. Then the factor $e^{i\eta t^2}$ decays as fast as $e^{-c\eta|t|^2}$, and one can easily justify differentiation under the integral sign as long as $\eta \geq \varepsilon$, for some positive ε.

We shall now show that $\hat{D}_p(\xi,\eta)$ is continuous near $|\xi| = 1$. What we must show is that $\lim_{\eta\to 0} \hat{D}_p(\xi,\eta)$ exists for ξ near ± 1. We shall show

41) $$\lim_{\eta\to 0} \int_{-\infty}^{\infty} e^{i\xi t} e^{i\eta t^2} \frac{dt}{t} = \int_{-\infty}^{\infty} e^{i\xi t} \frac{dt}{t}.$$

Assume for simplicity that $\eta > 0$.

Of course

42) $$\lim_{\eta\to 0} \int_{-\frac{1}{\eta^{1/3}}}^{\frac{1}{\eta^{1/3}}} e^{i\xi t}\, dt = \int_{-\infty}^{\infty} e^{i\xi t} \frac{dt}{t}$$

and

43) $$\left| \int_{\frac{1}{\eta^{1/3}}}^{\frac{1}{\eta^{1/3}}} e^{i\xi t}(e^{i\eta t^2} - 1) \frac{dt}{t} \right|$$

$$\leq 2\eta \int_0^{\frac{1}{\eta^{1/3}}} t\,dt \leq \eta^{1/3} \to 0 \quad \text{as} \quad \eta \to 0.$$

In view of 42) and 43) we can show 41) by showing

44) $$\int_{\eta^{-1/3} \leq |t|} e^{i\xi t} e^{i\eta t^2} \frac{dt}{t} \to 0.$$

We shall prove 44) by using Van Der Dorput's lemma. By using Van Der Corput's lemma with $j = 2$, we see

$$\int_{\eta^{-2/3}}^{t} e^{i\xi s} e^{i\eta s^2} ds \leq \frac{c}{\sqrt{\eta}} .$$

(Let $h(s) = \xi s + \eta s^2$, then $h''(s) = 2\eta$.) So an integration by parts shows

45) $$\int_{\eta^{-2/3}}^{\infty} e^{i\xi t} e^{i\eta t^2} \frac{dt}{t}$$

$$\leq \eta^{2/3} + \frac{1}{\eta^{1/2}} \int_{\eta^{-2/3}}^{\infty} \frac{dt}{t^2}$$

$$\leq C\, \eta^{2/3}/\eta^{1/2} < C\eta^{1/6} .$$

Note that if $t < \eta^{-2/3}$

$$|h'(\cdot t)| = |\xi + 2\eta t| > |\xi| - 2\eta^{1/3} .$$

So

$$\int_{\eta^{-1/3}}^{\eta^{-2/3}} e^{i\xi s} e^{i\eta s^2} ds$$

is bounded if ξ is close to minus one. Hence an integration by parts shows

46) $$\left| \int_{\frac{1}{\eta^{1/3}}}^{\frac{1}{\eta^{2/3}}} e^{i\xi t} e^{i\eta t^2} \frac{dt}{t} \right| \leq C \int_{\frac{1}{\eta^{1/3}}}^{\infty} \frac{dt}{t^2} + \eta^{1/3} \leq C\, \eta^{1/3} .$$

45) and 46) together with similar estimates for negative t prove 44) and hence the continuity of $\hat{D}_p(\xi,\eta)$ away from the origin. If one is a little more careful in the above argument, one can prove that $\hat{D}_p(\xi,\eta)$ satisfies a Lipschitz condition away from the origin. One may then use Riviere's [R] version of Hermander's multiplier theorem [H] to obtain some L^p results for $p \neq 2$. In fact if $\gamma(t) = (t,t^2)$ one obtains

$$47)\qquad \|H_\gamma f\|_{L^p} \subseteq C_p \|f\|_{L^p}, \quad \frac{4}{3} < p < 4 .$$

For quite a while we tried to prove the range of p, $\frac{4}{3} < p < 4$, in 47) was optimal — with no success. Also, there was a suspicion that the use of the Hörmander Riviere theorem lost something. In the Hörmander argument, one wishes to estimate an expression of the form

$$J = \int_{R \leq |x| \leq 2R} |K(x+h) - K(x)|dx$$

where K is a kernel, in terms of $\hat{K}$. One does this by using Schwartz's inequality and Plancherel's theorem.

$$J = \int_{R \leq |x| \leq 2R} \frac{1}{|x|^\alpha} \cdot |x|^\alpha |K(x+h) - K(x)|dx$$

$$\leq \left(\int_{R \leq |x| \leq 2R} \frac{1}{|x|^{2\alpha}} dx \right)^{1/2} \cdot \left(\int |x|^{2\alpha} |(x+h) - K(x)|^2 dx \right)$$

$$< \frac{1}{R^{\alpha - \frac{n}{2}}} \cdot \sum_{|j|=\alpha} \int |1 - e^{i\xi \cdot h}|^2 | \frac{\partial^j}{\partial \xi^j} \hat{K}(\xi)|^2 d\xi$$

where the sum is over all α'th derivatives of $\hat{K}$.

Now there was the feeling that a use of Schwartz inequality like that above lost too much, and that more careful estimates for J for particular kernels might lead to better results.

To get an idea of what to do we calculated $\hat{D}(\xi,\eta)$ very precisely by the method of steepest descents. We found

48) $$\hat{D}_p(\xi,\eta) = \operatorname{sgn}\xi + C\,\frac{|\eta|^{1/2}}{\xi}\, e^{i\xi^2/\eta}\psi\left(\frac{\eta}{\xi^2}\right)$$

$$+ \text{ Better terms },$$

where ψ is a C_0^∞ function on R^1 which is one near the origin. Hence, the crux of the matter was to study the transformation Tf given by

49) $$\widehat{Tf}(\xi,\eta) = m(\xi,\eta)\hat{f}(\xi,\eta) ,$$

where

50) $$m(\xi,\eta) = \frac{|\eta|^{1/2}}{\xi}\, e^{i\frac{\xi^2}{\eta}}\,\psi\left(\frac{\eta}{\xi^2}\right) .$$

This suggests introducing an analytic family of operators in the sense of [SI] as follows:

51) $$\widehat{T_z f}(\xi,\eta) = m_z(\xi,\eta)\hat{f}(\xi,\eta) ,$$

where

52) $$m_z(\xi,\eta) = \left(\frac{|\eta|^{1/2}}{\xi}\right)^z m(\xi,\nu) .$$

$T_z f$ is clearly bounded on L^2 if $\operatorname{Re} z = -1/2$. So if we could prove that the kernel K_z corresponding to T_z for $\operatorname{Re} z$ positive satisfied a condition of the Calderon-Zygmund type, we could prove T_z was bounded in each L^p if $\operatorname{Re} z > 0$. Hence by Stein's interpolation theorem we would know that $T_0 = T$ was bounded in all L^p, $p \leq 2$.

Then by a duality argument T would be bounded in all L^p, $1 < p < \infty$. It turns out that one can show by a messy calculation that T_z is of Calderon-Zygmund type if $\mathrm{Re}\, z > 0$.

Let us try to understand why the kernel for T_z, $\mathrm{Re}\, z > 0$, might be a little better than the kernel for T_0. T is essentially H_γ $\gamma = (t,t^2)$, and so the kernel K_0 of T_0 is essentially

$$K_0(x,y) = \frac{1}{x^3}\,\delta\left(1 - \frac{y}{x^2}\right)$$

from 22) and 22A). If we introduce "parabolic polar coordinates" $y = r^2 \sin\theta$ $x = r\cos\theta$, we see

$$K_0(x,y) = \frac{1}{r^3}\,\delta(\theta-\theta_0)$$

where

$$\frac{\sin\theta_0}{\cos^2\theta_0} = 1\,.$$

We might expect if $\mathrm{Re}\, z = \varepsilon > 0$ K_z to be ε better than K_0. So we might expect

$$K_z(x,y) \cong \frac{1}{r^3}\,\frac{1}{|\theta-\theta_0|^{1-\varepsilon}}\,.$$

Now we would like to explain why a $\dfrac{1}{|\theta-\theta_0|^{1-\varepsilon}}$ singularity is better than a $\delta(\theta-\theta_0)$ singularity. To see the situation more clearly, let us examine the analogous situation for the standard polar coordinates with $\theta_0 = 0$. Suppose

53) $$K_0(x,y) = \frac{\delta(\theta)}{r^2}$$

where $x = r\cos\theta$ and $y = r\sin\theta$, and

54) $$K_\varepsilon(x,y) = \frac{1}{r^2|\theta|^{1-\varepsilon}} .$$

(The factor $\frac{1}{r^2}$ for ordinary polar coordinates plays the same role as $\frac{1}{r^3}$ for parabolic polar coordinates.) We are trying to see whether

55) $$\int_{x^2+y^2>c(h^2+k^2)} |K(x,y)-K(x-h,y-k)| < C .$$

For either $K = K_0$ or $K = K_\varepsilon$. Let us take $h = 0$ and $k = 1$ and consider first $K = K_0$. Let us look at the contribution from y's which are very close to 0. We have

$$\int_{r>C} \int_{\text{near} o} \left|\frac{\delta(\theta)}{r^2} - \frac{\delta(\theta')}{r^2}\right| d\theta \, r dr .$$

If y is very close to 0 $|\theta'| \sim \frac{1}{r}$

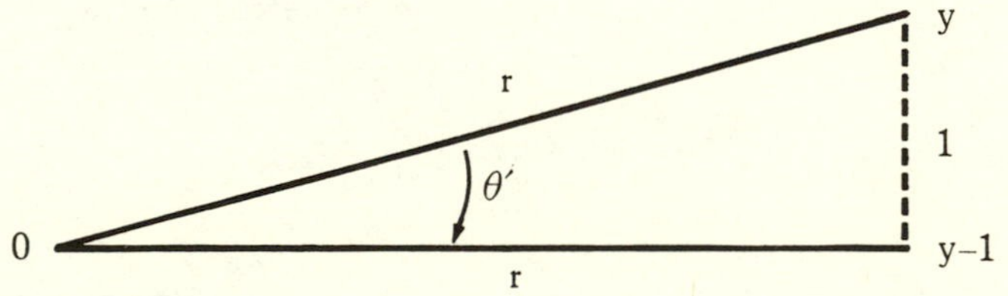

So the left-hand side of 55) is at least $\int_C^\infty \frac{1}{r} dr = \infty$. So 55) can't hold. Let us put the matter a little differently. If we consider the θ's with $\theta \geq 0$ where the difference

$$K_0(r,\theta) - K_0(r',\theta')$$

offers no cancellation, we find there is only one bad θ, $\theta = 0$. But still

$$\int_0^{\pi/4} |K_0(r,\theta) - K_0(r',\theta')|\, d\theta = 1 \,.$$

Let us consider now what happens with K_ε. We should expect no help from the difference $K(x,y) - K(x,y-1)$ when $y \geq 0$, if $y-1 \leq 0$. But this can only happen if $y \leq 1$ or $\theta < 1/r$. But over this set $K_\varepsilon(x,y)$ is integrable at infinity

$$\int_5^{\infty} \frac{1}{r^2} \int_0^{1/r} K_\varepsilon(r,\theta)\, d\theta\, r\, dr$$

$$\leq \int_5^{\infty} \frac{1}{r^2} \int_0^{1/r} \frac{1}{\theta^{1-\varepsilon}}\, d\theta\, r\, dr$$

$$\leq \int_5^{\infty} \frac{1}{r^{1+\varepsilon}}\, dr \,.$$

It is not difficult to complete the argument and to show 55).

After a laborious calculation one could prove that the kernel K_z corresponding to T_z of 51) was for $\operatorname{Re} z > 0$ an operator of Calderon-Zygmund type. This proved that H_γ was bounded in L^p $1 < \rho < \infty$ if $\gamma = (t,t^2)$. However it would be extremely difficult to carry over this proof to a three dimensional curve. For example it would be hard to derive an analogue of 48) for the curve (t,t^2,t^3). Essentially one needed a way to define a suitable analytic family T_z without using the asymptotic formula 48). Recall that $H_\gamma f = D_p * f$ where

56) $$\hat{D}_p(\xi,\eta) = \int e^{i\xi t} e^{i\eta t^2} \frac{dt}{t} \,.$$

So one might be tempted to define

57) $$\tilde{H}^z_\gamma f = \tilde{D}^z_p * f ,$$

where

58) $$\hat{\tilde{D}}^z_p(\xi,\eta) = \int \frac{e^{i\xi t}e^{i\eta t^2}}{(1+t^2)^{z/2}} dt .$$

It turns out that 58) is not a good idea for a very important reason. By changing variables in formula 56) we see that $\hat{D}_p(\lambda\xi, \lambda^2\eta) = \hat{D}_p(\xi,\eta)$, for any $\lambda > 0$. Note that also the function $m_z(\xi,\eta)$ defined in 52) also has this type of homogeneity, namely $m_z(\lambda\xi, \lambda^2\eta) = m_z(\xi,\eta)$, for $\lambda > 0$.

Now experience has shown that homobeneity is a powerful friend not to be tossed away lightly. However $\hat{\tilde{D}}_p$ does not have this homogeneity.

This situation can be remedied by defining

59) $$H^z_\gamma f = D^z_p \times f$$

where

60) $$\hat{D}^z_p(\xi,\eta) = \int_{-\infty}^{\infty} (1+\eta^2t^4)^{-z/4} e^{i\xi t}e^{i\eta t^2} \frac{dt}{t} .$$

Note that for $\lambda > 0$

61) $$\hat{D}^z_p(\lambda\xi, \lambda^2\eta) = \hat{D}^z_p(\xi,\eta) .$$

Let us see how formula 61) can help us. We would like to show

62) $$|\hat{D}^z_p(\xi,\eta)| \leq C(z)$$

if $\mathrm{Re}\ z > -\frac{1}{2}$. By formula 61) we may assume $\eta = \pm 1$, let us say $\eta = 1$. Then by Van Der Carput's lemma with $j = 2$, we see that

$$\left|\int_1^t e^{i\xi s} e^{i\eta s^2}\, ds\right|$$

$$= \int_1^t e^{i\xi s} e^{is^2}\, ds\Big| \leq C \;.$$

So an integration by parts shows

$$\left|\int_1^\infty (1+t^4)^{-z/4} e^{i\xi t} e^{i\eta t^2} \frac{dt}{t}\right| \leq C(z)$$

$$(\cdot\, \eta = 1)\;.$$

Now

$$\Bigg| \int_{-1}^1 (1+t^4)^{-z/4} e^{i\xi t} e^{i\eta t^2} \frac{dt}{t}$$

$$- \int_{-1}^1 e^{i\xi t} e^{i\eta t^2} \frac{dt}{t}\Bigg|$$

$$\leq C(z) \int_{-1}^1 t^2 \frac{dt}{|t|} \leq C(z)\;.$$

But we already know that

$$\left| \int_{-1}^1 e^{i\xi t} e^{i\eta t^2} \frac{dt}{t}\right| \leq C\;.$$

One may finally deal with $-\infty < t \leq -1$ in the same manner that one treated $1 \leq t < \infty$. Hence 62) is proved.

Let us calculate the kernel, K_z, corresponding to H_γ^z if $z = \varepsilon > 0$. Then

$$K_\varepsilon(x,y) = \iint e^{i\xi x}\, e^{-iny}\, \hat{D}_p^z(\xi,\eta)\, d\xi\, d\eta$$

$$= \int_{-\infty}^{\infty} \frac{dt}{t} \int_{-\infty}^{\infty} e^{-i\eta y}\,(1+\eta^2 t^4)^{-\varepsilon/4}\, e^{i\eta t^2}$$

$$\cdot \int_{-\infty}^{\infty} e^{-i\xi(x-t)}\,dx = \int_{-\infty}^{\infty} \frac{dt}{t} \int_{-\infty}^{\infty} e^{i\eta(t^2-y)}(1+\eta^2 t^4)^{-\varepsilon/4}\, e^{i\eta t^2}\,\delta(x-t)\,dt$$

$$= \frac{1}{x} \int_{-\infty}^{\infty} e^{i\eta(x^2-y)}(1+\eta^2 x^4)^{-\varepsilon/4}\, d\eta$$

$$= \frac{1}{x^3} \int_{-\infty}^{\infty} e^{i\eta\left(\frac{x^2-y}{x^2}\right)} \frac{1}{(1+\eta^2)^{\varepsilon/4}}\, d\eta$$

$$= \frac{1}{x^3}\, P_{\varepsilon/2}\left(\frac{x^2-y^2}{x^2}\right)$$

where $P_{\varepsilon/2}$ is a modified Poisson kernel. $P_{\varepsilon/2}$ decays exponentially fast at ∞ and $P_{\varepsilon/2}(u) \sim \dfrac{C}{|u|^{1-\varepsilon/2}}$ as $u \to 0$. See [SWE].

Thus $K_\varepsilon(x,y)$ has a singularity near the curve (t,t^2) of the form $\dfrac{1}{|\theta-\theta_0|^{1-\varepsilon/2}}$ which is just the improvement over the δ-function that we seek.

A modification of these ideas worked for curves

$$\gamma(t) = (t^{a_1}, t^{a_2}, \cdots, t^{a_n})$$

$a_1 < a_2, \cdots, < a_n$. See [NRW].

However, there is a natural generalization of these curves. All of these curves satisfy an equation of the form

63) $$\gamma'(t) = \frac{A}{t}\gamma(t)$$

where A is a real $n \times n$ matrix such that the real parts of the eigenvalues of A are positive. For example if

$$\gamma(t) = (t, t^2)$$

$$A = \begin{pmatrix} 1 & 0 \\ 0 & 2 \end{pmatrix}.$$

A curve satisfying 63), where all the eigenvalues of A have positive real part is called a homogeneous curve.

A will generate a group of transformations

$$T_\lambda = \exp(A \log \lambda) .$$

Then

64) $$H_\gamma f = D_\gamma * f$$

where

65) $$\hat{D}_\gamma(T_\lambda \xi) = \hat{D}_\gamma(\xi)$$

Moreover there is a distance $\rho_A(x)$ defined on R^n such that

$$\rho(T_\lambda x) = \lambda \rho(x) .$$

In the case of the cure (t, t^2) we may, as we said before take

$$A = \begin{pmatrix} 1 & 0 \\ 0 & 2 \end{pmatrix}.$$

Then

$$T = \begin{pmatrix} \lambda & 0 \\ 0 & \lambda^2 \end{pmatrix},$$

and

$$\rho(x,y) = (x^4+y^2)^{1/4} .$$

It turns out that in the case of a general homogeneous cure, we can obtain a satisfactory analytic family of operators by defining

66) $$H^z_\gamma f = D^z_\gamma * f$$

where

67) $$\hat{D}^z_\gamma(\xi) = \rho^{-z}_{A*}(\xi) \int |t|^{-z} \exp(i\xi\cdot\gamma(t)) \frac{dt}{t} .$$

ρ_{A*} is the distance function corresponding to A^*, the adjoint of A.

For a detailed description of the argument see [SW]. Here we shall just make a comment. If some of the eigenvalues of A have non-zero imaginary part, $\gamma(t)$ can be an infinite spiral. For example the curve

$$\gamma(t) = (t^\alpha \cos(\beta \log t), t^\alpha \sin(\beta \log t))$$

is an example. So one could believe it might be rather messy to prove integrals involving $\exp(i\xi\cdot\gamma(t))$ to be bounded. It might be difficult to show that at each t some derivative of $\xi\cdot\gamma(t)$ would be non-zero. However, if one makes a change of variables $t = e^u$, we would be led to consideration of integrals involving $\xi\cdot\eta(u)$ where $\eta(u) = \gamma(e^u)$. If $\gamma'(t) = \frac{A}{t}\gamma(t)$, $\eta(u)$ satisfies

68) $$\eta'(u) = A\eta(u) .$$

We shall show that if $\eta(u)$ is a curve in R^n satisfying 68) where the eigenvalues of A have positive real part, then either $\eta(u)$ lies in a proper subspace of R^n or for every $\xi \neq 0$ and u there is a j, $1 \le j \le n$ such that

69) $$\frac{d^j}{du^j}\, \xi \cdot \eta(u) \neq 0 .$$

From 68) we see that

$$\frac{d^{j+1}\eta(u)}{du^{j+1}} = A^j \eta'(u) .$$

By the Cayley-Hamilton theorem, we can find numbers a_j, $0 \le j \le n$, such that

$$\sum_{j=0}^{n} a_j A^j = 0 .$$

So

$$\sum_{j=0}^{n} a_j \frac{d^{j+1}}{du^{j+1}}\, \eta(u) = 0 ,$$

and

$$\sum_{j=0}^{n} a_j \frac{d^j}{du^j}\, \eta'(u) \cdot \xi = 0 .$$

In other words $\eta'(u) \cdot \xi$ satisfies an nth order constant coefficient differential equation. So if for some u and ξ

$$\frac{d^j}{du^j}\, n'(u) \cdot \xi = 0 \qquad j = 1,2,\cdots,n-1$$

$\eta'(u) \cdot \xi = 0$ for all u. Thus $\eta(u) \cdot \xi$ is a constant. But $\eta(u) \cdot \xi \to 0$

as $u \to -\infty$ since the eigenvalues of A have positive real part. Hence $\eta(u)$ is in the subspace of R^n orthogonal to ξ.

We shall conclude this section with the statement of some theorems that follow from the reasoning discussed above.

THEOREM 1. *Let* $\gamma(t)$ *satisfy*

$$\gamma'(t) = \frac{A}{t}\,\gamma(t)\,.$$

Suppose the span in R^n of $\gamma(t)$ for positive t and the span in R^n of $\gamma(t)$ for negative t agree. Then

$$\|H_\gamma f\|_{L^p} \leq C_{p,\gamma}\,\|f\|_{L^p}\,, \qquad 1 < p < \infty\,.$$

We say that a curve $\gamma(t)$ in R^n is well curved if

$$\left.\frac{d^j\gamma(t)}{d\gamma^j(t)}\right|_{t=0}, \qquad j = 1,2,\cdots$$

span R^n. It turns out that well curved curves can be approximated by homogeneous curves. We can then prove

THEOREM 2. *Let* $\gamma(t)$ *be well curved then,*

$$\left\| \int_{-1}^{1} f(x-\gamma(t))\,\frac{dt}{t} \right\|_{L^p} \leq A_{p,\gamma}\,\|f\|_{L^p}\,, \qquad 1 < p < \infty\,.$$

A general theorem in L^2 for curves which are approximately homogeneous was obtained by Weinberg [We].

III. *Maximal functions and* g-*functions*

We turn now to a discussion of maximal functions. We are especially interested in how the Fourier transforms and g-functions may be used as a tool to relate our maximal functions to more classical ones. The story began with the study of maximal functions along the curve (t,t^2). Thus we wish to consider averages

70) $$M_h f(x,y) = \frac{1}{h}\int_0^h f(x-t, y-t^2)\,dt\ .$$

After much frustration it was decided to take Fourier Transforms and try to see if anything could be learned. It is easy to see that

$$\widehat{M_h f}(\xi,\eta) = m_h(\xi,\eta)\hat{f}(\xi,\eta)$$

where

$$m_h(\xi,\eta) = \frac{1}{h}\int_0^h e^{it\xi}e^{it^2\eta}\,dt\ .$$

Now $m_h(\xi,\eta)$ cannot be evaluated explicitly. If one hopes to gain some insight by staring at a formula, one should have a formula that is as explicit as possible. Now a similar situation arose in the path integral approach to Quantum Mechanics. See [FH]. Feynman and Hibbs wished to have an explicit expression for the probability amplitude that a particle lies in a sphere of radius t. In essence, they had to consider an integral in R^n of the form

$$\int_{|x|\le\varepsilon} \exp Q(x)\,dx$$

where $Q(x)$ was a quadratic function of x. Instead they considered

$$\frac{1}{\varepsilon}\int_R e^{-\frac{|x|^2}{\varepsilon^2}+Q(x)}\,dx\ ,$$

which could be calculated explicitly. This suggests to consider instead of 70)

71) $$\nu_h * f(x,y) = \frac{1}{h} \int_{-\infty}^{\infty} \exp\left(-\frac{t^2}{h^2}\right) f(x-t, y-t^2)\,dt \,.$$

Then

72) $$\widehat{\nu_h * f}(\xi,\eta) = \hat{\nu}(h\xi, h^2\eta)\hat{f}(\xi,\eta)$$

where ν is the measure considered in 23). In particular

$$\hat{\nu}(h\xi, h^2\eta) = \int_{-\infty}^{\infty} e^{-t^2} e^{ih\xi t} e^{ih^2\eta t^2}\,dt \,,$$

and this integral can be computed explicitly by completing the square. We find

73) $$\hat{\nu}(h,h^2) = (\text{nice smoothly decaying function}) \cdot \exp i \frac{h^4\xi^2\eta}{1+h^4\eta^2} \,.$$

One might guess that the appearance of the oscillatory factor $\exp i \dfrac{h^4\xi^2\eta}{1+h^4\eta^2}$ is a reflection of the fact that ν_h is a singular measure. On the other hand we see that if $h^2\eta$ is large

$$\exp i \frac{h^4\xi^2\eta}{1+h^4\eta^2} \sim \exp i \frac{\xi^2}{\eta} \,,$$

which is independent of h. Thus one might try to write (from 72)

$$\widehat{\nu_h * f}(\xi,\eta) = \left(\hat{\nu}(h\xi, h^2\eta) \exp\left(-i\frac{\xi^2}{\eta}\right)\right) \cdot \left(\left(\exp i^{\xi^2}\right) f(\xi,\eta)\right).$$

One might now hope that if one defines a measure $\tilde{\nu}_h$ by the formula

$$\hat{\tilde{\nu}}_h(\xi,\eta) = \hat{\nu}(h\xi,h^2\eta)\exp -i\frac{\xi^2}{\eta}$$

ν_h could be dealt with by classical arguments, while

$$\left\{\exp\left(i\frac{\xi^2}{\eta}\right)f(\xi,\eta)\right\} = \hat{g}(\xi,\eta)$$

where g is another L^2 function having the same norm as f. So one could hope that

74)
$$\|\sup \nu_h * f\|_{L^2} = \|\sup \tilde{\nu}_h * g\|_{L^2}$$
$$\leq C\|g\|_{L^2} = C\|f\|_{L^2} .$$

Roughly speaking this works out. See [NRWM].

The proof of 74) was a hint on how to proceed. However, it depended (as had happened before) on very special computations. What was needed was a way to compare averages like ν_h to more classical averages by using only the decay of $\hat{\nu}_h$ and not so much the explicit expression of $\hat{\nu}_h$ as was used above.

Stein [Ssp] and [SH] succeeded in doing this by introducting appropriate g-functions. Stein's first argument with g-functions dealt with the averages $M_{B_r}f$ of equation 4). Recall

75)
$$M_{B_r}f(x) = \int_{B_r} f(x-y)\,d\mu_r(y)$$

where B_r is the ball of radius r centered at the origin, and $d\mu_r$ is the unit rotationally invariant measure on B_r. We set, as before,

76)
$$\mathfrak{M}f(x) = \sup_{r>0} |M_{B_r}f(x)| .$$

Stein used g-functions to prove

THEOREM 3:

77) $$\|\mathfrak{M}f\|_{L^p} \leq C_{p,n} \|f\|_{L^p}$$

if $p > \frac{n}{n-1}$ *and* $n \geq 3$.

Simple examples of the form

$$f(x) = \frac{\psi(x)}{|x|^{\frac{n-1}{n}} \log^{\alpha} \frac{1}{|x|}}$$

where $\psi = 1$ near the origin and has compact support show that $p > \frac{n}{n-1}$ is necessary in order that 77) hold. The situation for $n = 2$, $p > 2$ is unknown at this time.

I would like to present here Stein's original argument which proved 77) for $p = 2$ and $n = 4$. We define

78) $$g(f)(x) = \left\{\int_0^{\infty} t\left|\frac{d}{dt} M_t f(x)\right|^2 dt\right\}^{1/2}.$$

Assume that we could prove

79) $$\|g(f)\|_{L^2} \leq C(n) \|f\|_{L^2},$$

and let us see how 77) would follow. Now

$$r^n M_r f(x) = \int_0^r \frac{d}{ds} s^n M_s f(x) ds$$

$$= n \int_0^r s^{n-1} M_s f(x) ds + \int_0^r s^n \frac{d}{ds} M_s f(x) ds .$$

Thus

$$M_r f(x) \le \frac{C}{r^n} \int_0^r s^{n-1} M_s f(x)\,ds + \frac{1}{r^n} \int_0^r s^n \frac{d}{ds} M_s f(x)\,ds$$

$$= I(r) + II(r) \; .$$

Now $I(r)$ is dominated by the Hardy-Littlewood Maximal function and

$$II(r) \le \frac{1}{r^n} \int_0^r s^{n-1/2} s^{1/2} M_s f(x)\,ds$$

$$\le \frac{1}{r^n} \left(\int_0^r s^{2n-1}\,ds \right)^{1/2} g(f)(x)$$

$$\le C g(f)(x) \; .$$

So if we assume 29), we have

$$\|\sup M_r f(x)\|_{L^2} \le C \|f\|_{L^2} \; .$$

We turn now to the proof of 79).

$$\int_{R^n} |g(f)(x)|^2\,dx = \int_0^\infty t \int_{R^n} \left| \frac{d}{dt} M_t f(x) \right|^2 dx\,dt$$

$$= \int_0^\infty t \int_{R^n} \left| \widehat{\frac{d}{dt} M_t f}(\xi) \right|^2 d\xi\,dt$$

$$= \int_0^\infty t \int_{R^n} \left| \frac{d}{dt} \widehat{M_t f}(\xi) \right|^2 d\xi\,dt \; .$$

But $M_t f(\xi) = \hat{f}(\xi)\, m(t|\xi|)$, where

$$m(r) = \frac{1}{r^{\frac{n-2}{2}}} J_{\frac{n-2}{2}}(r) .$$

Here $J_{\frac{n-2}{2}}$ is the usual Bessel function. We shall need to know

$$\left|\frac{dm(r)}{dr}\right| \leq C \frac{1}{r^{\frac{n-2}{2}} r^{1/2}} = C \frac{1}{r^{\frac{n-1}{2}}} ,$$

and $\frac{dm}{dr}$ is bounded. See [SWE]. Thus

$$\int |g(f)(x)|^2 d\lambda \leq \int_0^\infty t \int_{R^n} \left|\frac{d}{dt}\{m(t|\xi|)\, |\hat{f}(\xi)|\}\right|^2 d\xi$$

$$\leq \int_{R^n} |\hat{f}(\xi)|^2 \int_0^\infty t \left|\frac{d}{dt} m(t\,|\xi|)\right|^2 dt$$

Thus to obtain 79) we need to show

$$\int_0^\infty t \left|\frac{d}{dt}\, m(t|\xi|)\right|^2 dt \leq C .$$

First since $\frac{dm}{dr}$ is bounded,

$$\left|\int_0^{1/|\xi|} dt\right| < |\xi|^2 \int_0^{1/|\xi|} t\, dt \leq C .$$

Next, since $|m'(t)| \leq \frac{C}{t^{\frac{n-1}{2}}}$,

$$\int_{1/|\xi|}^{\infty} t \left|\frac{d}{dt} m(t|\xi|)\right|^2 dt$$

$$\leq |\xi|^2 \int_{1/|\xi|}^{\infty} t \frac{1}{(t|\xi|)^{n-1}} dt \leq C ,$$

if $n \geq 4$.

Let us be more precise about the counterexample in 2-dimensions. We take

$$f(x) = \begin{cases} \dfrac{1}{|x| \log \frac{1}{|x|}} & x \text{ very near } 0 \\ \\ \text{in } C_0^\infty & \text{away from } 0 . \end{cases}$$

We can disprove 77) for $p = 2$, $n = 2$ by showing

80) $$M_{B_{|x|}} f(x) = \infty$$

for all small x.

Because of rotational symmetry it suffices to prove 80) for points $(a,0)$ with a small. In that case

$$M_{|x|} f(x) = \int_{-\pi}^{\pi} \frac{d\theta}{\{a^2(1-\cos)^2 + a^2 \sin^2\theta\}^{1/2} \log \dfrac{1}{a^2[(1-\cos)^2 + \sin^2\theta]}}$$

$$\sim \int_{-\varepsilon}^{\varepsilon} \frac{d\theta}{|\theta| \ln \frac{1}{|\theta|}} = \infty .$$

A similar argument shows

81) $$\sup_{1\leq r\leq 2} |M_{B_r} f(x)| = \infty$$

a.e. for an appropriate f.

We wish now to prove a theorem indicating how bad 81) fails in $L^2(R^2)$. We set

82) $$\mathfrak{M}_k f(x,y) = \sup_{2^k \leq j \leq 2^{k+1}} |M_{B_{\frac{j}{2^k}}} f(x)y| ,$$

where f is in $L^2(R^2)$. We shall show

83) $$\|\mathfrak{M}_k f\|_{L^2} \leq Ck\|f\|_{L^2} .$$

To prove 83), it suffices to show that for each function $j(x)$ taking the values $0,1,2,\cdots,2^k$,

84) $$\|M_{B_{1+j(x)2^{-k}}} f(x)\| < Ck\|f\|_{L^2} ,$$

with C independent of the function $j(x)$. We show 85) by induction on k. That is given a $j(x)$ taking the values $0,1,2,\cdots,2^k$, we shall define a function $j^*(x)$ so that $j^*(x)$ takes values in the set $0,1,2,\cdots,2^{k-1}$ and

85) $$\|M_{B_{1+j(x)2^{-k}}} f - M_{B_{1+j^*(x)2^{-k+1}}} f\|_{L^2} \leq C\|f\|_{L^2} .$$

85) provides the inductive step to prove 86). If $j(x)$ is given define

$$j^*(x) = \begin{cases} \frac{j(x)}{2} & \text{if } j(x) \text{ is even} \\ \frac{j(x)-1}{2} & \text{if } j(x) \text{ is odd .} \end{cases}$$

Then, if we set

86) $$gf(x) = \left\{ \sum_{j=0}^{2^k} |M_{B_{1+\frac{j}{2^k}}} f(x) - M_{B_{1+\frac{j-1}{2^k}}} f(x)|^2 \right\}^{1/2},$$

we see

$$|M_{B_{1+j(x)2^{-k}}} f - M_{B_{1+j^*(x)2^{-k+1}}} f| \le gf(x) .$$

Thus to prove 86) and hence 85) it suffices to prove

87) $$\|g(f)\|_{L^2} \le C\|f\|_{L^2} .$$

We shall prove 87) by using the Fourier Transform.

$$\int |gf(x)|^2\,dx = \sum_{j=0}^{2^k} \int |M_{B_{1+\frac{j}{2^k}}} f(x) - M_{B_{1+\frac{j-1}{2^k}}} f(x)|^2\,dx$$

$$= \sum_{j=0}^{2^k} |\widehat{M_{B_{1+\frac{j}{2^k}}} f}(\xi) - \widehat{M_{B_{1+\frac{j-1}{2^k}}} f}(\xi))^2\,d\xi$$

$$= \int |\hat{f}(\xi)|^2 \sum_{j=0}^{2^k} |m\left(\left(1+\frac{j}{2^k}\right)|\xi|\right) - m\left(\left(1+\frac{j-1}{2^k}\right)|\xi|\right)|^2\,d\xi ,$$

where $m(r) = J_0(r)$.

So to prove 87) it suffices to show

88) $$\sum_{j=0}^{2^k} |J_0\left(\left(1+\frac{j}{2^k}\right)|\xi|\right) - J_0\left(\left(1+\frac{j-1}{2^k}\right)|\xi|\right)|^2 \le C .$$

But 88) follows because for s positive and r positive

89)
$$|J_0(r)| \le C/\sqrt{r}$$

and

90)
$$|J_0(r+s) - J_0(r)| \le C\frac{s}{\sqrt{r}} .$$

To prove 88) we use 89) if $|\xi| \ge 2^k$ and 90) if $|\xi| \le 2^k$.

Finally we will show how Stein [SH] proved the maximal function along the parabola (t,t^2) is bounded by using g-functions.

We start with the measure $d\mu$ defined in 24)

$$d\mu(\phi) = \int_1^2 \phi(t,t^2)\,dt .$$

We set

91)
$$d\mu_h(\phi) = \int_1^2 \phi(ht,ht^2)\,dt .$$

Then

$$d\mu_h * f(x,y) = \int_1^2 f(x-ht,y-ht^2)\,dt ,$$

or

92)
$$d\mu_h * f(x,y) = \frac{1}{h}\int_h^{2h} f(x-t,y-t^2)\,dt .$$

We choose a function $\psi(x,y) \in C_0^\infty(R^2)$ with $\hat{\psi}(0) = 1$. We set

93) $$\psi_h(x,y) = \frac{1}{h^3}\,\psi\left(\frac{x}{h}, \frac{y}{h^2}\right),$$

and

94) $$g(f)(x,y) = \left\{\int_0^\infty |d\mu_h * f(x,y) - \psi_h * f(x,y)|^2 \,\frac{dh}{h}\right\}^{1/2}.$$

Let us first assume

95) $$\|g(f)\|_{L^2} \le C\|f\|_{L^2}.$$

Note that

$$\sup_{\varepsilon>0} \frac{1}{\varepsilon}\int_0^\varepsilon |d\mu_h * f(x,y) - \psi_h * f(x,y)|\,dh$$

$$\le \sup_{\varepsilon>0} \frac{1}{\varepsilon^{1/2}}\left\{\int_0^\varepsilon |d\mu_h * f(x,y) - \psi_h * f(x,y)|^2 dh\right\}^{1/2}$$

$$< \left\{\int_0^\varepsilon |d\mu_h * f(x,y) - \psi_h * f(x,y)|^2 \,\frac{dh}{h}\right\}^{1/2}$$

$$\le g(f)(x,y).$$

So

96) $$\sup_{t>0} \left|\frac{1}{\varepsilon}\int_0^\varepsilon d\mu_h * f(x,y)\,dh\right|$$

$$\le g(f)(x,y) + \sup_{h>0} |\psi_h * f(x,y)|.$$

A classical argument (see [R]) shows

$$\|\sup_{h>0} \psi_h * f\|_{L^p} \leq C_p \|f\|_{L^p} .$$

Thus by 95), we see

$$\left\| \sup_{\varepsilon>0} \left| \frac{1}{\varepsilon} \int_0^\varepsilon d\mu_h * f(x,y)\, dh \right| \right\|_{L^2} \leq C \|f\|_{L^2} .$$

If $f \geq 0$,

$$\frac{1}{\varepsilon} \int_0^\varepsilon d\mu_h * f(x,y)\, dh$$

$$= \frac{1}{\varepsilon} \int_0^\varepsilon \frac{1}{h} \int_h^{2h} f(x-t, y-t^2)\, dt\, dh$$

$$\geq \frac{1}{\varepsilon} \int_0^\varepsilon f(x-t, y-t^2) \int_{\varepsilon/2}^\varepsilon \frac{1}{h}\, dh$$

$$\geq \frac{1}{\varepsilon} \int_0^\varepsilon f(x-t, y-t^2)\, dt .$$

So from 96) we infer

$$\left\| \sup_{\varepsilon>0} \frac{1}{\varepsilon} \int_0^\varepsilon f(x-y, y-t^2)\, dt \right\|_{L^2} \leq A \|f\|_{L^2} .$$

It remains to prove 96).

$$\int |(gf)(x,y)|^2\,dx\,dy = \int_0^\infty \frac{dh}{h}\int |d\mu_h * f - \psi_h * f|^2\,dx\,dy$$

$$= \int_0^\infty \frac{dh}{h}\int |\widehat{d\mu_h}(\xi,\eta) - \hat{\psi}_h(\xi,\eta)|^2\,|\hat{f}(\xi,\eta)|^2\,d\xi\,d\eta$$

$$= \int_0^\infty |f(\xi,\eta)|^2 \int_0^\infty \frac{dh}{h}\,|\widehat{d\mu_h}(\xi,\eta) - \hat{\psi}_h(\xi,\eta)|^2\,d\xi\,d\eta\,.$$

So to prove 96) it suffices to prove

97) $$\int_0^\infty \frac{dh}{h}\,|\widehat{d\mu_h}(\xi,\eta) - \hat{\psi}_h(\xi,\eta)|^2 \le C\,.$$

The integral on the left side of 97) is

98) $$\int_0^\infty \frac{dh}{h}\,|\widehat{d\mu}\,(h\xi,h^2\eta) - \hat{\psi}\,(h\xi,h^2\eta)|^2\,.$$

Thus by replacing h by λh $\lambda > 0$ we may write the expression in 98) as

$$\int_0^\infty \frac{dh}{h}\,|\widehat{d\mu}(\lambda h\xi,\lambda^2h^2\eta) - \hat{\psi}\,(\lambda h\xi,\lambda^2h^2\eta)|^2\,.$$

By choosing λ so that $\lambda^2\xi^2+\lambda^4\eta^2 = 1$, we see that it suffices to estimate 99) when $\xi^2+\eta^2 = 1$. In this case we see

99) $$\int_0^1 \frac{dh}{h}\,d\mu(h\xi,h^2\eta) - \hat{\psi}(h\xi,h^2\eta)|^2 \le C\int_0^1 \frac{h^2}{h}\,dh \le C$$

since $\widehat{d\mu}(0) = \hat{\psi}(0) = 1$.

Then from 27)

$$|\widehat{d\mu}(h\xi, h^2\eta)| \leq Ch^{-\delta}$$

for some $\delta > 0$. So

100) $$\int_1^\infty \frac{dh}{h} |\widehat{d\mu}(h\xi, h^2\eta)| \leq \int_1^\infty \frac{dh}{h^{1+2\delta}} \leq C \, .$$

Also $\hat{\psi}(\xi,\eta) \leq \dfrac{C_N}{(1+\xi^2+\eta^2)^N}$ for any N, so

101) $$\int_1^\infty \frac{dh}{h} |\hat{\psi}(h\xi, h^2\eta)| \leq C \, .$$

Now we obtain 98) and hence 96) by combining 100), 101) and 102).

In this section we have emphasized L^2 methods. L^p results for $p > 1$, can be obtained by combining the L^2 estimates presented here with the techniques of section 2. Altogether one can prove the following theorems:

THEOREM 4. *If* $\gamma(t)$ *satisfies* $\gamma' = \frac{A}{t}\gamma(t)$ *where all the eigenvalues of* A *have positive real part,*

$$\| \sup_{0<h<\infty} \frac{1}{h} \int_0^h |f(x-\gamma(t))| dt \|_{L^p} \leq C \|f\|_{L^p}, \quad 1 < p \leq \infty \, .$$

THEOREM 5. *Let* $\gamma(t)$ *be a curve in* R^n. *If the vectors* $\gamma'(0), \gamma''(0), \gamma^{(3)}(0) \cdots$ span R^n,

$$\| \sup_{0<h\leq 1} \frac{1}{h} \int_0^h |f(x-\gamma(t))| dt \|_{L^p} \leq C_p \|f\|_{L^p} \quad 1 < p \leq \infty \, .$$

IV. *Vector field problems*

We turn now to problems III and III′. To study problems III and III′ it is convenient to make a change of variables. It is possible to make a change of variables so that the integral curves of the vector field v(x) become lines parallel to the x-axis. Under this change of variables the vectors v(x) transform into curves γ which vary from point to point. So we are led to studying problems IV and IV′. We facilitate the statement of these problems with 3 definitions.

If $\gamma(x,t)$ is for each x a smooth curve in t with $\gamma(x,0) = 0$, let

102) $$H_\gamma f(x) = \int_{-1}^{1} f(x-\gamma(t,x))\,\frac{dt}{t}\,,$$

let

103) $$M_\gamma^h f(x) = \frac{1}{h}\int_0^h f(x-\gamma(t,x))\,dt\,,$$

and

104) $$\mathfrak{M}_\gamma f(x) = \sup_{1\geq h>0} |M_\gamma^h f(x)|\,.$$

We are now ready for the statement of problem IV and IV′.

Problem IV: When do we have

$$\|H_\gamma f(x)\|_{L^p} \leq C_p\|f\|_{L^p}\,?$$

Problem IV′: When do we have

$$\|\mathfrak{M}_\gamma f\|_{L^p} \leq C_p\|f\|_{L^p}\,?$$

The change of variables described above preserves tangency and curvature conditions. So for example $\left.\frac{\partial\gamma}{\partial t}\right|_{t=0}$ will be parallel to the x axis, and if the curvature of the integral curves of v(x) never vanishes $\left.\frac{\partial^2\gamma}{\partial t^2}\right|_{t=0}$ will not be zero. It turns out that one can prove the following theorems:

THEOREM 6. *Let* $\gamma(t,x) = (t,\Gamma(t,x))$ *be a smooth curve in* R^2 *satisfying*

$$\left.\frac{\partial\Gamma(t,x)}{\partial t}\right|_{t=0} = 0 \quad \textit{and} \quad \left.\frac{\partial^2\Gamma}{\partial t}\right|_{t=0} \neq 0 .$$

Then

105) $$\|H_\gamma f\|_{L^2(R^2)} \leq C\|f\|_{L^2(R^2)} ,$$

and

106) $$\|\mathfrak{M}_\gamma f\|_{L^2(R^2)} \leq C\|f\|_{L^2(R^2)} .$$

A consequence of Theorem 6 for vector fields will then be

THEOREM 7. *If* v(x) *is a smooth vector field in* R^2 *such that the integral curves of* v *have nowhere vanishing curvature,*

$$\|H_v f\|_{L^2(R^2)} \leq \|f\|_{L^2(R^2)}$$

and

$$\lim_{h\to 0} \frac{1}{h} \int_0^h f(x-tv(x))\,dt = f(x) \qquad \text{a.e.}$$

for f *in* $L^2(R^2)$.

These theorems are announced in NSWV. Here we shall give some discussion of the ideas. Because $\gamma(t,x)$ depends on x, the Fourier Transform no longer seems like a good tool to study H_γ and $\mathfrak{M}_\gamma$. We must find a different way to employ orthogonality. Here we're motivated by work of Kolmogorov and Silveristov [Z] on the partial sums of Fourier Series and an approach to the Poisson Integral by Paley [P]. The idea is to consider $H_\gamma \cdot H_\gamma^*$ and $M_\gamma^{h(x)} \cdot (M_\gamma^{h(x)})^*$ where $\cdot$ denotes composition of operators and $*$ signifies Hilbert space adjoint. In order to gain some insight into this method we shall just discuss it in the case $\gamma(t) = (t,t^2)$ where, of course, we already know the results by a different method.

Let us consider $K = H_\gamma \cdot H_\gamma^*$. The kernel of K will have support on $\{(u,v) | u=t-s, v=t^2-s^2\}$. Hence one might hope that K would have a much smoother kernel than H_γ. (Of course if K were bounded on L^2 so would H_γ.) One might hope then, that K would be a Calderon-Zygmund operator. Let us see if this could be the case. One can see that

$$H_\gamma^* f(x,y) = \int f(x+t, y+\gamma(t)) \frac{dt}{t},$$

and hence that

$$H_\gamma H_\gamma^* f(x,y) = \iint f(x+t-s, y+t^2-s^2) \frac{dt\,ds}{ts}.$$

Let $u = s-t$, $v = s^2-t^2$, and we find

$$H_\gamma H_\gamma^* f(x,y) = \iint f(x-u, y-v)\, k(u,v)\, du\, dv$$

where

$$k(u,v) = \frac{C}{|u| \left(u - \frac{v}{u}\right) \left(u + \frac{v}{u}\right)}.$$

Now $k(u,v)$ does have its support spread out. However the singularities of k across the curves $v = \pm u^2$ are not locally integrable away from the origin. Hence $H_\gamma H_\gamma^*$ cannot be a Calderon-Zygmund operator. Let us try to see what goes wrong on the level of the Fourier Transform. Recall

$$H_\gamma f = Q_p * f$$

where

$$\hat{Q}_p(\xi,\eta) = \int_{-\infty}^{\infty} e^{(i\xi t + i\eta t^2)} \frac{dt}{t}.$$

We know from 48) that

$$Q_p(\xi,\eta) = \text{sgn} + C\,\frac{|\eta|^{1/2}}{\xi}\, e^{i\frac{\xi^2}{\eta}}$$

$$+ \text{ better terms.}$$

The multiplier for $H_\gamma \cdot H_\gamma^*$ would be essentially $|\hat{Q}_p|^2$. Notice that $|\hat{Q}_p|^2$ doesn't look any nicer than $\hat{Q}_p$. However $|\hat{Q}_p(\xi,\eta) - \text{sgn}\,\xi|^2$ is much nicer than $\hat{Q}_p(\xi,\eta)$ or $\hat{Q}_p(\xi,\eta) - \text{sgn}\,\xi$. Now $\text{sgn}\,\xi$ corresponds to the operator

$$Lf = \int f(x-t,y)\,\frac{dt}{t}.$$

L is known to be bounded in L^2. This suggests that we try to consider

$$M = (L - H_\gamma)(L^* - H_\gamma^*).$$

If M is bounded in L^2 so will be $L - H_\gamma$. Hence so will be H_γ. This actually works and is the basic idea in the proof of Theorem 4. We turn now to the idea of the proof of Theorem 5. Paley showed that

107) $$P_{h(x)} \cdot (P_{h(x)})^* f(x) \le C[P_{h(x)} f(x) + P_{h(x)}^* f(x)]$$

where P_h is the Poisson Kernel. It follows from 107) that

$$\|P_{h(x)}f\|^2_{L^2} \le C\|P_{h(x)}f(x)\|_{L^2} ,$$

and hence

$$\|P_{h(x)}f\|_{L^2} \le C .$$

Since the function $h(x)$ is arbitrary we have

$$\|\sup_h P_h * f\|_{L^2} \le C\|f\|_{L^2} .$$

The fact that we had to modify H_γ suggests that we should not expect 107) to hold, but we might expect a variant to hold – perhaps involving an operator

$$R_{h(x,y)}f(x,y) = \frac{1}{h(x,y)} \int_0^{h(x,y)} f(x-t,y)\,dt .$$

(We know

$$\|R_{h(x,y)}f(x,y)\|_{L^2} \le C\|f(x,y)\|_{L^2} .)$$

We might hope to prove

108) $M_{h(x)}(M_{h(x)})^* f(x) \le C(R_{h(x)}M^*_{h(x)}f(x) + M_{h(x)}R^*_{h(x)}f(x) + B_{h(x)}f(x))$

where $B_{h(x)}$ is some bounded operator. Even 107) is not quite right. We refer the reader to NSWV for the correct technical modification of 107). This concludes our discussion of the vector field problem.

V. *Recent developments*

The positive results of Theorem 2 and Theorem 5 assume that the curve $\gamma(t)$ has some curvature at the origin. There have been a number of papers trying to understand what happens when this curvature condition is dropped. See [C], [CNVWW], [NVWW] 1), 2), 2), and [NE]. Let us note that we don't have positive results for all C^∞ curves.

Suppose for example $\gamma(t)$ is odd and $\gamma(t) = (t,\Gamma(t))$ where

$$\Gamma(t) = \begin{cases} 0 & \text{for } 0 \le t \le 1 \\ t-1 & \text{for } t > 1 \end{cases} .$$

Then

$$\widehat{H_r f}(\xi,\eta) = m_\gamma(\xi,\eta)\hat{f}(\xi,\eta)$$

where

$$m_\gamma(\xi,\eta) = \int_{-\infty}^{\infty} e^{i\xi t} e^{i\eta\Gamma(t)} \frac{dt}{t}$$

$$= \int_0^1 \sin\xi t \frac{dt}{t} + \int_1^\infty \sin(\xi t + \eta t - \eta) \frac{dt}{t}$$

which is easily seen to be unbounded if $\xi = -\eta$. In fact it is easy to see the following: Suppose $\gamma(t)$ is odd and linear on a sequence of intervals $a_i \le t \le b_i \le 1$ and assume that the linear extension of γ on $[a_i,b_i]$ does not pass through the origin. Then if the ratios $\frac{b_i}{a_i}$ are unbounded,

$$\lim_{\varepsilon \to 0} \int_{|t| \le \varepsilon} f(x-\gamma(t)) \frac{dt}{t}$$

cannot exist in the L^2 sense for every f in L^2.

It is somewhat more difficult to produce counterexamples for the operator $\mathfrak{M}_\gamma$. For example in the case of the two dimensional curve $(t,\Gamma(t))$ described above, it is easy to see that $\mathfrak{M}_\gamma$ is bounded in every L^p, because it is dominated by

$$\sup_h \frac{1}{h} \int_0^h |f(x-t,y)|\,dt + \sup_h \frac{1}{h} \int_0^h f(x-t,y-t+1)\,dt\ .$$

Both of these operators are bounded by the classical theory. We refer to [SW] for an example to show that the maximal function along a C^∞ curve has no non-trivial positive results.

The above authors have been investigating the behavior of the Hilbert Transform and maximal functions related to convex curves. Let $\gamma(t) = (t, \Gamma(t))$ a plane curve with $\Gamma(t)$ convex for positive t. Let $h(t) = t\Gamma'(t) - \Gamma(t)$. $-h(t)$ represents the y-intercept of the line tangent to the curve γ at $\gamma(t)$. We then have

THEOREM 8. *Let* $\gamma(t) = (t, \Gamma(t))$ *with* $\Gamma(t)$ *convex and increasing for* $t > 0$. *If* $\Gamma(t)$ *is even*

$$\|H_\gamma f\|_{L^2} \leq C\|f\|_{L^2}$$

if and only if

$$|\Gamma'(Ct)| \geq 2|\Gamma'(t)| \qquad t > 0$$

for some $C > 0$.

If $\gamma(t)$ *is odd*

$$\|H_\gamma f\|_{L^2} \leq C\|f\|_{L^2}$$

if and only if

$$*) \qquad h(Ct) \geq 2h(t)\ , \qquad t > 0$$

for some $C > 0$.

There are generalizations of Theorem 8 to higher dimensions, and an investigation of the L^p theory has begun. We know that

$$\|\mathfrak{M}_\gamma f\|_{L^2} \le C\|f\|_{L^2}$$

if $*$ holds for some $C > 0$, however the maximal functions can be bounded on L^2 (and in fact in L^p for any $p > 1$) for some convex curves even if $*$ fails.

Recently Phong and Stein [PS] introduced a general problem of which our problems are special cases. They consider at each point P in R^n a submanifold M_p of dimension say ℓ and an ℓ dimensional Calderon-Zygmund kernel $K(P,Q)$. Then they consider

$$Tf(P) = \int f(Q)\,K(P,Q)\,dm(Q)$$

where $dm(Q)$ is a measure on M_p. They show that if $n \geq 3$ and $k = n-1$

$$\|Tf\|_{L^p} \le C\|f\|_{L^p}$$

if M_p satisfies a kind of generalized curvature condition. See [PS].

Various authors have also considered multiple parameter problems which are essentially multiple Hilbert transforms on surfaces and multi-parameter maximal functions on surfaces. See [NW2], [V], [STR1], and [CSS].

Appendix 1. *An introduction to the method of steepest descents.*

Here we shall try to give an explanation of the main ideas of the method of steepest descent. The interested reader can find a more detailed description in [B].

Let us first consider the behavior of the integral

$$\text{A-1)} \qquad I(\lambda) = \int_{-\infty}^{\infty} e^{-\lambda x^2}\,\frac{dx}{x}\,,$$

for large λ. Of course we can make a change of variables

$$t = x\sqrt{\lambda}$$

and observe

A–2)
$$I(\lambda) = \frac{B}{\sqrt{\lambda}}$$

where

A–3)
$$B = \int_{-\infty}^{\infty} e^{-t^2}\, dt .$$

The point we wish to make here is that if λ is large most of the contribution to the integral $I(\lambda)$ comes from a small neighborhood of the origin. In fact

$$\int_{|t|>\frac{1}{\lambda^{2/5}}} e^{-\lambda t^2} dt \le \int_{\frac{1}{\lambda^{2/5}}\le t\le 1} e^{-\lambda t^2} dt + \int_{t\ge 1} e^{-\frac{\lambda}{2}-\frac{t^2}{2}} dt$$

$$\le e^{-\lambda^{1/5}} + e^{-\lambda/2}$$

$$\to 0 \text{ very fast as } \lambda \to \infty .$$

Thus the main contribution to the integral $I(\lambda)$ comes from the small interval $-\frac{1}{\lambda^{2/5}} \le t \le \frac{1}{\lambda^{2/5}}$. Now if we perturb the integrand in $I(\lambda)$, we can expand the integrand in a power series in that little interval. For example we might consider

$$J(\lambda) = \int_{-\infty}^{\infty} e^{-\lambda h(t)} dt$$

where $h(t) \geq t^2$ for large t, $h(t) = t^2 + \mathcal{O}(t^3)$ for small t and $h(t) > 0$ for $t > 0$.

Then one can easily see that

$$\int_{|t| \geq \lambda^{2/5}} dt$$

is exponentially small as before. Now

$$\int_{|t|<\lambda^{-2/5}} e^{-\lambda h(t)} dt = \int_{|t|<\lambda} e^{-\lambda t^2} \cdot (1 + \mathcal{O}(\lambda t^3)$$

$$= \int_{|t| \leq \lambda^{-2/5}} e^{-\lambda t^2} dt + \mathcal{O}\lambda^{-1/5} \int_{-\infty}^{\infty} e^{-\lambda t^2} dt$$

$$= \frac{B}{\sqrt{\lambda}} + \mathcal{O}(e^{-\lambda^{1/5}}) + \mathcal{O}\left(\frac{1}{\lambda^{1/2+1/5}}\right).$$

In the method of steepest descents we try to choose a contour of integration so that on the new contour the situation would be essentially that of J. Thus for example, if we had

$$K = \int_{-\infty}^{\infty} e^{i\lambda t^2 + \lambda P(t)} dt$$

and $P(t)$ were very negative at infinity and $\mathcal{O}(t^3)$ near $t = 0$ we would try to write $t = \sigma + i\tau$ and integrate on the line $\sigma = \tau$ for $|t| < 1/\lambda^{2/5}$, and we would expand $e^{\lambda P(t)}$ in a power series in this small interval.

More, generally, if we were concerned with the asymptotic behavior for large values of λ of an integral of the form

$$\int g(z)e^{\lambda h(z)}dz$$

$g(z)$ and $h(z)$ are holomorphic, we would try to choose a contour on which $\operatorname{Re} h(z)$ had only a finite number of maxima, and argue that the main contribution to the integral should come from a small neighborhood of the largest maximum or perhaps an endpoint and we then expand g and h in a Taylor series at such points. At such a maximum, ξ, $h'(\zeta) = 0$. Thus the main contribution to our integral should come from a point where $h'(\zeta) = 0$ or an endpoint. This is also reflected in Van Der Corput's lemma. A variant of this principle for non-analytic functions is called the principle of stationary phase and is discussed in Professor Stein's lectures in these proceedings.

Appendix 2. *The method of stationary phase and quantum mechanics*

The method of stationary phase lends itself to a formulation of quantum mechanics that is very appealing to at least some mathematicians. The principle of stationary phase asserts that the integral

$$I = \int_a^b e^{i\lambda f(t)}dt$$

with $f(t)$ real gets most of its contributions for large λ, near a, b, or a zero of $f'(t)$.

If we had no endpoints for example if f were periodic with period $b-a$ or if the interval of integration was from $-\infty$ to ∞ and f oscillated very rapidly for large t, we would expect the main contribution to the integral to come from small neighborhoods of a zero of f'.

Let us now turn to quantum mechanics. In particular let us consider a particle moving from a point x_a to x_b as time evolves from time t_a to time t_b. According to classical physics, the particle will follow a path for which the classical action is stationary. If $x(t)$ is any path, and our particle has mass m and is moving under the influence of a potential $V(x,t)$, the classical Lagrangian is defined by

$$L = \frac{m}{2}[\dot{x}(t)]^2 - V(x(t),t) .$$

The action along a path x is defined by

$$S(x) = \int_{t_a}^{t_b} L(x(t))dt .$$

The path on which the classical particle moves will be a path $\overline{x}(t)$ such that $\overline{x}(t_a) = x_a$, $\overline{x}(t_b) = x_b$, and such that

$$\frac{d}{d\delta} S(x+\delta x)\Big|_{\delta=0} = 0 .$$

One of the most important principles of quantum mechanics asserts that motion on a classical scale must be essentially described by the laws of classical mechanics. In our example it means the only paths that are important are paths near the path $\overline{x}$ of $*$. This principle is expressed in terms of a small number h. The principle says that the only classical paths that should be important are paths which differ from $\overline{x}$ only on an h scale of measuring. Now the Feynman path integral formulation is in terms of probability amplitudes of events. In our case the probability amplitude of passing from x_a at time t_a to x_b at time t_b is given by an integral

$$F = \int e^{\frac{i}{h}S(x)} \mathcal{D}x$$

where the integration is an integral over all paths $x(t)$ such that $x(t_a) = x_a$ and $x(t_b) = x_b$. Leaving aside the question of how such an integral can be precisely defined, let us try to guess what paths contribute the most to the integral F. Since h is very small, the principle of stationary phase would indicate that the main contribution to the integral F should come from a small neighborhood of a path of which some kind of a derivative of S was zero. Thus we might expect the main contribution to the integral F to come from paths which are very close (on a scale of h) to the path $\bar{x}$ defined by $*$.

There have been many papers in the mathematical and physical literature dealing with the problem of making sense out of the definition F. See for example [CS] and references cited there. However, the original definition in [FH] serves the purpose of making many formal calculations. We may imagine dividing the t interval into 2^j subintervals, I_j, of equal length

$$I_k = \left[[t_b - t_a]\frac{k}{2^j} + t_a, [t_b - t_a]\frac{k+1}{2^{\cdot n1}} + t_a \right].$$

We consider only paths which are linear on I_k.

Such paths are determined by

$$x_k = X\left(t_a + \frac{k}{2^j}[t_b - t_a]\right)$$

$k = 1,2,\cdots,2^j - 1$. We then take

$$F = \lim_{j\to\infty} A_j \int_{R^{2^j-1}} e^{\frac{i}{h}S(x_1,\cdots,x_{2^j-1})} dx_1 dx_2,\cdots,dx_k$$

where $S(x_1,\cdots,x_{2^j-1})$ is the action along the polygonal path determined by $x_1,\cdots,x_{2^j-1}$, and A_j is a normalizing factor. For more details see [FH]. We would like to make one final remark about the book [FH]. It's great. It explains quantum mechanics in terms of mechanics and does not use notions of atomic physics as many of the standard books do. Thus it is accessible to many more mathematicians than standard quantum mechanics texts. The book also elucidates the differences between the nature of physicists and mathematicians. If you don't want to know h to 3 significant figures in ergs/sec, whatever they are, you probably would rather be a mathematician than a physicist. Finally, many mathematicians could probably learn a great deal to improve themselves as mathematicians by reading the book.

STEPHEN WAINGER
DEPARTMENT OF MATHEMATICS
UNIVERSITY OF WISCONSIN
MADISON, WISCONSIN

REFERENCES

[B] N. DeBruijn, *Asymptotic Methods in Analysis*, North Holland Publishing Co., Amsterdam, 1958.

[BF] H. Busemann and W. Feller, "Zur Differentiation des Lebesguesche Integrale," Fund. Math, Vol. 22, 1934, pp. 226-256.

[CS] R. Cameron and D. Storvick, *A simple definition of the Feynman integral with applications, Amer. Math. Soc.*, Providence, 1983.

[CSS] H. Carlsson, P. Sjogren, and J. Stromberg, "Multiparameter maximal functions along dilation-invariant hypersurfaces" to appear in Trans. of the A.M.S.

[CW] H. Carlsson and S. Wainger, "Maximal functions related to convex polygonal lines," to appear.

[C] M. Christ, preprint.

[CS] J. L. Clerc and E. M. Stein, "L^p multipliers for non-compact symmetric spaces," Proc. Nat. Acad. Sci., U.S.A., Vol. 71, 1974, pp. 3911-3912.

[COR1] A. Cordoba, "The Kekeya maximal function and the spherical summation multipliers," Amer. J. of Math., Vol. 99, 1977, p. 1-22.

[COR2] ________"Maximal functions, covering lemmas and Fourier multipliers," Proc. Symp. in Pure Math., Vol. XXXV, Part I, 1979, pp. 29-50.

[CF1] A. Cordoba and R. Fefferman, "A geometric proof of the strong maximal theorem," Annals of Math., Vol. 102, 1975, pp. 95-100.

[CF2] ———, "On differentiation of integrals," Proc. Nat. Acad. Sci., U.S.A., Vol. 74, 1977, pp. 2211-2213.

[CF3] ———, "On the equivalence between the boundedness of certain classes of maximal and multiplier operators in Fourier Analysis," Proc. Nat. Acad. Sci., U.S.A., Vol. 74, 1977, pp. 423-425.

[CNVWW] A. Cordoba, A. Nagel, J. Vance, S. Wainger, and D. Weinberg, "L^p bounds for Hilbert Transforms along convex curves," preprint.

[F] E. B. Fabes, "Singular integrals and partial differential equations of parabolic type," Studia Math., Vol. 28, 1966, pp. 81-131.

[FEF] R. Fefferman, "Covering lemmas, maximal functions, and multiplier operators in Fourier Analysis," Proc. Symp. in Pure Math., Vol. XXXV, Part 1, pp. 51-60.

[FH] R. Feynman and A. Hibbs, *Quantum Mechanics and Path Integrals*, McGraw Hill, New York.

[H] L. Hörmander, "Estimates for translation invariant operators in L^p spaces," Acta Math., Vol. 104, 1960, pp. 93-139.

[KS] R. Kunze and E. Stein, "Uniformly bounded representations and harmonic analysis of the 2×2 unimodular group," Amer. J. of Math., Vol. 82, 1960, pp. 1-62.

[NRW] A. Nagel, N. Riviere, and S. Wainger, "On Hilbert transforms along curves, II, "Amer. J. Math., Vol. 98, 1976, pp. 395-403.

[NRWM] A. Nagel, N. Riviere and S. Wainger, "A maximal function associated to the curve (t,t^2)," Proc. Nat. Acad. of Sci., U.S.A., Vol. 73, 1976, pp. 1416-1417.

[NSWB] A. Nagel, E. Stein, and S. Wainger, "Balls and metrics defined by vector fields I; Basic Properties," to appear in Acta Mathematica.

[NSW] A. Nagel, E. Stein, and S. Wainger, "Boundary behavior of functions holomorphic in domains of finite type," Proc. Nat. Acad. Sci. U.S.A., Vol. 78, 1981, pp. 6595-6599.

[NSWV] A. Nagel, E. Stein, S. Wainger, "Hilbert transforms and maximal functions related to variable curves," Proc. of Symposia in Pure Math., Vol. XXXV, part I, 1979, pp. 95-98.

[NVWW1] A. Nagel, J. Vance, S. Wainger, and D. Weinberg, "Hilbert transforms for convex curves," Duke Math. J., Vol. 50, 1983, pp. 735-744.

[NVWW2] A. Nagel, J. Vance, S. Wainger, and D. Weinberg, "The Hilbert transform for convex curves in R^n," to appear in *Amer. J. of Math.*

[NVWW3] ———, "Maximal functions for convex curves," Preprint.

[NW] A. Nagel and S. Wainger, "Hilbert transforms associated with plane curves," Trans. Amer. Math. Soc., Vol. 223, 1976, pp. 235-252.

[NW2] ———, "L^2 boundedness of Hilbert transforms along surfaces and convolution operators homogeneous with respect to a multi-parameter group," Amer. J. of Math., Vol. 99, 1977, pp. 761-785.

[NE] W. Nestlerode, "Singular integrals and maximal functions associated with highly monotone curves," Trans. Amer. Math. Soc., Vol. 267, 1981, pp. 435-444.

[P] R. Paley, "A proof of a theorem on averages," Proc. Lond. Math. Soc., Vol. 31, 1930, pp. 289-300.

[R] N. Riviere, "Singular integrals and multiplier operators," Ark. Mat., Vol. 9, 1971, pp. 243-278.

[PS] D. Phong and E. Stein, "Singular integrals related to the Radon transform and boundary value problems," Proc. Nat. Acad. Sci., U.S.A., Vol. 80, 1983, pp. 7697-7701.

[S] E. Stein, *Singular Integrals and Differentiability Properties of Functions*, Princeton University Press, 1970.

[SBC] ———, *Boundary Behaviour of Holomorphic Functions of Several Complex Variables*, Princeton University Press, Princeton, 1972.

[SH] ———, "Maximal functions: Homogeneous curves," Proc. Nat. Acad. Sci., U.S.A., Vol. 73, 1976, pp. 2176-2177.

[Ssp] ———, "Maximal functions: Spherical means," Proc. Nat. Acad. Sci., U.S.A., Vol. 73, 1976, pp. 2174-2175.

[SPL] ———, *Topics in Harmonic Analysis related to the Littlewood-Paley Theory*, Princeton University Press, Princeton, 1970.

[SI] ———, "Interpolation of linear operators," Trans. Amer. Math. Soc., Vol. 88, 1958, pp. 359-376.

[SWA] E. Stein and S. Wainger, "The estimation of an integral arising in multiplier transformations," Studia Math., Vol. 35, 1970, pp. 101-104.

[SW] ———, "Problems in harmonic analysis related to curvature," Bulletin of the A.M.S., Vol. 84, 1978, pp. 1239-1295.

[SWE] E. Stein and G. Weiss, *Introduction to Fourier Analysis on Euclidean Spaces*, Princeton University Press, Princeton.

[STR1] R. Strichartz, "Singular integrals supported on submanifolds," Studia Math., Vol. 74, 1982, pp. 137-151.

[STR] J. Stromberg, "Weak estimates on maximal functions with rectangles in certain directions," Ark. Mat., Vol. 15, 1977, pp. 229-240.

[STRO] J. Stromberg, "Maximal functions associated to rectangles with uniformly distributed directions," Ann. of Math., Vol. 107, 1978, pp. 399-402.

[V] J. Vance, "L^p boundedness of the multiple Hilbert transform along a surface," Pacific J. of Math., Vol. 108, 1983, pp. 221-241.

[WE] D. Weinberg, "The Hilbert transform and maximal function for approximately homogeneous curves," Trans. Amer. Math. Soc., Vol. 267, 1981, pp. 295-306.

[Z] A. Zygmund, *Trigonometric Series*, Vols. I & II, Cambridge University Press, London, 1959.

INDEX

Library of Congress Cataloging-in-Publication Data

Beijing lectures in harmonic analysis.

(Annals of mathematics studies ; no. 112)
Bibliography: p.
Includes index.
1. Harmonic analysis. I. Stein, Elias M.,
1931- . II. Series.
QA403.B34 1986 515'.2433 86-91452
ISBN 0-691-08418-1
ISBN 0-691-08419-X (pbk.)

Elias M. Stein is Professor of Mathematics at
Princeton University